HOW TO DESIGN AND BUILD YOUR OWN CUSTOM ROBOT

BY DAVID L. HEISERMAN

TAB BOOKS Inc.
Blue Ridge Summit, PA 17294

FIRST EDITION

SIXTH PRINTING

Printed in the United States of America

Library of Congress Cataloging in Publication Data

Heiserman, David L 1940-
    How to design and build your own custom robot.

    "TAB book # 1341."
    Includes index.
    1. Automata.   I. Title.
TJ211.H37   629.8'92   80-28207
ISBN 0-8306-9629-6
ISBN 0-8306-1341-3 (pbk.)

Questions regarding the content of this book
should be addressed to:

> Reader Inquiry Branch
> Editorial Department
> TAB BOOKS Inc.
> Blue Ridge Summit, PA 17294

Preface

Electronic technology has developed to such a point that nearly anyone has access to ideas and hardware that can transform the dreams and aspirations of science fiction into reality. Such is the case of robots and robot-like machines.

Through the opening years of popularized, personal robot building (say, between 1976 and 1980), a number of do-it-yourself robot construction books appeared on the market. I had the privilege of contributing two of them. For the most part, however, those first books (including mine) described how to build specific machine systems. And while the designs often allowed some latitude for modifying and extending the machines to suit a reader's own ideas, most hobbyists found it very difficult to do so.

Indeed, it can be interesting and instructive to build a robot that duplicates the ideas of an author/experimenter. But there comes a time in an experimenter's experience when it is far more desirable to construct a custom robot machine—one that expresses the experimenter's own dreams and aspirations.

The purpose of this book is to provide the background information and some general guidelines for developing a custom robot machine from the ground up. The level of presentation assumes a fairly sound knowledge of basic electronics, algebra and machine language programming for microprocessor devices.

That is not to say, however, that the book is necessarily too deep for anyone lacking a good background in any one of these special disciplines. Throughout the book, you will find commentaries suggesting ways to make up for any lack of knowledge in special technical areas. So in addition to being a guide to building a robot of your own design, this is also a guide to learning what you must know to get the job done in your own fashion.

I would like to express my appreciation to the staff of The Ohio Institute of Technology in Columbus, Ohio. Without the encouragement and special effort of the adminstrative staff of Ohio Tech, and Bell & Howell Schools in general, I could not have hoped to complete a project of this scope and magnitude.

David L. Heiserman

Contents

Taking Hold
of a Dream

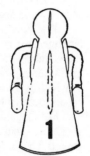

1

Maybe there is no accounting for the reason why so many people dream of building a machine that behaves in a fashion that reminds us of living creatures. Some say robots represent the next logical step in the development of machine technology, but that, in itself, doesn't explain why so many want to get involved in the matter. I don't think we are all that concerned about world technology and its development.

The notion of building a robot, or at least a robot-like machine, is a personal thing. It is simply something that some of us want to do. It is a bit of fantasy, drawn from science fiction, that can be transformed into reality. We cannot build a personal spaceship that will carry us to Mars, we cannot build a time machine for exploring the past and future, and we cannot set up conditions for conversing with alien beings about the purpose and meaning of the Universe. So by default, we are left with the notion of building robots. That, we can do at home and in our spare time.

I cannot really explain why you want to build a robot, or even read a book like this one. I can't even explain why I like to build robots and write such books. All I know for sure is that I've wanted to build robots since I got some inkling how it could be done in the early 1960s. I have found that half the fun is sharing what I've done and learned.

My first working mechanism was a radio-controlled gadget that was built around a vacuum-tube technology. It wasn't very sophisticated, and it had a lot of bugs in it. But that first machine gave me my earliest glimpse at something that can be described as machine personality.

Being a radio-controlled machine that used a commercial AM radio link, it was susceptible to outside radio interference, especially lightning. The poor little thing was deathly afraid of lightning responding to lightning strokes by going through a series of terrible spasms. Certainly it would

7

have been possible to iron out that little bug, making the machine immune to outside electrical interference, but that would have destroyed an essential element of its personality.

That first machine whetted my interest in robot machines and gave me a comfortable feeling about living with machines that might one day rival the mechanical and intellectual capabilities of living creatures. I have never lost the impression that such machines can, and should be, regarded as pets and companions.

That impression colors all my thinking about robots and machine intelligence in general. Unlike so many other experimenters, I have a hard time thinking about these machines as potential slaves that are programmed to do my bidding. I am convinced that thinking about machines as slaves clamps an impenetratable lid on an experimenter's imagination. It seems to straightjacket one's thinking.

Just play this little game in your head for a couple of weeks. Force yourself to think of robots as pets and companions, instead of slaves. Dream about a machine that can tell you what *it* wants to do. Censor any thought about your playing a direct role in dictating the machine's behavior.

As you work through this exercise, you will find your impression of robots blossoming into new and exciting dimensions. Don't let "reality" mess up your dreaming! Let yourself go and you'll begin painting a remarkable mental image. Practical "reality" will come crashing into your dreams later on, but that's a terrible place to start.

My second major robot project took shape in the early 1970s. I named the machine Buster and wrote a book describing exactly how to build him. It is TAB book No. 841, *Build Your Own Working Robot*. Buster was a fascinating machine pet that taught me a lot about digital control mechanisms. Built around a TTL technology, Buster was quite an ambitious project, but it was fun to play with.

A number of experimenters have built Busters over the years, and I still have a bulletin board in my office that displays photos of Buster machines that reader/experimenters were kind enough to send me. That is one of the great pleasures of writing this sort of project book.

I could control Buster whenever I wanted, making him do my bidding in a slave-like fashion. That wasn't the point of the project, though. It was far more fun and satisfying to turn Buster loose in a room and watch him deal with the environment in his own way. That little machine had an engaging personality; if you choose to do so, you can read some anecdotes about Buster in the book just cited.

Watching my 4-year-old son and the family cats interact with Buster greatly reinforced my thinking about such machines as creatures—creatures that can, and will, be worked into the fabric of daily living. This certainly was not a dumb slave machine whose behavior was strictly controlled by some predetermined programs, at least not in the traditional sense. How, for instance, can a cat be expected to interact socially with a machine that must be told exactly how it is supposed to behave at any given moment?

Even before completing the finishing touches on Buster and the book describing how anyone can build him, I began learning some other things about robots and intelligent machines in general. I sensed that it was possible to carry the notion of a creature/pet machine even further than Buster. And as I worked over the idea, a machine I later called Rodney began to take shape.

Rodney is featured in my second robot book, TAB book No. 1241, *How to Build Your Own Self-Programming Robot*. The purpose of that machine was to demonstrate some seemingly farfetched philosophical notions about machines and machine intelligence.

Rodney retains the essential character of Buster, exhibiting a sort of unique personality and an ability to interact with the environment in some intriguing ways. Rodney carried things a lot further, though.

Rodney comes close to being what I can only describe as a *machine creature*. That machine programs itself to deal with the environment in a unique fashion. And every time you wipe out Rodney's memory banks and start him over again from scratch, he takes on a slightly different personality and way of dealing with problems and situations he encounters.

By making the machine self-programming, human biases and prescriptions for behavior are divorced from the mechanism. This action breeds forms of behavior that are unique to machines, instead of those deemed proper by human beings. Rodney can be a little nasty at times, striking out at things that interrupt the processes going on in his microprocessor "brain." At other times he is incredibly shy, and sometimes he does some pretty stupid things.

One never knows for sure how Rodney's personality is going to develop. The general idea, however, is to erase his self-programming memory, then turn him loose in a room by himself for a couple of days. I cannot fully express the feeling of hearing that machine blundering around in a room by itself. I wonder, upon hearing every little bump and click, what is going on in that silicon brain. I wonder what memory patterns are developing and I worry about how well Rodney is generalizing his first-hand experiences to situations not yet encountered in a direct way.

I can still recount quite vividly every instance when I stepped into that room after a couple of days. Rodney always looked the same, but he always responded differently to my presence. Usually he ignored me, going about his business of probing the room or snuggling up to his battery charger "nest."

Sometimes he would respond to my voice, stopping in his tracks for a moment as some memory circuits responded to the unique sound experience. Often he would simply process the new information without changing any of his well-developed patterns of behavior. Other times, though, he would dart away from me or, as in one instance, make a headlong rush toward me.

Generally, Rodney would already know to deal with sudden sounds picked up by his microphone "ear." He had learned to deal with the sounds he generated himself through his initial self-programming phases of life. What was most intriguing were those instances when he found something different in the sound of my voice.

Rodney was especially suceptible to bright flashing lights. His playpen rarely contained such a light. So when I flashed a light at his phototransistor "eye," he had to process the event as a truly new situation.

It is possible to write some rather lengthy scenarios about Rodney, but this is not an appropriate place for that. Instead, I invite you to build Rodney as described in my second robot book, or use the information in this book to develop a Rodney-like machine of your own.

What have I learned from building and working with Rodney? I have to be careful about answering that question because there is the ever present risk of sounding like a crank or mad scientist.

I am convinced that Rodney represents, or at least points the way toward, an entirely new class of machine. It is an autonomous machine that interacts with its environment in a fashion that is almost wholly dictated by its own view of things.

Machines must not be forced to operate according to patterns of behavior dictated by human operators—unless, of course, you want an electromechanical puppet. Machines are machines, and humans are humans. If the two are to meet and interact in the most dynamic fashion, they must meet on their individual terms. People should not be forced to think like machines, and machines should not be forced to think as people do. Understanding that principle is crucial to building true robots.

As my Rodney project drew to its conclusion, I began to see that robotics is suffering from a serious lack of development at the machine-intelligence level. We have some remarkable mechanics available to us today, but only passing fads when it comes to dynamic machine intelligence.

With this thought in mind, I began developing personal computer simulations of Rodney-like intelligence. Setting aside the expense and labor involved in constructing mechanical systems, I concentrated on the machine intelligence.

The first results of this work appear in my book, TAB book No. 1191, *Robot Intelligence...with experiments*. This book shows how to simulate self-programmed machine behavior on a home computer.

That is my present niche in the world of experimental robotics, and there seems to be no end to the variety and quantity of exciting work to be done. I will be sharing much of this in future books on the subject.

I am, however, fully aware of the sort of intrigue and excitement connected with building a real machine—one that buzzes around the floor and, perhaps, picks at objects with an arm-and-hand manipulator. Watching little spots of light bounce around on the monitor for a home computer does not appeal to many robot experimenters, and that's where this book fits into the picture.

My first two robot books, as well as virtually all the robot books written by other experimenters, describe how to build one particular machine. We robot experimenters, as a group, are outgrowing this sort of thing, wanting to move ahead with machines that satisfy each of our own imaginations.

The main purpose of this book is to outline some of the principles involved in building robots and robot-like machines. There are no particular machines described here, but you will find the information necessary for designing and building one of your own—one of your very own conception.

Unfortunately, it isn't technically possible to offer every single bit of information and know-how regarding robotics within a single volume. It's going to take several books, prepared over a period of a couple of years, to do that.

My mental image of you, the reader, is of someone who has a background in basic electronics. I thus assume you have a good understanding of some fundamental points. I will therefore dwell on the basics only to show you how they fit into the context of robots and robot-like machines.

As you read through this book, be on the alert for suggestions and principles that you do not clearly understand. Not everything can be explained in great detail, so it is up to you to see what you must learn in order to make significant progress. In a sense, I suppose this book offers the sort of challenge that many experimenters need in order to discipline themselves for a self-teaching program.

A case in point concerns the machine-language programming that occupies the later chapters of the book. It is wholly inappropriate to spell out the fundamentals of microprocessor programming in the book; you will have to dig out those basic principles for yourself. All I can do here is show you how they are implemented in the context of these machines.

WHAT IS A PARABOT?

In the world of robotics, it is important to distinguish certain classes of projects and machines. Every experimenter should have a clear notion of how his or her work fits into the overall picture. Otherwise, there is the risk of confusion, and that's something none of us can afford at this stage of the game.

A *parabot* is a machine that can behave in a fashion similar to that of a robot, but is not a robot in the truest sense. A parabot is a machine that lacks an element of autonomy or self-determination. It is generally a machine that is either controlled directly by a human operator or indirectly by means of a prescribed program.

Sometimes it is easy to define a machine as a parabot. If its essential modes of behavior call for direct human control, either via a radio link or umbilical cord, it is definitely a parabot—a robot-like machine. Generally, such machines are mechanical extensions of human activity. The operator supplies the most critical elements of brainpower, telegraphing the results of the parabot.

There is nothing wrong about working with such parabot machines. In fact, that sort of work is the first in the business to show practical payoffs—most notably, prosthetic limbs for amputees. Such extensors have been used for decades in the nuclear business, allowing humans to control mechanical hands built into extremely hazardous environments.

In other cases, however, it is not quite so easy to see that a machine is a parabot and not a true robot. A case in point is the class of so-called *industrial robots*. Such machines operate according to some rather elaborate computer programs, doing useful tasks as prescribed by a human operator or programmer.

The element of human intervention is indirect in this case. The machine appears to be working on its own. It isn't. When it is time to change the task, for instance, a human operator must intervene to change the program. Such machines do what they are told to do—sometimes less, but rarely more.

The popular media often hypes industrial "robots" as examples of what we can expect of robotics in the future. It will be quite unfortunate if robotics is limited to general applications of such parabots. Those machines are incapable of dealing with a dynamic environment on the level demanded by daily living. It is difficult to imagine an industrial "robot" coming up with an imaginative solution to a problem not anticipated by its designer and programmer.

And here is why it is important to distinguish parabots and robots. If your thinking is limited to that spelled out for conventional parabot machines, you run the risk of eventually running headlong into an unsurmountable barrier. You will reach a point where further progress toward realizing your dreams of making a mechanical companion becomes virtually impossible.

That barrier is one of principle, and not of technological achievement. Experimenters who cannot see the difference between parabots and robots generally think only in terms of parabots. As a result, they live in that land of *Someday*. Someday we will be able to do this or that. Someday we will have the technology that is necessary for making a machine that does thus and so. Someday ... Someday ... Someday.

I recently read a magazine article written by a superbly gifted experimenter who, unfortunately, is trapped in Somedayland. In that article, he cries out for help from computer programmers—help with the ghastly task of preparing a program estimated to take 24 man-years of time. That experimenter is attempting to fulfill a wonderful dream within an inappropriate context: that of a parabot machine.

I also think about little R2-D2 in the motion picture, *Star Wars*. I think about the scene where the robot is wandering through a desert on some unknown planet, dealing with situations no human programmer could have anticipated. R2-D2 was not a parabot. He/she/it portrayed a real robot.

I am not suggesting there is anything inherently wrong with parabot experiments and machines. I am simply saying that every experimenter should have a good perspective on the project at hand. And if it is a parabot, there should be no confusion regarding its limitations. There was a time—not long ago—when I was clearly prejudiced against parabot experimenters. That must have been an overreaction to what I had learned about the significance of true robots.

I have since mellowed on the point. While I still prefer to work with robot machines, I am fully aware of the value of experiments and developments in parabot technology. In fact, you will find that the vast majority of the material in this book refers specifically to parabot applications.

If you set up a parabot project carefully, you can always retrofit it for robot experiments. And if you are the sort of individual who needs the acclaim and well-wishes of others to keep you going, you will find parabots reaping such rewards rather quickly. (Real robots aren't often very showy machines. People have a hard time understanding what they are doing and seeing any point to it.)

WHAT IS A REAL ROBOT?

It is not easy to define a real robot. It is generally easier to define a parabot. Then if the machine is something more than that, it might be a robot by default. Does that seem to be an unsatisfactory definition of a robot? Indeed, it is. But it's a start.

One key word in the definition of a real robot is *autonomy*. A real robot responds to its environment in its own fashion. A human programmer can set limits on the range of responses, but the machine is free to work out matters on its own within those boundaries.

Limits will always be imposed upon the robot's ability to perceive elements of its environment, too. But the robot must be allowed to interpret those sensory elements in its own fashion. Give it an ability to sense light but let it determine for itself how it will *interpret* light.

Another key word in defining robots is *adaptability*. A robot must be able to adapt its behavior to unforeseen circumstances. Let me illustrate this point with a story about robot Rodney.

I once let Rodney explore a room for a couple of days. At the end of that time, he knew that room "like the back of his own hand" (which he didn't have, but you get the idea). He had no trouble getting away from obstacles of any sort, even those I later placed into the room.

I then shook his confidence by disconnecting the wiring to one of his two drive/steer wheels. That meant I crippled him, making it impossible to move in a straight line. Whereas he could once get away from an obstacle by zooming straight backward, he now blundered around with crazy, circular motions. Such an "accident" would devastate a parabot preprogrammed with important, straight-line motions.

Rodney gradually adapted to his disability, however, by learning to cope with the environment in a different way. One-wheeled Rodney was soon as competent as he was with two wheels.

You can blind a robot, stuff cotton in its "ears," damage sections of memory, reverse the connections to motors and scramble some of its programming. Still, it works! Working under such handicaps, the robot might not be as effective or as efficient as before but it adapts. No parabot can do that.

On the positive side, you can add new sensory and response mechanisms. If the robot is properly designed, it will begin taking new mechanisms into account right away, learning to exhibit wider ranges of responses and modes of behavior that indicate a larger amount of information is available for making decisions.

The matter of robot adaptability, as I see it, absolutely precludes a capacity for self-programming. If a machine is to adapt to any situation, it must be able to alter its own programming. That 24 man-years of programming required for a sophisticated parabot now becomes the task of the robot, itself. Let the thing do all of its own programming. You have better things to do with your spare time.

If you don't like the way your robot is responding to a certain set of circumstances, you can work to change its mind. Robots are trainable. But you have to study its patterns of behavior before you can work out an effective training program.

Perhaps it is needless to say that real robots are full of surprises. They often work out solutions to problems that alternately humble and delight programmers who would attempt to work out solutions in advance.

Autonomy and *adaptability* are hardly terms that are fully adequate for a good technical definition of a real robot. The terms are applicable and, indeed, essential; but at the same time, they are too general for conventional scientific and technical thinking.

I am not sure that a precise, clear-cut definition really exists. It is the nature of the beast to defy conventional technical definition. It is a flavor of defiance that seems to pervade the whole business working with real robots.

A sad mistake for anyone would be to disregard the possibility of working with autonomous, self-programming (adaptable) robots on the grounds that a precise definition has not yet evolved. That would be like a trained physician refusing to practice medicine until given a precise definition of life.

It is far easier to weigh, measure and define in great detail the nature of a dead creature than it is to define that special something possessed by a live animal. If you can live with that notion, you will be comfortable with the differences between parabots and robots.

AI, MI AND EMAI

I am not enamored with acronyms, but one can hardly do a great deal of work in psychology and computer programming without feeling the need to make up acronyms. I am especially turned off by an overly self-conscious effort to make up expressions whose acronyms turn out to spell cute, jazzy little sounds. Forget it; I am digressing.

I do want to impose a certain bias on you concerning AI (artificial intelligence) and MI (machine intelligence), however. I once found myself in the position of explaining my work to a sophisticated, middle-aged lady. Upon hearing me mention the expression, "artificial intelligence," she shuddered, "Oooh, I just hate anything artificial, especially artificial intelligence."

Normally, one writes off such comments as being of no real significance. But the experience stuck with me for some reason. And having the opportunity to work for several years with genuine robot intelligence, I, too, began disliking the term, *artificial intelligence.* There seemed to be nothing at all artificial about it. It is real intelligence at work. I'm not saying these machines think—that always starts an argument that never gets anywhere. I'm simply stating there is nothing artificial about machine intelligence. It is intelligence in its own right.

If you build a parabot or write a computer game program that mimics the workings of intelligence, *that* is artificial intelligence. These robots don't mimic anything. They are electromechanical creatures in their own right, and the sort of intelligence that is reflected by their behavior is genuine and unique to them. It is anything but artificial intelligence. What separates it from human intelligence or animal intelligence of any other sort is the fact that it is generated or exhibited by a machine; thus, it is *machine intelligence.*

I don't think you will see the term *artificial intelligence* used anywhere else in this book. It is an inappropriate expression when used in the context of robot machines.

So what's EAMI? That's my concession to coining an acronym. It refers to a particular process for developing a hierarchy of machine intelligence: *e*volutionary *a*daptive *m*achine *i*ntelligence.

At one time, I proclaimed the process as the one and only true road to developing honest-to-goodness, real-life robots. I have recently mellowed on that point, too. There are other ways to manage the job, but I still hold that EAMI is the most suitable approach for budget-minded amateur robot experimenters.

The basic motivation behind using an EAMI scheme is to give the experimenter a chance to build an intelligent machine gradually. It is a building-block approach that lets the experimenter become comfortable working at one level of intelligence before advancing to the next. This is the *evolutionary* aspect of the job.

I have just described my views of *machine intelligence*, so that takes care of the *MI* part of the acronym. And in the previous section of this chapter, I pointed out the importance of adaptability as it applies to real robot mechanisms. Thus you have all the background information for understanding the meaning of EAMI. Now for some more details about the evolutionary side of things.

Alpha-Class Intelligence

The first step in an EAMI program is to develop a machine that responds and adapts to its environment in a purely random fashion. This *Alpha-Class* machine has absolutely no capacity for remembering past experiences; it exists only in a very narrow time frame of the moment.

Such a machine makes a random response to any condition in the environment that it can sense and to which it can react. Such a machine might seem to be sort of stupid, useless, inefficient and a whole lot of other negative adjectives. Until you've actually played with one of them, however, reserve judgment.

It might be surprising to note how sophisticated such a machine can be and how well it can adapt to changing circumstances in its environment. In fact, an Alpha-Class machine is superbly adaptable to drastic events. A great deal of space in my book on simulating robot intelligence on a personal computer is devoted to justifying and demonstrating the adaptive character of Alpha-Class machines. You will find some flowcharts and machine-language programs for an Alpha-Class creature in Chapter 13 of this book.

Beta-Class Intelligence

The second step in the evolution of adaptive machine intelligence is to add some memory to the basic Alpha-Class version. The idea is to give the machine an ability to remember responses from past encounters with the environment, and then call up those responses when similar conditions arise later.

A Beta-Class machine actually works on an Alpha-Class level through the early part of its life when it is still exploring things for the first time. But as the Beta's hierarchy of experience grows, it exhibits smaller amounts of purely random behavior in favor of "habits" it has learned.

A *Beta* always deals with a brand-new situation in an Alpha-like, random fashion at first. It refines its responses with time and experience, however, eventually programming itself to cope with virtually any condition it can perceive. Chapter 13 shows how to set up a Beta-Class mechanism, too.

Gamma-Class Intelligence

An Alpha-Class machine responds to its environment in a purely random fashion. It learns and remembers nothing. A Beta-Class machine does remember responses to conditions it experienced directly at some time in the past. It responds according to remembered experiences, but it cannot anticipate situations never dealt with on a first-hand basis. The shortcoming of the Beta-Class machine is covered by the third, and final, step in the EAMI program: the Gamma-Class machine.

A Gamma-Class machine starts its life as an Alpha, but it soon begins exhibiting the memory-related programming of a Beta. Then, as the hierarchy of remembered responses grows and becomes refined with further experience, a Gamma begins generalizing that knowledge to situations not yet encountered on a first-hand basis. Ideally, a Gamma has at least some vague notions about how to deal with new circumstances when they arise in the future.

My Rodney robot can include Gamma behavior, and the programs necessary for engendering that sort of behavior are spelled out in TAB book No. 1241, *How to Build Your Own Self-Programming Robot.* Some Gamma simulations for home computers are listed in TAB book No. 1191, *Robot Intelligence...with experiments.* I have, however, omitted Gamma schemes from this book.

Perhaps it is unfortunate that I am not encouraging Gamma-Class experiments here. One reason, however, is a very practical one: There simply isn't enough space available between the two covers of this book. The subject of Gamma-Class intelligence is a very extensive one, dealing essentially with self-determining heuristics. Explaining programs that write programs for writing programs is a topic worthy of its own book, space here must be left for more basic topics.

A second reason for passing over gamma intelligence here is that its behavior is difficult to appreciate with an actual working mechanism. Gamma-Class behavior is not easily detected in a real-world setting, and the experimenter has to devise intricate tests for the machine to see any real elements of that behavior at work. It all shows up much better through a *faster-than-real-time computer simulation.*

That doesn't mean the door is closed for Gamma-Class machines, though. The beauty of an EAMI system is that you can extend it by adding little or no hardware. In fact, you can insert a Gamma program into an existing Beta program without having to change an iota of hardware.

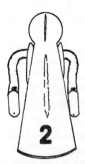

The Electrical And Mechanical Subsystems

2

Every new parabot/robot project must begin with a careful consideration of the electrical and mechanical subsystems it will eventually require. Knowing what you are going to do ahead of time is tantamount to getting it done, and the more precisely the system is defined from the beginning, the easier the task is and the greater the odds of success are.

Of course it is possible, and sometimes workable, to dig into the construction phase before giving full attention to how things should turn out. More often than not, though, such a haphazard approach spells a great waste of time and money; there's too much backtracking and duplication of effort involved.

This chapter is a general, overall summary of the mechanical and electrical elements, or subsystems, of just about any sort of parabot/robot project. The model is representative of a rather complete system, but modifying it for simpler systems is a matter of leaving out some of the subsystems.

The purpose of the chapter is to help you define exactly what you want in terms of a parabot or robot system. And once you know what basic subsystems you want, you can begin working out the details contained in most of the remaining material in this book.

A SYSTEM BLOCK DIAGRAM

Figure 2-1 illustrates a complete working parabot or robot machine. From the most general viewpoint, there are two different environments: and internal and an external environment.

The *internal environment* includes the working machine, itself. It includes all the electrical parts and mechanical schemes that characterized the machine, setting it apart from anything else in the world.

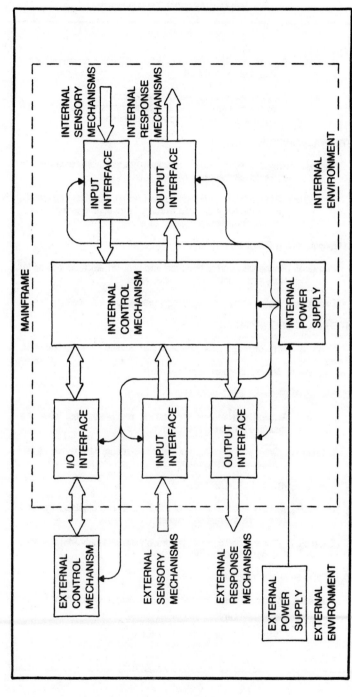

Fig. 2-1. An overall system block diagram.

19

Table 2-1. Summary of Parabot/Robot Subsystems.

MAINFRAME

Purpose: Support and house all mechanical and electrical systems within the internal environment.

Examples: Aluminum, plastic or wooden framework
Structures for external response mechanisms
Essential and cosmetic panels and coverings

INTERNAL POWER SUPPLY

Purpose: Provide electrical power directly to all internal electrical circuits and external response mechanisms.

Examples: Storage battery for high-current motors (lead-acid, gel-cell)
Batteries for electronic circuitry (lead-acid, gel-cell, ni-cad, carbon-zinc, etc.)

EXTERNAL POWER SUPPLY

Purpose: Recharge internal batteries and provide electrical power for tests and troubleshooting procedures.

Examples: Battery charger, regulated power supply, solar cells

INTERNAL RESPONSE MECHANISMS

Purpose: Provide responses relevant to the machine's internal operation

Examples: Motors, lamps, relays, solenoids

EXTERNAL RESPONSE MECHANISMS

Purpose: Provide the means for making responses that alter the machine's external environment.

Examples: Drive and steering motors, manipulators, lamps, relays and solenoids, loudspeaker

INTERNAL SENSORY MECHANISMS

Purpose: Monitor internal status of mechanical and electrical systems relevant to the machine's internal operation.

Examples: Voltage and current sensors, phototransistors, limit switches, temperature sensors

EXTERNAL SENSORY MECHANISMS

Purpose: Monitor relevant conditions in the external environment.

Examples: Microswitches, phototransistors, microphone, ultrasonic transducer, temperature sensor

INTERNAL CONTROL MECHANISM

Purpose: Coordinate all ongoing activity, evaluate the status of all internal and external sensory mechanisms, make decisions, and determine the status of all internal and external response mechanisms.

Examples: TTL or CMOS logic circuitry, bus-oriented microprocessor system, program and control memory

EXTERNAL CONTROL MECHANISM

Purpose: Give the user some measure of control over the function of the machine. The greater the amount of external control, the less likely the machine will qualify as a true robot.

Examples: Control panel, cassette tape player, home computer interface

INPUT INTERFACE

Purpose: Condition the mechanical or electrical signals from internal and external sensory mechanisms, making them compatible with the internal control mechanism

Examples: Switching amplifiers, pull-up or pull-down resistors, A/D converters, tachometers, digital counters, one-shot multivibrators

OUTPUT INTERFACE

Purpose: Condition digital signals from the internal control mechanism, making them compatible with the requirements of all internal and external response mechanisms.

Examples: DC power amplifiers, audio amplifiers, D/A converters, digital latches, counter circuits, timers

I/O INTERFACE

Purpose: Serve as both an input and output interface. See INPUT INTERFACE and OUTPUT INTERFACE for more complete definitions and examples

The external environment is anything else in the universe except the machine. Certainly, that's a bit general. For practical purposes, the scope of the external environment can be narrowed down to any environment the machine is likely to encounter and with which it will interact.

Table 2-1 summarizes the main elements of a complete parabot/robot system. That table defines the blocks and offers a few examples of the circuits or mechanisms used in each case.

The *mainframe*, for example, is the mechanical housing and supporting framework for the machine. Much of the machine's physical appearance is dictated by the nature of the mainframe assembly.

Then there are a couple of power supplies. If the machine moves by its own locomotion, it should include a self-contained, internal power source, such as a storage battery. Quite often it is desirable to have one high-capacity battery for running the motors and another battery of high reliability, but lower current capacity, for operating the electronic circuits.

In all but the very smallest machines, it is likely that at least one of the internal power supplies will be a rechargeable battery. Consequently, an external source of energy for recharging that battery is needed. A battery charger of some sort is thus on the bill of materials for most parabot/robot projects.

Whether the battery charger is carried on-board or set apart from the main machine isn't relevant. If the charger happens to be an integral part of the main machine, there is still a need for an external power source—a source of 120 VAC, 60-Hz utility power, for instance.

Even the notion of using solar cells to recharge the on-board power supply of the machine doesn't alter the general format of the block diagram. In that case, the block labeled "external power supply" is simply relabeled "sun."

Internal response mechanisms are those devices that are largely responsible for making things happen within the internal environment of the machine. Such mechanisms do not have any direct influence on the external environment, although whatever might happen internally can well have some secondary influence on the world outside the machine.

Consider relays as internal response mechanisms. The opening and closing of on-board relay contacts has no direct bearing on the outside environment. The fact that the relays might control the main drive motors of the machine is a secondary, indirect means of manipulating the external conditions.

Motors are a prime example of external response mechanisms. Motors make the machine move in some fashion, and that motion affects the external environment in a direct way.

Internal sensory mechanisms monitor the status of any number of activities taking place within the machine. One of the most critical internal sensing circuits is one that monitors the output power of the main batteries. Whenever these sensors detect a low-power condition developing, they must generate a signal that tells the machine or operator to do something about it.

External sensory mechanisms monitor events taking place outside the machine, itself. Touch or contact sensors, microphone "ears" and photosensitive "eyes" fall into this category.

Some of the most important and difficult subsystems in any parabot or robot are the interfacing schemes. You will probably invest more time, money and effort into working out interfacing systems than anything else.

It would be nice if the entire environment of the machine were composed of nothing but digital electrical signals. That can only happen when simulating machine environments via a digital computer, however (see TAB book No. 1191, *Robot Intelligence . . . with experiments*). The "real" world is largely nondigital and nonelectrical, and virtually all parameters the machine is to sense in the external environment must be translated into a digital, electrical form that is compatible with its primary internal control mechanism. That is the all-important task of the input interface subsystems.

Interfacing the machine with the outside world is, in some respects, simpler than input interfacing. With the notable exception of mechanical manipulators and speech synthesizers, the transformation of digital, electrical energy into some other form of energy that influences the external environment is a relatively simple task. It is fairly easy to turn motors on and off, blink lights and create buzzing sounds through the use of some simple and inexpensive output interface circuits.

The *external control mechanisn* is any subsystem used for getting information into or out of the machine in a direct fashion—one that bypasses the usual sensory and response mechanisms. With a purely parabot-type system, the external control mechanism might be the only control mechanism of any real significance. Made up of switches and dials, dials, the operator uses the control mechanism to control most—if not all—operations of the machine.

The external control mechanism might also include a cassette tape player that directs the activity of the machine, or it can be a connection to the (input/output) I/O plug on a home computer. The more essential this external control mechanism might be to the function of the machine, the more parabot-like the machine will be. The need for an external control mechanism might not disappear altogether in the case of true robots, but its role is minimized to the point where it is used only for routine tests and troubleshooting.

The *internal control mechanism* is central to virtually all events taking place in the system. This mechanism controls the operation of all input and output devices, sensory and response mechanism; and makes logical decisions affecting both the internal environment of the machine and the generating signals that ultimately alter the external environment.

The closer a machine approaches the quality of a true robot, the more complex and critical the internal control mechanism becomes. In the ideal robot, the internal control mechanism is the sole source of behavioral activity. In a parabot system, however, much of the responsibility of the internal control mechanism is turned over to its external counterpart.

To see how the general scheme blocked out in Fig. 2-1 can be modified to suit a particular parabot/robot scheme, consider what might well be the simplest sort of parabot. You want a little machine that you can steer around the floor by operating some switches on an external control panel. The external control mechanism is an input-only device; there are no provisions for monitoring the action of the machine through that interface. There are no external sensory mechanisms, because you, the parabot operator, are keeping an eye on the little machine yourself.

This simple scheme requires some external response mechanisms and an appropriate output interface, however. The purpose is to set some little drive wheels into motion and adjust the steering. The internal power supply might be nothing more complicated than a couple of C-cells, so there is no need for an external power supply.

You are doing all the thinking for the machine, so there is no need for an internal control mechanism. And the thing is so simple that there is no need to monitor and adjust any of the internal workings in an automatic fashion. If you follow this description and redraw the block diagram to include only those subsystems that are used, you will end up with a much simpler block diagram.

At the other extreme, you might dream up a very large and complicated robot system. If you are on the right track as far as your subsystem planning is concerned, you will find that every conceivable subsystem fits the diagram in Fig. 2-1.

PRELIMINARY MAINFRAME DESIGN CONSIDERATIONS

Designing a mainframe assembly for a parabot/robot project can be tricky. The multitude of things to be considered in advance range in importance from the critical to the trivial. Some parts of the mainframe have to be assembled a certain way, simply because the machine cannot do its intended tasks properly if these things aren't done that certain way. On the other hand, you will end up doing a lot of things simply because you feel like doing them.

I suppose the amount of agony involved in designing a mainframe depends on the experimenter's level of patience and how much of a perfectionist he is. To date, I haven't met a robot experimenter who is totally satisfied with his or her mainframe design. It's often a matter of learning to live with a less-than-perfect design for the sake of getting anything done at all.

To illustrate some of the difficulty involved in an initial mainframe design, consider that the largest and heaviest single object in most parabots and robots is the battery for the main drive motors. Much of the mainframe must be built around that particular object.

But then you cannot do a really good job of selecting the battery until you know how much power the motors will consume from it. Generally speaking, the higher the power rating of the motors, the larger and heavier the battery must be. But then the size and weight of the battery has a lot to do with the motor requirements, and you end up in a *Catch-22* situation: The size of the motors determines the size of the battery, the size of the battery determines the size of the motors, and both have a lot of influence on the size, weight and configuration of the mainframe assembly.

The way around this sort of dilemma is to be content with an initial, educated guess about the mainframe specifications. Use that initial design as you plan the rest of the system on paper. The closer you come to a final set of specifications for the other subsystems, the more apparent any mistakes in your initial mainframe design will become.

Revise the physical design as many times as necessary, bearing in mind that it is better to err on the side of being too generous with dimensions than too stingy. That is why my own robot designs tend to be a bit larger than necessary. In each case, if I ever choose to rebuild them from scratch, they will turn out to be a lot smaller.

Mechanical Loading Considerations

The subsystems having the greatest influence on the mainframe configuration are the main drive and steering assemblies and any manipulators you might want to attach to the machine. You will find some general discussions of drive/steer and manipulator assemblies in the two major sections of this chapter that follow this one.

Motors and manipulator assemblies can be relatively bulky in larger machines, and you must take into account the influence their operation will have on the distribution of weight and stresses on the mainframe. Some manipulator designs can complicate the whole matter of weight and force distribution. If you have any plans for having your machine pick up relatively heavy objects with a long arm, you must be prepared to counterbalance the torque that sort of situation creates. Working with manipulators can make the machine's center of gravity shift all over the place; and if you want the whole thing to be fairly stable, you probably have to make some trade-offs concerning the placement of batteries, drive motors, steering and driving mechanisms as well as the manipulator, itself.

Also remember that the main batteries are going to determine the center of gravity of the system in most instances, so count on mounting the batteries at the very bottom of everything. Whether the batteries should be mounted toward the rear, center or front of the machine depends on its steering configuration. See the section of this chapter, "Preliminary Drive and Steering Considerations," for further details.

This matter of working out a mainframe design in advance might sound terribly complicated, but the situation isn't hopeless. Just give it a lot of serious thought in the early going, and be fully prepared to give up some of your pet ideas in favor of expediency. You might even go so far as to try a cardboard scale model and simulate the placement of heavier parts on a first-hand basis.

Some Hints About the Placement of Smaller Parts

In a practical sense, the subsystems of lesser size and weight can be tacked onto the mainframe assembly just about anywhere you choose. Once you have made provisions for the bulkier parts, there will always be some room for a lot of smaller parts.

Just keep in mind a couple of facts about electronic systems in general:

☐ The fumes from lead-acid batteries can corrode the conductors and contacts on printed circuit (PC) boards. Keep all circuit boards at least a couple of inches away from such a battery. Use a blank aluminum panel to separate them, if necessary.

☐ Heat sinks with parallel fins must be mounted so that the fins are vertical most of the time. That arrangement ensures a free flow of convection currents.

☐ Circuit boards must be fixed so that they cannot shake loose. Simply suspending them by their sockets is not sufficient in most cases. Screw or clamp the free ends of circuit boards securely to a fixed frame.

☐ While your machine design might call for a proximity-type obstacle sensor, consider that it might fail under certain circumstances. Arrange all the delicate components so that they are not subjected to direct contact with any object in the outside world.

☐ Make provisions for clamping long cables to a fixed frame. An exposed loop of wire can get caught on all sorts of objects.

☐ Keep cosmetic items or unnecessary razzle-dazzle to a minimum. Such gimmicks often cause more trouble than they are worth.

☐ Leave room for the placement of plugs and barrier strips. You should be using a lot of these solderless connectors in the machine.

☐ Things are going to go wrong with your system, and you are going to have to fix it from time to time. So save yourself a lot of grief in the future by laying out all parts so that they are accessible without much trouble.

Selecting Mainframe Building Materials

Another phase of the mainframe design concerns the choice of building materials. It is tempting to use wood or a wood substitute, such as particle board or pressed wood. Wood is easy to use and doesn't require a lot of special tools. But wood has some serious disadvantages in the context of parabots and robots.

One problem with wood is that it is heavier than aluminum or plastic of equivalent strength and durability. Every ounce of weight cuts minutes off the running time of battery-operated equipment, so using wood will reduce the effectiveness of the machine.

Wood is also a poor conductor of heat. Getting rid of excess heat is vital for power transistors and motors.

Wood tends to expand and contract a significant amount with changes in ambient temperature and humidity. Who wants to risk cracking the conductive tracks on a $300 microprocessor board every time the weather changes? Of course, it is possible to seal the wood with a good shellac. Even then, though, the other disadvantages remain.

Avoid using steel in the mainframe construction. It is relatively heavy, but worse yet, it can pick up magnetic fields that can play havoc with certain kinds of delicate sensing devices.

Plastic can be used for secondary items on the mainframe, but it should not be a primary construction material. Commercial-grade plastic tends to be brittle, so there is a chance it will crack the first time your machine collides with an immovable object. Plastic can also be very expensive, and it's poor conductor of heat.

So plan to use aluminum wherever possible. It takes some talent to work with it, but a knowledgeable experimenter can do wonders with it. Aluminum extrusions now sold in hardware stores offer a lot of possibilities for making some neat-looking and functional mainframe assemblies.

Aluminum is the antithesis of wood. It is strong and lightweight, it is a very good conductor of heat (better than steel, too) and it does not expand and contract a significant amount with changes in temperature and humidity.

The only problem with aluminum, as compared to wood and plastic, is that it is a very good conductor of electricity. That means you have to insulate all electrical contacts (others than those that are supposed to be connected to system ground) from the mainframe.

PRELIMINARY DRIVE AND STEERING CONSIDERATIONS

As mentioned earlier, one of the primary considerations in the design of a mainframe is the drive and steering format you want to use. And farther down the line, the drive/steer format also dictates the kinds of response and sensory interface subsystems required for operating the motors and sensing their operational status.

Taking it for granted that you will be using a wheeled machine (the walking alternative is generally impractical), there are just two fundamental formats for driving and steering the machine. It will appear that there are three designs, but I think that after you study them, you will agree that two of them use the same basic format.

Separate Drive and Steering Motors

Figure 2-2 illustrates one drive/steer format that uses separate motors for driving the machine and steering it. It is the format most suitable when working with a machine that has a rectangular, top-view configuration.

The steering gear motor turns the entire front-wheel assembly a number of degrees to the left or right of straight ahead. The turning radius of the system is limited only by the maximum steering angle; that, of course, is limited by the physical construction of the steering assembly and the load balance near the front end of the mainframe.

If this steering scheme is to be used with anything but a purely operator-controlled system, there must be provisions for sensing at least

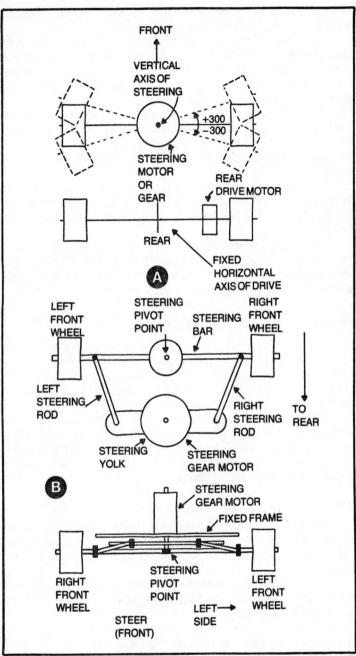

Fig. 2-2. A scheme for front-wheel steering and rear-wheel drive. (A) An overall top view. (B) Front steering detail.

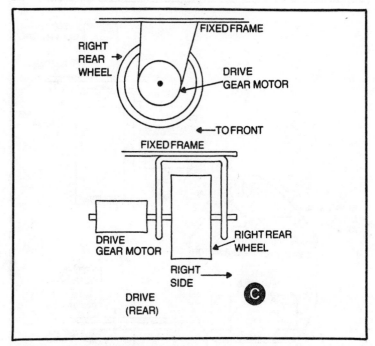

Fig. 2-2. C). Rear drive wheel detail.

the two extreme steering limits as well as the on-center position. Achieving finer, automatic steering calls for additional position sensors at, say, 5-degree or 10-degree increments along the steering arc.

The drive motor is mounted to one or both of the rear wheels. For simplicity, that gear motor can power just one of the two rear wheels, leaving the other as a idle wheel for balance. But of course both rear wheels can be fixed to the rear axle, and the gear motor can drive both of them at the same time.

I used this drive/steer scheme with my first major robot project, Buster. It worked out quite well with the rectangular configuration of the machines, and Buster rarely found a tight corner which prevented the steering mechanism from finding an angle that wouldn't let him get away.

My only regret in that instance was I mounted the heavy lead-acid batteries directly over the rear axle. Consequently, a sudden surge of drive power raised the front steering wheels from the floor and temporarily rendered the steering useless. So plan to mount the batteries a bit forward of the rear axle.

The drive and steering system illustrated in Fig. 2-3 also uses separate motors for steering and driving operations. Here, however, the two motors run just one wheel, and the other two in the tricycle arrangement are simple idle wheels.

29

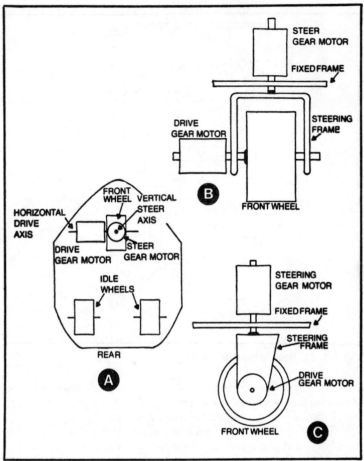

Fig. 2-3. A scheme for front-wheel drive and steering. (A) An overall top view. (B) End view of the front wheel mechanism. (C) Side view of the front wheel mechanism.

This tricycle arrangement is suitable for fairly small machines that have an ovoid, top-view configuration. It is the one used in the classic cybernetic demonstrator, Turtle, invented by Dr. Grey Walter.

The two motors drive the same wheel arrangement along two different axes. The steering motor is fixed to the mainframe and is allowed to rotate the rest of the assembly through some angle left and right of dead center. Using the ovoid configuration and a thoughtful placement of the center of gravity of the machine, the front wheel can be turned nearly 180 degrees in either direction from straight ahead. In fact, the only thing preventing continuous, 360-degree steering is the wiring that runs to the drive motor. If you care to run that wiring to the drive motor through a set of slip rings, you can achieve that full-rotation steering effect.

In any case, controlling this sort of steering arrangement in a fully automatic fashion calls for placing steering-position sensors along the arc of rotation. The more sensors you use, the finer the automatic steering can be.

The drive gear motor is fixed to the steerable frame, and directly to the front wheel axis of rotation. The drive motor thus determines the running speed of the machine and whether it is running forward or in reverse. The two rear wheels are both idle, serving only as the two extra supports for the tricycle arrangement.

Actually, the definitions of front and rear used in Fig. 2-2 are arbitrary. There is nothing wrong with considering the drive/steer wheel as the rear wheel, and the two idling wheels as being at the front of the assembly.

While the drive/steer arrangements in Figs. 2-2 and 2-3 are vastly different in a mechanical sense, they are virtually indentical in terms of the

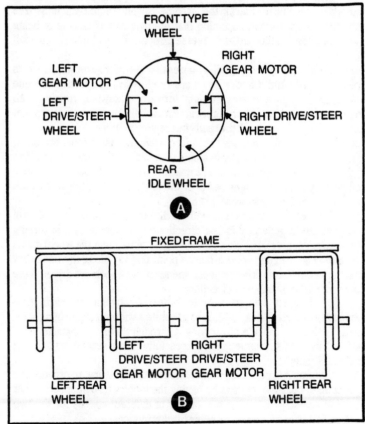

Fig. 2-4. A scheme for independent drive/steer motors. (A) An overall top view. (B) End view of the drive/steer motor assembly.

control mechanisms and interface circuitry required for operating them. The feature that separates them from the alternate scheme suggested in the following section is the fact that they use motors fully dedicated to different operations—one for steering and one for driving the machine.

Two Motors with Combined Drive/Steer Operations

The diagram in Fig. 2-4 represents the most popular drive and steering formats used by amateur roboticists today. Its popularity probably stems from its relative simplicity and, perhaps, the fact that it can create spinning motions that are unique to it.

The two powered wheels are arranged at opposite ends of an imaginary drive/steer axis. Considered independently, the motors perform as simple drive motors, being capable only of spinning in a forward or reverse motion.

Considered together, however, the two, independently-driven motors are capable of steering the machine as well. Table 2-2 summarizes this drive/steer format, assuming that the motors are capable of being stopped, or run in forward and reverse directions at just one speed—full speed.

If both motors are stopped, it figures that the machine is stopped in its tracks as well. And if both motors are run forward at exactly the same speed, the machine moves straight forward. Likewise, running both motors at the same speed and in reverse makes the machine move straight backward. But those aren't the really interesting motions.

The interesting motions take place when the two motors are doing different things. Suppose, for example, the left motor is stopped and the right motor is running forward. The machine will move forward, but with a left turning component centered over the left motor. In other words, the machine moves with a forward, left turn.

Whenever one motor is stopped and the other is turning, the motion of the machine is generally in the direction of the motor that is turning (forward or reverse), but in a circle having its center over the wheel that is stopped. The machine spins around a point midway along the imaginary axis between the two driven wheels whenever the wheels are driven at the same speed, but in opposite directions.

So with just two gear Motor/wheel arrangements, each independently operated, it is possible to generate a wide spectrum of machine motions that are capable of carrying it through the most obstacle-laden paths. I used this scheme quite successfully in my second major robot project, Rodney.

Of course it is possible, but not altogether necessary, to broaden the spectrum of possible motions by having the motors turn at speeds other than full speed. Working with a variety of different speeds for the two motors, it is possible to get spiralling turning motions.

The simplicity of the scheme becomes apparent when you consider that the two motors can be operated from identical circuitry. There is no

Table 2-2. Function Table for the Two-Motor Drive/Steer Operations.

Left Motor	Right Motor	System Response
STOP	STOP	STOP
STOP	FORWARD	FORD, LEFT TURN
STOP	REVERSE	REVERSE, LEFT TURN
FORWARD	STOP	FORWARD, RIGHT TURN
FORWARD	FORWARD	FORWARD
FORWARD	REVERSE	CLOCKWISE SPIN
REVERSE	STOP	REVERSE, RIGHT TURN
REVERSE	FORWARD	COUNTERCLOCKWISE, SPIN
REVERSE	REVERSE	REVERSE

need to work with separate drive and steering logic, motor drivers and motion-sensing circuits.

Incidentally, one idle wheel will suffice if you are careful about balancing the overall center of gravity and are willing to trade off purely circular motions by setting the axis of the two motors behind the center axis of the machine. If you have any doubts about the feasibility of this drive/steer scheme, you can test your ideas using some pieces of cardboard and some wheels from toy cars. Build a simple scale model and try it for yourself.

PRELIMINARY MANIPULATOR CONSIDERATIONS

Most parabot/robot systems have *motion* as one of the primary modes of expressing their behavior. If that mode of motion happens to be limited to drive and steering operations, the big hassle involved in devising the mainframe assembly is over by now.

A good many experimenters, however, want to tackle the challenge of making a robot-like system that can reach out and touch an object or, better yet, reach out and pick up an object. That sort of behavior falls into the realm of manipulator design.

You will find an entire chapter devoted exclusively to the mechanical and electrical details of manipulators. The following discussion merely sets the stage, providing some insight that is indispensable to working out a preliminary mechanical design.

Figure 2-5 illustrates the evolution of a manipulator scheme in a schematic fashion. The first—and simplest—is said to offer one degree of freedom. Just one motor rotates a shaft and arm fixed at right angles to the shaft.

Having just one degree of freedom, the end point of the manipulator (point E) touches a point around a circle of radius 1. The orientation of the manipulator is not really relevant to the drawing: It doesn't have to be oriented so that the "touching" plane of E is parallel to the world horizon. It can be fixed at any desired angle.

The main feature of the manipulator having just one degree of freedom is that the end point can reach only those points at the radius of 1. It cannot touch any point within that circle.

The diagram in Fig. 2-5B begins with the system having just one degree of freedom, and adds to it another axis of motion. This arrangement, having two degrees of freedom, is capable of making end-point contact anywhere along the surface of a geometric torus, or donut shape. The second motor in this case, M2, moves right along with the frame of motor M1.

Again, you are free to orient the schematic in any way you choose. The main feature is that the end point (E) can contact only those points laying on the surface of the torus. Reducing the length of member 1_1 closes up the torus, creating a figure that looks something like a pumpkin, or a donut having just a couple of depressions, rather than a hole all the way through it.

The diagram in Fig. 2-5C shows three distinctly different degrees of freedom. It is merely a two-degree system extended to include a third motor and extensor, M3 and 13.

The nice feature of a manipulator having three degrees of freedom is that its end point can be directed at any point on the surface or within the volume of the torus. It has unlimited contact-making capability within that three-dimensional torus figure. Reducing 1_1 to zero (attaching M2 directly to the shaft of M1) gives the manipulator a contact volume that resembles a sphere. The machine can touch or pick up any object within that space.

The mechanism with three degrees of freedom is clearly superior to the other two when it comes to making contact with objects in a three-dimensional space. However, the two simpler versions are not completely useless.

A manipulator having just one degree of freedom, for example, can be used as the driving mechanism for an optical scanner. If the plane of its rotation is parallel to the floor, the machine can scan the path ahead, behind and on both sides. Using the scheme with two degrees of freedom gives the optical scanner the ability to "see" above and below itself as well. Using a scheme of three degrees of freedom just for optical scanning would be a waste of effort and hardware.

Bear in mind that these are very general diagrams that are intended to illustrate the capabilities of several fundamental modes of manipulator motion. There are countless variations possible from each one of them.

To demonstrate one kind of variation, consider the manipulator scheme illustrated in Fig. 2-6. Here, two manipulators are ganged together, each having three degrees of freedom. The first section, composed of M1, M2, M3 and their respective extensors, can drive motor M4 to any point within a major torus figure. For many practical applications, the dimensions of this first section are much greater than those of the second section.

The second section is schematically identical to the first, being composed of M4, M5 and M6. This section, usually scaled to smaller dimensions, works within its own little toroidal-shaped space.

Putting these two sections together, the first and larger section

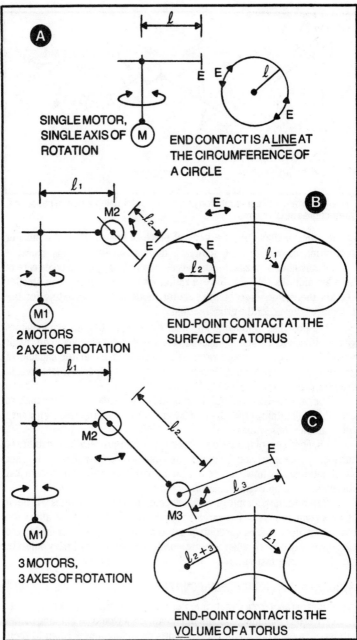

Fig. 2-5. Highly schematic views of the three basic manipulator configurations. (A) One degree of freedom. (B) Two degrees of freedom. (C) Three degrees of freedom.

35

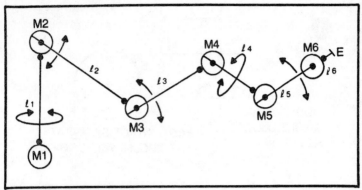

Fig. 2-6. Schematic representation of a manipulator using a pair of cascaded, three-degree mechanisms.

carries the smaller one to a general piece of space. Then the smaller section makes smaller manipulations within its own little toroidal space.

In a manner of speaking, the larger manipulator section takes care of broader motions that can have a rather sloppy tolerance. The smaller section then takes care of finer motor motions, compensating for the tolerance allowed for the gross motions.

It is possible to imagine any number of three-degree manipulators ganged in sequence and in successively smaller dimensions. The result would be a highly sophisticated and versatile manipulator capable of grasping very small objects. Paralleling some of the finer elements would create something akin to the human hand.

The only mechanical limitation on this idea is the size of the motors available. Something other than rotary motion becomes necessary for very tiny manipulator sections.

A second sort of limitation concerns the data-processing capability of the logic used for driving the motors and monitoring the behavior. It is not at all unthinkable, however, to use an entire microprocessor system just for running a complex manipulator.

No matter how simple or complex your manipulator designs will be, it is important to give some careful thought to how they might affect the mainframe design. You certainly want to minimize the risk of having the parabot/robot upset itself when attempting to manipulate a heavy object or one that is firmly fastened to something else that is heavy.

A LOOK AT THE ELECTRONIC ENVIRONMENT

Most parabot/robot experimenters are mechanical tinkerers at heart. If they weren't building parabots or robots, they would be building model airplanes, modifying motorcycle engines or adding a new room to their homes. The prospect of building a parabot or robot tends to attract people who already have some background and experience building mechanical things.

So very few robot experimenters are totally inept when it comes to planning and building the mechanical parts of the systems. Given time, most people can figure out how to mount a wheel, screw together two pieces of metal, cut some aluminum and fasten down a battery. It is highly unlikely that the prospect of building one of these systems would ever appeal to someone having absolutely no confidence and experience with mechanical matters.

The picture is totally different when it comes to the electronics. I have seen so many people jump headlong into a sophisticated parabot/robot project while knowing practically nothing about the electronics. It is only through a combination of good fortune and dogged determination that the job gets done at all.

Maybe some electronics writers, including myself, have oversold the notion that building electronic projects is a task that is possible for anyone. Indeed, "anyone" can do it, but there is a catch: The individual must have *some* background in the subject. Certainly it is possible to learn the necessary principles as the job progresses, but I am speaking of people who don't even know how to make a good solder joint, count the pins on an IC or solve Ohm's law problems with anything but the *magic triangle*.

On a different level of the same problem, some prospective experimenters who know the fundamentals of electronics have no heart for mastering the *technical* details. Such people are willing to build a power amplifier without knowing—or really caring in some cases—how it works. They will invest a lot of money in a microprocessor system, hoping that the "magic" of microprocessors will somehow make up for their lack of know-how concerning its programming language.

And then there are would-be experimenters who have no feeling at all for testing and troubleshooting electronic circuitry. Even the most competent experimenters often end up spending more time troubleshooting a circuit than building it in the first place. There is no way around Murphy's law: If something can go wrong, it will.

As a writer of a number of how-to electronics books, my severest critics are those who are unprepared fot the task of working out the circuitry. People who would not dream of building a purely mechanical system without thinking it through carefully ahead of time, will jump into an electronic system with the faith that it will all work out somehow.

The point is that you should work through all the mechanical and electronic subsystems of your intended project long before starting to build them. If you are honest with yourself, you will find the points of weakness in your own understanding and skills. And that is the time to start rounding up some reference materials and textbooks dealing with your own areas of weakness.

Most of the technical material in this book is written for experimenters having some electronics background that is roughly equivalent to that of a graduate from a post-high school vocational-technical (vo-tech) school. That doesn't mean you must be a vo-tech graduate to do the job, but when you find some of the theory going over your head, that's the time to start digging through more basic sources of information.

Here is another survey of the electronics subsystems for just about any conceivable sort of parabot/robot project. Give the notions special attention, being on the lookout for principles you might have to study further.

Some External Response Mechanisms

The most visible feature of just about any sort of parabot or robot is the thing it does. The way it moves, sounds and displays blinking lights attract the attention of the least sophisticated observers. These features fall into the general category of *external response mechanisms*.

Most robots and parabots move in some way, which means you will be working with motors. And since the system requires a portable power source in most cases, a battery of some sort must be used. Batteries are sources of DC electrical energy, so the most commonly used motors are DC motors. The logical conclusion is that you can't do much with a parabot/robot unless you know something about the nature of DC motors.

For that reason, an entire chapter in this book is devoted to the principles of DC motors and their gearing arrangements. See Chapter 4 when you're ready to take on that particular subject.

Maybe you want your system to do more than move around on the floor and move objects from one place to another. You might want the system to make noises—some as simple as a wailing or beeping sound or as complex as synthesized human speech.

Making sounds of any sort calls for a loudspeaker and audio amplifier. You have to know how such things work, how to build them and how to hook them up. You have to know something about audio oscillators, too. And if you plan to get into speech synthesis, you are bound to get yourself into a lot of trouble if you don't take the time to figure out how your chosen speech synthesizer works.

Are you going to use some lights as external response mechanisms? You will have to know how those lights work and their general specifications. Do you know exactly how to use an LED? Are you aware of the fact that a good many IC devices do not have the current-handling capacity for lighting even very small LEDs? If so, do you know how to make a current driver circuit that will light up the LED when it's supposed to light?

Motion, sound and light are the "big-three" external response mechanisms for parabot/robot machines. Emitting radio energy runs a distant fourth place. I don't know anyone who has used heating or cooling devices as response mechanisms, but it is possible to make that part of the hierarchy of external responses of the system. In fact, anything your machine does to alter the external world in the most general sense qualifies as an external response mechanism.

So define what you want your robot to do and select the external response mechanisms accordingly. At that point in your analysis, you should be getting a fairly clear picture of the sort of homework you'll have to do in order to get the job done right.

Some External Sensing Mechanisms

Generally speaking, the role of external sensing devices is just the opposite of external response mechanisms. Rather than doing something to alter the external environment, the sensing devices monitor relevant conditions.

As mentioned earlier, most parabot/robot systems include a motion-generating response system. If the system is to be classified as "intelligent" in the barest sense of the word, there must be some sensing mechanisms associated with the motion.

For example, the system must be able to know when it makes contact with another object. The contact-sensing mechanism can be as simple as a normally open pushbutton that closes by the force of the contact. Sometimes it is more desirable to sense an impending contact before it actually occurs. The sensing devices in that case can be an ultrasonic or optical sensor coupled, of course, with the appropriate source of energy—ultrasound or light.

The exact mechanism you use for sensing physical contact with objects in the external environment depends a lot on how critical the contact situation is and how sophisticated you want the system to be. You will find a variety of examples throughout this book, but you have to be the final judge in the matter. You have to understand the merits and limitations of the ideas in the context of the machine you are planning.

Maybe you have just spent a week studying external response mechanisms, and now here you go again with the sensors. Sorry, but that's the way it has to go if you have any hope of doing something out of the ordinary, such as building a working robot.

Some systems call for external sensing mechanisms other than those related to sensing contact with other objects. Position sensing can be important, especially in the case of manipulators. Microswitches, tip switches, potentiometers, optical and acoustical sensors enter the scheme at this point.

Then there is the matter of optical sensing for purposes other than sensing physical contact or the position of manipulators. On the simplest side of the matter, you might want the system to monitor the existence of light and dark. All you need there is a suitably biased phototransistor. The phototransistor conducts whenever an external light source reaches a certain level of intensity, and the phototransistor stops conducting whenever that light is not present.

On the other hand, you might want the system to be able to pick out one particular light from other, equally bright sources of light in the external environment. How can the sensing system tell one light source from another? Certainly some color filters could do the job, but a more satisfactory procedure is to have the system responsive to light that is blinking on and off at a certain frequency. And that means running the output of the photosensor to a tuned circuit, which is a circuit that responds only to signals of a predetermined frequency.

At the extreme end of optical complexity is the notion of optical pattern recognition. TV cameras can play a vital role as external sensory mechanisms here, but if you can live with a low level of resolution or you are working from a tight budget, it's possible to substitute a rectangular array of phototransistors. The larger the number of tightly packed phototransistors is in such an array, the finer the resolution of the system will be. In a practical sense, however, it cannot approach the high quality of TV-like optical scanning. *Optical pattern recognition* (OPR) is a big topic, and the sensory part of the matter is only the tip of the iceberg—a lot of computer "number crunching" has to take place in the background. If you are seriously considering a system built around the principles of OPR, make that the main thrust of all your work. It is a subject worthy of intense study in its own right.

On the other hand, if you are dreaming of building a working system that does other things not directly related to OPR, I suggest you set aside the OPR work until the basic system is up and running. The bottom line of the matter is that you might never get around to finishing the parabot or robot if its function is contingent on a sophisticated OPR scheme.

Acoustical sensing runs about the same spectrum of complexity as optical sensing. If you want the system to sense merely some changes in overall sound levels in the environment, all you need is a microphone and suitable audio amplifier. If you want the system to detect certain audio frequencies and reject others, the output of the audio amplifier will have to be coupled to an audio tuned circuit.

The most sophisticated audio sensing scheme available today is one that responds to human speech, generating patterns of computers words for various speech qualities. Just a few years ago, the notion of talking to your machine and having it understand you was a bit farfetched. It's a viable idea these days, though. Whether or not you use a speech interpreter in your system depends mainly on how important it is to the operation of the system. Speech interpretation isn't quite as involved as optical pattern recognition, but it doesn't miss by much. So before making a commitment to speech recognition, find out what is involved in terms of your own knowledge, skills and financial resources.

The "big-three" sensing mechanisms are contact or position sensors, optical sensors and acoustical sensors. There can be useful roles for temperature and radio-frequency sensors as well. You could add taste and odor sensors to the list, but unless you are thinking about the most rudimentary forms of taste and smell, these sensing mechanisms are either impractical or generally unreliable at this time.

Internal Response and Sensory Mechanisms

As described earlier in this chapter, the purpose of the internal response and sensory mechanisms is to take care of the internal needs of the machine. It can be argued that any internal mechanism ultimately has some interaction with the external environment, but that's a case of missing the point. The point of setting up this separation between external and internal response/sensory mechanisms is to help you grasp the real roles played by all the circuits outside the internal control mechanism.

A very simple parabot system might not have any internal sensing and response circuits at all. In such cases, it is up to the operator to keep a close eye on the internal well being of the machine and make any adjustments, such as changing the batteries, that are necessary for keeping the system running properly.

One good example of an internal sensing mechanism is one that monitors the voltage from the main battery. Such a system can be critical to fully automatic machines, making the difference between one that can run indefinitely and one that "dies" everytime the battery runs down.

Another kind of internal sensing mechanism monitors the status of certain motors, detecting critical stall conditions. And on an entirely different level, a high-performance system might be on the lookout for control situations that lead to self-destructive or self-defeating behavior—going crazy, in other words.

Before any brief discussion of internal response mechanisms can be meaningful, it is necessary to understand a particular feature of internal sensory/response mechanisms that further sets them apart from their external counterparts. Unlike external response and sensing mechanisms, the internal versions deal more directly with an electrical environment. One of the troublesome features of external sensing mechanisms is that they most often have to transform one kind of energy into electrical energy. The same notion, only in reverse, applies to external response mechanisms; they generally translate electrical energy into some other form of energy.

By contrast, the internal mechanisms can be wholly electrical. They monitor electrical parameters within the machine and respond with changes of an electrical nature. That, if nothing else, accounts for their relative simplicity.

So when you are working out your preliminary machine designs, make up a list of internal electrical conditions worthy of monitoring. Ask yourself what could possibly happen to the machine that would cause critical changes in its innards. How could those changes be detected? What should be done about them? Answer those questions as thoroughly as possible and you have pretty well designed the internal sensory and response mechanisms your machine will have.

More About the Internal Control Mechanism

If your machine is to have any automatic features, the success of the project rests heavily on the working of the internal control mechanism. The closer your machine approaches the definition of a true robot, the more important the internal control mechanism becomes.

Perhaps the first point to be resolved in this regard is whether or not the internal control mechanism has to be a microprocessor scheme. Strange as it may seem at first, the more complex the machine becomes, the fewer options you have about setting up the internal control.

Given a simple system that is largely under the control of a human operator, the internal control mechanism can be built around a wide variety of technologies. You can use TTL, CMOS or MOS ICs, a lot of discrete transistors, and combinations of linear and digital circuits. Having such a broad selection of device technologies, you have an equally wide choice of power-supply voltages and current levels.

As you study the jobs your internal control mechanism must do, keeping track of all the on-going activities of the machine, there comes a point where the simpler notions of logic design stop looking so simple. When you begin getting the impression that ordinary ICs cannot handle the job without creating a whole lot of problems in terms of cost, complexity and reliablity, it is time to begin thinking about using a microprocessor-based system.

And once you make the decision to use a microprocessor, you will find a lot of other options slipping away. Microprocessor devices, for example, work most effectively with low-power Schottky TTL devices. Suddenly your power-supply options come down to a single 5V supply, or a +5V supply with a low-current, +12V auxiliary. Whether or not you need the +12V supply depends on the microprocessor IC you choose. In any event, you will need the +5V for the TTL devices that support it.

Choosing to use a microprocessor in the internal control mechanism dictates the use of a bus-oriented control scheme, and that further narrows the choices as far as the nature of the interface devices are concerned. All the interface subsystems must be compatible with that microprocessor and its bus configuration.

It is very easy these days to say, "Sure. Of course I will use a microprocessor in my system." You ought to have some compelling reasons for holding that view, however. Some popular reasons don't stand up very well in the final analysis.

It is possible these days to buy a good microprocessor chip on the surplus market for less than $20. That's pretty cheap, considering you might need $50 worth of conventional ICs to do the job you want to do. What is often overlooked, though, is the fact that a microprocessor chip cannot stand alone. It requires support devices and, more significantly, outboard memory ICs. You should also have some provisions for preserving the programs in the event of a shutdown catastrophe that wipes out the RAM programming. That means you will need a cassette tape machine and an acoustical interface for getting the information swapped around between the tape and memory.

On top of that, every microprocessor system should be supported by a *firmware bootstrap*. In the event you aren't acquainted with this bit of jargon, it means there isn't any way to get a program into the system unless you have a program-loading program. The bootstrap is a little program that makes it possible to program a microprocessor, but where does the bootstrap come from, and how does it get into the system in the first place? The answer to that puzzle is some firmware in the form of a ROM (read-only memory).

Unless you are prepared to spend many hours entering a tedious bootstrap by hand from a switch panel, your system will require a ROM. And how many people do you know who have the facilities for "burning" a custom ROM for you? For every reader who can answer that question in the affirmative, I'll bet there are at least 25 who cannot.

To be sure, microprocessor chips don't cost much these days, but the necessary memories, support devices and ROMs do. Justifying the use of a microprocessor system on the basis of economy, alone, rarely stands up to close scrutiny.

It isn't my intention here to discourage the use of microprocessors in parabot/robot systems. Quite the contrary. What I am hoping you will guard against is a casual and uniformed choice that will prove discouraging in the long run.

I think the best criterion for selecting a microprocessor-oriented system is whether or not the whole project is possible without it. Maybe "feasible" is a better term to use here. Despite the cost and special problems involved, my Rodney robot could not have met its goals successfully without the use of a microprocessor. In fact, I wanted to avoid using a microprocessor, but I finally had to give in for the sake of getting the job done.

If you have now made an intelligent and impartial assessment of your system and find that you must use a microprocessor, it's time to do some more serious reckoning. Should you build the microprocessor system from scratch or use one of the popular one-board computers? If you have any misgivings about deciding to use a microprocessor in the first place, the answer to that question will settle the matter, once and for all.

Engineering and building your own microprocessor system from scratch has to be one of the most demanding and frustrating experiences in this whole business of hobby electronics. What might appear rather straightforward to a skilled technician turns out to be a monster in reality. One of the big problems is that a microprocessor system is really a *digital communications system*. The little ones and zeros that plunk out of simple digital circuits so easily have a tendency to get lost while flying around at radio frequencies in a microprocessor system. To work out a microprocessor system of your own, you have to be skilled in two distinctly different disciplines: radio technology and digital technology.

Route a couple of signal-carrying conductors just the right way, and the whole system will go haywire. If you can live with that, be my guest. Build the system from scratch.

The alternative is to buy a one-board computer, either in kit form or ready-made. Shop around a little bit to see what you can find. Some are fairly inexpensive (about $200), but don't do much in the way of helping you get the system running and maintaining it. Others are fairly expensive (between $300 and $400) and do a lot for you.

You can judge the merits of the inexpensive ones on your own. Here are some of the advantages of the more expensive models. See which ones are common to the less expensive model you might be considering.

☐ A built-in ROM bootstrap that makes it possible to start entering programs for your system right away.

☐ A built-in hexadecimal keypad that makes it possible for you to enter both addresses and data in a convenient and conventional format. There is no need for a complicated panel of toggle switches and cumbersome binary notation.

☐ An on-board, ready-to-go serial acoustical interface that permits programs to be stored and retrieved easily from a conventional cassette tape machine. It is possible to swap tapes with other experimenters in instances where the boards are identical or, in some instances, where different boards use the same acoustical format, such as the *Kansas City standard*.

☐ A built-in LED character readout that allows you to see the actual addresses and data at any point in the memory.

☐ At least 1K of program RAM—usually expandable to 16K—in the system. You won't need much, if any, additional and expensive memory for your system.

☐ Bus compatibility with many home computer systems, which means you can interface your machine with a home computer, making it possible to do the programming and even control the machine from a home computer system.

☐ A built-in ROM burner that lets you make custom ROMs for your system.

If spending something in the neighborhood of $350 for your robot project sounds a bit too rich, at least give some thought to how nice things would be if you were able to make the investment. Maybe one day you can do it.

If $200 for a basic one-board computer also sounds too expensive, you should ask yourself whether or not the effort and frustration involved in building one from scratch is worth $200 to you.

Try this suggestion. Buy just the user's manual for the ready-made microprocessor board you want to use. Study the book, learning all the ins and outs of programming the thing. By the time you know the system well enough to use it on your own, you have had a chance to save up the money for buying one of them.

Whether you build a procèssor board of your own or buy a ready-made version, you still have some say about which microprocessor chip you use. At the time of this writing, the processor marketplace is undergoing some convulsions with the introduction of 16-bit microprocessors. There are some distinct advantages in using a 16-bit data bus for Beta-Class and Gamma-Class robot machines. Parabot-Class and Alpha-Class machines do not require the same high level of number crunching, so they do not often justify using a larger chip.

Of all the available microprocessor devices, the Zilog Z80 offers the widest range of software features. For the machine-language programmer, this means fewer steps in the programs, and for the one who has to pay for the memory chips, it means less memory space for getting any job done.

The Intel 8080 series, notably the 8085, runs a close second in terms of programming features and the number of internal registers. Then there is the 6800 series from Motorola, the 6502 from Mostec, and the latecomers from Texas Instruments. Hardware simplicity is the name of the game here, and their programming procedures are fairly easy to understand. The only problem is that it often takes six or eight instructions to do a job that can be carried out with just one or two instructions when using Z80 or 8085 device.

In the final analysis, however, it doesn't make much difference which microprocessor device you choose. Like most other people in the business, you probably want to use the chip with which you are most familiar. There is nothing wrong with that—not really.

Most of the programming examples presented later in this book are built around the Z80. Some are for the 8085, though.

The matter of selecting a particular microprocessor device for your project is far less critical than the decision regarding whether you should use one or not. This discussion about preliminary designs for the internal control mechanism has been a rather extensive one, and a brief summary of comments is in order.

☐ Should you use standard logic devices or a microprocessor?

—Give the matter serious thought and don't use a microprocessor just because it's the "thing to do."

—Work out some preliminary logic designs, using standard logic elements. If the complexity of the scheme begins getting out of hand, you have at least one good reason for considering a microprocessor.

—Bear in mind that there are rarely any economic advantages in using a microprocessor. While the chips, themselves, might be fairly inexpensive, the cost of the support devices, including memories, can get out of hand.

—A microprocessor can greatly simplify the hardware, but you have to pay for that advantage with writing a lot of programs.

—In the long run, using a microprocessor cannot make up for your personal lack of understanding concerning the design, construction and maintenance of ordinary logic circuitry.

—Using a microprocessor-oriented system allows you to modify and expand the capability of your system by simply rewriting some programs, rather than by making extensive hardware modifications.

□ If you find it is necessary to use a microprocessor system, should you build it up from scratch yourself or buy a ready-made "one-board" computer?

—Building a microprocessor system up from scratch, even from someone else's plans, is a drawn-out and exceedingly tedious task. In a manner of speaking, the extra money you pay for a ready-made version is the fee you pay someone else to take care of all the hassle for you. Is it worth it? Only you can answer that question.

—Ready-made microprocessor boards generally include a lot of special features that will make your project more fun and easier to use. Incorporating such extras in a from-scratch operation might exceed your own level of skill.

—Using standard, ready-made micro boards offers a better opportunity to swap programs and parts with other experimenters.

□ If you find it necessary to use a microprocessor, which one should you use?

—The Z80 is the most powerful processor chip available. This means a considerable savings in programming effort and RAM space.

—The Intel 8080 series, especially the 8085, runs a close second to the Z80 in overall performance.

—The lines of microprocessors represented by Mostec, Motorola and Texas Instruments have fewer software options and working registers, but that does not eliminate them from the race.

—The newer 16-bit processors can speed up operations for Beta-Class and Gamma-Class robot machines, but of course they cost more than their 8-bit counterparts.

Internal and External Power Supplies

It is unwise to select the power-supply subsystems until you have made the first decisions concerning the nature of all the other subsystems. The specifications for the motors, lamps, semiconductors, relays and so on all dictate the voltage and current ratings of the power supplies.

Unless you are building a fairly small system, it is generally a good idea to count on at least two internal power supplies, one for the motors and another for the rest of the circuitry. Motors, you see, have a nasty habit of loading their power source with a lot of electrical noise, and that noise can play havoc with ICs elsewhere in the system. Microprocessors are especially vulnerable, and memories aren't much better when it comes to noise immunity.

The voltage rating of the main motor battery must match that of the motors it drives. That usually means 6V or 12V for larger motors, and 3V, 6V, 9V or 12V for smaller ones. You can certainly simplify matters by selecting motors that have the same voltage rating.

As far as the current rating for the motor battery is concerned, it should be able to deliver full stall current to all motors at the same time. The chances of having all the motors stall or start at precisely the same

46

time are remote, but you certainly don't want the system to bog down if that event should ever occur.

The ampere-hour rating of the main battery supply should be selected so that the system can run at full power for a reasonable length of time. What is a reasonable length of time? My own rule-of-thumb on that matter happens to be four hours. That might sound like a terribly short time to you, but look around at the size and weight of the battery necessary for running your system any longer than that. The detailed discussions about batteries in Chapter 3 will help you in this regard.

You must give the same sort of consideration to a power supply for all the logic and control circuitry. This power supply will usually be smaller than the one needed for the motors, but if you are new to this business of building parabot/robot machines, you might be surprised at the large amount of power necessary for operating all those innocent-looking IC devices.

The logic power supply should have a voltage rating equal to that of the ICs. Sometimes that means using a 12V battery, followed by a 5V regulator. The regulated 5V portion is a must for TTL circuits, and a lot of situations call for the 12V output as well.

To get a realistic figure for the current rating, assume that every logic IC will draw about 10 mA from the power supply. That's a ballpark figure, of course, but it works out in most cases. If your system has a microprocessor, a couple of support chips and 8 memory ICs, you are already up to 1A of logic current—or pretty close to that much, anyway.

The ampere-hour rating of the logic power supply should at least match that of the battery for the motors. The smart thing to do is give the logic power supply a somewhat larger ampere-hour rating. That way, the machine will still have some decision-making capability in the event the motor power supply runs down.

And finally, you might want to include a highly reliable, separate power supply for the memory circuits of the system. The last thing you want to happen with a microprocessor-based system is to lose the memory.

With at least one, and as many as three, batteries carried on board the machine, there can be no question about the matter of an external power supply. Its first function is to recharge those on-board batteries, a matter that cannot be treated lightly. You must use a carefully planned battery-charging scheme, being especially careful to avoid recharging the batteries too quickly. If you have ever had the experience of cleaning electrolyte out of the delicate workings of a robot, you have learned your lesson about recharging batteries with too much current.

Another Note About the External Control Mechanism

The external control mechanism can be a blessing or a curse, depending on your objectives and how you handle this particular subsystem. If you are building a true, robot-class machine, the external control mechanism plays one kind of role—a secondary role, at that. On the other hand, the external control mechanism is a primary control device in the case of parabot machines or operations.

For true robot-class machines, the external control mechanism ought to do little more than the following:

☐ Provide an input point for the basic operating programs. Sources of these programs include manual switches, prerecorded programs on cassette tape and output connectors from a home computer system.

☐ Provide an input point for special test and troubleshooting operations.

☐ Provide an output point for monitoring or recording the internal behavior of the machine, including memory scans for Beta-Class and Gamma-Class robots.

☐ Provide an input point for carefully selected signals that override normal internal functions.

In the case of parabot machines, the external control mechanism is the main input point for all system controls. The information might come from the human operator, prerecorded sequences of operations, or commands from a small computer terminal. It is also the output point for monitoring and recording all ongoing activity.

Batteries and Power Supplies

3

While it is altogether possible to build some fine parabot/robot projects that are operated directly from standard utility power sources, not many experimenters care for the idea of having their machines limited in their motion by a long line cord. That, of course, means that most experimenters like to operate their machines from on-board batteries.

Just sticking some batteries into your unit doesn't end the matter, however. Along with the batteries comes the need for a battery recharging scheme. And then you must distribute the on-board battery power, sometimes reducing and regulating it, throughout the system. Finally, you should be made aware of the importance of having a backup power source, an auxiliary power supply.

A complete power source scheme for a parabot or robot project can thus be divided into four basic sections:

- ☐ One or more on-board batteries.
- ☐ A suitable battery charger.
- ☐ An on-board power distribution and control scheme.
- ☐ An auxiliary power supply.

The on-board battery (or batteries) supplies electrical power to all of the motors, relays, indicator lamps and electronic circuits of the machine. All electrical components in a self-contained machine will operate from the battery supply.

The simplest sort of battery system is one that has voltage and current specifications compatible with the voltage and current requirements of the machine. Deciding to operate the entire system from a single battery can bypass a lot of problems cited later in this chapter.

But it is also possible to make a good case for using more than one battery. Give some serious thought to having one battery for operating the higher-current electromechanical devices (namely motors and relays) and a second battery for powering the noise-sensitive electronic circuits.

DC motors and digital ICs, especially microprocessor devices and CMOS logic circuits, do not make very good bedfellows. Motors can draw a great deal of current from a battery, creating a low-voltage condition every time they start and stall. Indeed, the motors in parabot and robot machines undergo a lot of starting and stalling.

Power surges on the battery supply can upset digital ICs and, at worst, destroy them. Even isolating the electronic circuits from the battery with a voltage regulator is no guarantee that there will be no interference under worst-case conditions.

When running through your preliminary design, then, think about using one battery that matches the requirements of the motors, and at least one more for the electronic devices. That is a case of power supply isolation at its best.

As far as battery chargers are concerned, the matter is best handled by buying a commercial version, provided it matches the characteristics of the batteries being used. For some reason, many experimenters try to save a few dollars by using battery-charging scheme that is wholly inappropriate.

The battery charger must match the specifications of the batteries. All too often, experimenters see nothing wrong with recharging batteries from a standard DC power supply or any old battery charger they happen to have at hand. The result is that the batteries are often charged much faster than they should be, warping the plates, upsetting the chemical reaction and, in short, quickly destroying the batteries.

A power distribution system includes fuses, voltage regulators, terminal blocks and carefully selected wiring. The fuses, of course, protect the system in the event of a catastrophic short-circuit condition; a condition that would otherwise "smoke" a lot of circuitry or, worse, set the machine on fire. Some batteries can deliver an incredible amount of current to a short circuit, and the fuses in the power distribution system prevent subsequent disasters from occurring.

Most of the ICs you will be using in your project are rather fussy about power-supply regulation. Motors, relays and some less critical electronic circuits can operate directly from the nonregulated battery supply, but most of the digital circuits and precision linear devices must be powered through voltage regulators. For our immediate purposes, those regulators will be considered part of the power distribution system.

Selecting connectors, terminal blocks and wiring for the power distribution system is neither an arbitrary nor a trivial task. If you have any thought of using ordinary hookup wire for the power distribution system, you might be flirting with some disappointments. Ordinary hookup wire cannot handle the current surges for DC motors. The same can be said for solder connections. So when it comes to distributing battery power, you will need some heavier gage wire as well as solderless terminals and connectors of the appropriate sizes.

An auxiliary, bench-type power supply can be important while you are working on the machine. The idea is to provide a reliable source of DC power while the system is "on the bench." This source temporarily replaces any battery you might use while the system is fully operational.

With an auxiliary power supply, you can keep the electronics fired up indefinitely while working on the system. Otherwise, a gradually declining battery voltage can upset all tests and calibration procedures.

It is sometimes possible to use the battery charger as an auxiliary power supply. You have to be careful, however, because a few battery chargers have filtered outputs, and the better models regulate their output currents with an SCR or triac switching scheme. That makes for a nice battery charger but a lousy DC power supply.

The safest bet, as far as auxiliary power supplies are concerned is to buy or build a nice regulated model. That is at least one element of your robot system you can later use for other kinds of electronic ventures.

You are probably now impressed with the notion that the battery and power supply system for your prospective parabot/robot machine is not a matter to be taken lightly. It is just as important as the more romantic processor systems in your unit, and it deserves equal attention from the outset.

SELECTING BATTERIES

The world abounds with all sorts of batteries these days. Thomas Edison would be delighted with the recent resurgence of interest in DC power sources. There are miniature batteries for electronic wristwatches and hearing aids, medium-sized batteries for electronic calculators and toys, a bulky, hight-current batteries for golf carts and electric automobiles.

With such a wide selection of batteries on the market today, a robot experimenter might be justifiably confused at first. Which kind of battery is best for my system?

Here is a brief summary of the most common kinds of batteries and their main characteristics:

☐ **Carbon-zinc dry cells and batteries**—low cost, available in a number of different voltage ranges, readily available, relatively low current, moderate useful lifetime, nonrechargeable.

☐ **Alkaline dry cells**—similar to carbon-zinc batteries, but slightly more expensive, slightly longer life expectancy.

☐ **Mercury cells**—fairly high cost, very small selection of voltage ratings, relatively low current, extended useful life (if used properly), *must not be recharged*.

☐ **Nickel-cadmium cells and batteries**—fairly high cost, relatively narrow range of voltage specifications, fully rechargeable, moderately high current ratings, indefinite useful lifetime (if treated properly).

☐ **Lead-acid batteries**—fairly low cost, relatively narrow range of voltage specifications, fully rechargeable, available with very high current ratings, very long life expectancy (if treated properly), require routine maintenance.

☐ **Gel-cell batteries**—very high cost, relatively narrow range of voltage specifications, fully rechargeable, moderately high current ratings, indefinite useful lifetime, require no maintenance.

While this summary is hardly adequate for making an intelligence decision regarding the selection of batteries for a parabot/robot project, it does point out some features that can at least eliminate some possibilities.

Carbon-zinc batteries, for example, drop off the list of possibilities right away. Being not rechargeable kills them as far as parabots and robots are concerned. It is true that you can put a carbon-zinc battery into one of those little commercial battery rechargers and then get a little more "juice" from them. But you can do the same thing by warming them in the sun. Warming carbon-zinc batteries tends to reduce their internal resistance for a short time, leading one into thinking they have recharged the battery. And putting them into one of those "rechargers" doesn't recharge them; it merely warms them up a little bit. Besides, that trick works only a couple of times. Exit the idea of using carbon-zinc batteries, even if they are cheap.

Alkaline dry cells are quite similar in performance to carbon/zinc batteries. While they do have a longer life expectancy, they aren't truly rechargeable either. That, plus it would take lot of them connected in parallel to supply a decent level of motor current, eliminates them from the picture.

Mercury cells have the excellent ability to maintain their rated voltage almost to the moment they completely die. You might want to consider mercury cells for those special operations calling for a steady reference voltage, but they are wholly unsuitable for most parabot/robot applications.

Incidentally, you should *never attempt to recharge a mercury cell.* Recharge current heats them up rather dramatically, running the risk of an explosion. That's why you aren't supposed to throw them into a fire.

So what about mercury cells? They may be OK if you want to hold a very close tolerance on some reference voltages. In such instances, keep the current drain in the low-milliampere level, and be prepared to change them according to a fixed maintenance schedule. Otherwise, forget about them.

What about nickel-cadmium (nicad) cells and batteries? Now we are getting somewhere. These are very good batteries. They are clean (no acid mess), require no maintenance, and are built to be fully recharged as often as necessary. But the picture isn't altogether a perfect one.

For one thing, nicads tend to have relatively low current capacities. They are certainly better than carbon-zinc, alkaline and mercury cells in this respect, but you have to count on paralleling several of them to get enough current for running moderately sized motors. That, coupled with their relative expense, makes them a questionable choice.

You might consider a nicad or two when it comes to a power source for low-current electronic circuits. If you are planning to use separate batteries for the motors and electronic circuits, nicads certainly hold their own for the electronic circuit supply.

No matter where they are used in your system, nicads must be discharged and recharged within rather tight specifications. Discharging or recharging them too quickly sets up the wrong sort of chemical reaction within them. That reaction generates a gas, and since the cells are sealed, there is a chance of a nasty, little explosion.

It isn't unusual to find nicads that have discharge limits of 1 or 2 amperes. That's nice. But as a rule-of-thumb, a nicad should never be recharged at a rate greater than 10 percent of its maximum discharge rate. So a 1A nicad should not be recharged with a current exceeding 100 mA—and that's the catch. Try recharging a nicad battery from a battery charger that doesn't have built-in current limiting, and you're in for some trouble.

Have you ever inspected the "cheap" transformer in one of those recharging units for nicad-operated calculators and home appliances? It's a little transformer wound with fine wire in such a way that it has to be inefficient. It is made inefficient on purpose. That provides the automatic current limiting. The heat it generates is a sign that it is doing its job properly.

Nicads do not current-limit themselves. The current limiting must be provided by the battery charger system, and the level of current limiting must match the battery specifications.

If that seems all to complicated for you, you aren't alone. A lot of experimenters have given up on nicads for that reason. In short, a nicad battery is a prime candidate for a battery that supplies a couple of amperes of current for IC devices—provided it is matched with the right, current-limiting recharger.

Incidentally, there is another reason why nicads aren't suitable for the main motor batteries. Nicads, like mercury cells, tend to hold up their rated voltage level until they are just about exhausted and ready for a recharge. If you plan to use a low-voltage sensing scheme for your robot, the circuit will not detect the low-voltage condition until it is too late for the machine to find its way back to the battery charger.

Lead-acid batteries are the most widely used kind of battery in today's parabot/robot workshops. Certainly, the lead-acid battery is a crude battery compared to the nicad and gel-cell, but it can pack a real high-current punch and is a lot more forgiving when it comes to occasional abuse.

Lead-acid batteries require constant attention, making certain the electrolyte remains above the plates. On the other hand, though, they are relatively inexpensive and widely available.

I have built every one of my working robot projects around lead-acid motorcycle batteries, which are small and light enough for the job. They are very reliable if handled properly, and I have yet to burn a hole in the carpet with any spilled electrolyte (not to say that couldn't happen, though).

Lead-acid cells can be deep discharged (run with excessive discharge current) or deep charged (recharged with excessive current) on occasion, but you do not make a habit of it. If the current capacity of the battery is selected to suit your system's worst-case current demand, you won't have any trouble with deep discharging. It is up to you , however, to make sure the battery charger doesn't deep charge the battery very often.

The rule-of-thumb for recharging a lead-acid battery is to keep that recharge current about 10 percent of its rated discharge current. This means one of two things: Either you have to buy a battery charger that matches the current rating of your battery or use one of those adjustable current-limiting devices that are available at most stores that carry batteries.

To be sure, a service station might deep charge your auto battery if you are in a hurry. As mentioned earlier, that is all right on occasion. But anyone who knows what they are doing will recommend an overnight *trickle charge.* That's a case of recharging the battery at its rated recharge specification.

Give some thought to using a marine battery—a class of lead-acid battery that is specially constructed for routine deep charging. Such batteries are rather large, heavy and expensive. That might be a rather high price to pay for your impatience. The next section in this chapter is devoted exclusively to selecting lead-acid batteries for parabot/robot projects, so there is no need for further details at this point.

Finally, there is the gel-cell battery. It is a newcomer to the battery world, and some would say that it combines the positive features of nicad and lead-acid batteries, while retaining few of their negative features.

While gel-cell batteries have a great future in robotics, the right time hasn't arrived yet. The things are simply too costly for most experiment- er's budgets. The fact that gel-cells do not have the same high level of current capacity aggravates the cost problem; you might need several of them to do the job right.

Gel-cells are just beginning to appear on the surplus market at tolerable prices, but a closer inspection of the matter shows that you need a number of them to get a reasonable current capacity. Thus the cost, even at surplus prices, quickly exceeds that of a single lead-acid battery.

When the price is right, most robot experimenters will probably jump at the chance to convert from lead-acid to gel-cell batteries. In the meantime, we'll just have to keep an eye on electrolyte levels, take some precautions to avoid spilling acid on the floor, and put up with that little bit of hydrogen gas that bubbles out of the lead-acid batteries.

PICKING THE RIGHT LEAD-ACID BATTERY FOR YOUR PROJECT

The following discussions about selecting battery specifications apply specifically to lead-acid batteries. It will be possible to update the information rather easily when gel batteries become more economically feasible.

Lead-Acid Voltage and Capacity Specifications

Lead-acid batteries are specified according to two ratings: *full-charge output voltage* and *current-delivering capacity*. Determining the right voltage rating for your project is a rather simple task, but figuring out the necessary current capacity is something else.

As far as the voltage rating is concerned, most robot and parabot projects call for either 6V or 12V, depending mainly on the voltage rating of the motors and electronic power supplies. Simply select a 6V battery when using 6V motors, or a 12V battery when using 12V motors.

A lead-acid battery outputs its rated voltage whenever it is fully charged. A discharged battery, on the other hand, is one that outputs no more than 80 percent of its full-charge rating. A "dead" 6V battery, for instance, is one that shows only 4.8V when fully loaded (connected to the circuit), and a "dead" 12V battery shows only about 9.6V when fully loaded.

The battery voltage, by the way, should be checked only under full-load conditions. Even a "dead" battery can show a healthy output voltage when it isn't delivering current to an external load.

Batteries should never be routinely discharged below their 80-percent, dead voltage level. It can happen every once in a while; but routinely discharging them below that level causes the electrolyte levels to drop drastically, the batteries tend to heat up and generate excessive amounts of hydrogen gas when recharging, and the plates show a very rapid rate of corrosion.

So much for a look at battery voltage ratings. Now for the current capacity figures.

Most lead-acid batteries are specified according to their ampere-hour rating. This figure is an indication of their current-delivering capacity, expressed as the mathematical product of some amount of current and the amount of time the battery delivers that current before going "dead."

By way of a specific example, suppose you find the specifications for a 12V battery having an ampere-hour rating of 10 A-h. This means that battery can supply maybe 10A for 1 hour before its voltage drops to the 80 percent "dead" level, or 9.6V. Or the same battery can supply 1A for 10 hours, 2A for 5 hours, and so on. Of course this assumes the battery is fully charged at the beginning of the timing interval.

Auto batteries are sometimes specified in a slightly different fashion, showing how long they can provide 25A before going dead. These figures can be easily converted to a standard ampere-hour rating by converting the time into hours and multiplying the result by 25A.

Suppose, for instance, a catalog shows that a certain auto battery can supply 25A for 130 minutes. Converting the minutes to hours (by dividing by 60), we get about 2.2 hours. Multiplying by the 25A standard discharge rate, the ampere-hour rating turns out to be rather close to 54 a-h. This particular auto battery will supply 54A for 1 hour, 1A for 54 hours, 27A for 2 hours, etc. The *cold-cranking current* specified for auto batteries isn't really relevant to parabot and robot projects, unless you happen to think your machine will get stuck on some ice in −20°F weather.

Ampere-hour ratings of batteries are usually determined by the battery engineers when they try discharging the battery for a standard 10-hour interval. So when testing a 12 A-h battery, the engineer will drain 1.2A from it for 10 hours to see how close the output voltage is to the 80-percent, full-charge voltage level.

This 10-hour magic number is indicated in the battery catalogs as the *10-hour discharge rate*. The practical significance of this figure is that *the battery should not be routinely recharged at a current any higher than its 10-hour discharge rate*. When the specifications state a battery has a 12 A-h capacity at a 10-hour discharge rate, for example, that means it should not be recharged at a rate any higher than 1.2A—12 A-h divided by 10 hours. Routinely recharging a battery too rapidly can destroy it in a rather short time.

A few catalogs list 20-hour discharge rates. Determining the maximum recharge current in this case is a matter of dividing the ampere-hour rating by 20 hours, instead of 10.

Bearing in mind these facts about battery capacity and recharge ratings, the next step is to figure out an ampere-hour rating most appropriate for your own project. There are two main parts to this procedure: First, determine the average current demand of the robot; then settle for some maximum amount of continuous running time.

Estimating the Average Current Demand

Most robot systems today have two main sections that draw current from the battery: the DC motor section and the section composed of all the electronic circuitry. The average current drain on a single battery is the sum of the average motor current and the current consumed by the electronics. If you are planning to use two separate batteries, one for each major section of the system, you still have to calculate the average current drains for those two sections. You don't sum them together, however.

If you already have the motors at hand, you are in an excellent position to make some good estimates concerning average motor current. First, find the start/stall current (I_p) of the motors, either from the manufacturer's literature or by measuring the current while stalling the motors by hand. (More details are in Chapter 4.)

Then find the running current level (I_r) of the motor by loading the shaft with the normal torque force of the motor and measuring the motor current. If you have no idea what that torque force will be, simply load down the motor shaft by hand until it slows to about 75 percent of its no-load running speed. In any case, you must run these tests from a good voltage supply, preferably the battery that will ultimately operate them in the working machine.

If more than one motor is being used, add together the stall currents to get the maximum, total start/stall current level. Likewise, add together the average, nominal running currents to get the total running current.

These motors aren't going to be stalled all the time, but neither are they going to be running smoothly all the time. And this is where your first ballpark estimate enters the picture.

Under most parabot/robot operating conditions, the motors spend no more than 10 percent of the time starting up or in a stall condition. On the average, then, the motors draw stall-current levels about one minute in 10.

That means the motors must be spending 90 percent of the time either turned off or running at their normal load. To be conservative, forget about the time the motors are completely turned off and assume they are operating at their nominal running current 90 percent of the time.

The following equation shows how to use these estimates to come up with an overall, average motor current demand:

$$I_m = 0.1I_p + 0.9I_r \qquad \textbf{Equation 3-1}$$

where I_m = the average current required for the motors, I_p = the sum of the stall currents for the motors, and I_r = the sum of the nominal running currents for the motors.

Using a specific example, suppose you are working with a pair of identical motors that are to be used in a two-motor, drive/steer system. You find in both cases that the stall current is 12A, and the nominal running current is 3A. Summing the stall currents, you get 24A, and summing the running currents, you get 6A. Those are the figures needed for solving Equation 3-1: $I_m = 0.1(24A) + 0.9(6A) = 7.8A$.

Being conservative in your estimates, you can expect that drive/steer system to draw a long-term, average current of 7.8A. The peak demand might run as high as 24A, but the long-term figure is more important at this point in the analysis.

If you have not yet selected the motors for your project, you are going to have some trouble with this analysis. See Chapter 4 for some discussions that will help you out of the quandary.

So much for the motor current. Now for the electronic circuits that control everything the machine will do.

If your circuits are already finished, you can measure the current drain directly with an ammeter. Unfortunately, it isn't easy to make close estimates of the current drain if the circuits aren't built yet. It is possible, however, to play around with some ideas that lead to a ballpark figure.

Few parabot/robot projects that do anything significant draw any less than 1A from the power supply. You should thus figure at least 1A, even for the simplest systems. The more sophisticated the electronics you anticipate using, the larger that estimate should be, perhaps up to 6A or 8A for rather large systems.

Another way to approach the matter of estimating current drain for the electronics is to first estimate the number of IC devices the system will use. Just come up with an order of magnitude, anyway—10, 20, 50, or whatever seems reasonable to you. Then figure about 1A of supply current for every 15 IC devices, and toss in an ampere or two for special interfacing circuits.

Does that all sound a little crazy? Sure, it's very much like telling someone how to bake a cake by saying, "Put in a pinch of this, a dab of that, a handful of this and a smattering of that." Who in the world can work from a recipe like that? A highly experienced chef can. A beginner has a lot of trouble with it.

The same applies to estimating the current drain of a circuit you haven't even designed yet. The more experience you have, the more comfortable you become with rule-of-thumb estimates. And if you are comfortable with the estimates, it's because you've made them work before.

But if you're unsure about this, your alternative is to build the circuits first. Then come back to this section with a nice figure for circuit current drain.

No matter how you come up with an average circuit current drain, assign it to I_s in Equation 3-2. That equation provides the overall, average current demand for a single-battery power supply.

$$I_d = I_m + I_s \qquad \textbf{Equation 3-2}$$

where I_d = the overall, average current demand of a single-battery system, I_m = the average current required for the motors (see Equation 3-1), and I_s = the estimated current required for the electronic circuit.

If you are using separate batteries for the motors and electronics, I_m is the overall current demand for the battery that is to operate the motors, and I_s the overall current demand for the battery operating the electronic circuits. Applying Equation 3-2 to a specific example, suppose you have found the average motor current (from Equation 3-1) to be on the order of 4.5A. The current for the electronics is about 1.2A. Working these figures into Equation 3-2, the total battery current demand averages out to be on the order of 5.7A. Round up to 6A to cover yourself in the event one of your estimates was on the low side.

Estimating Minimum Battery Ampere-Hour Rating

Knowing the total average current drain of the system is not very helpful without some additional information concerning how long the battery should supply that current level before going dead. What you need to select a battery is a minimum ampere-hour rating. Multiply the total average current drain of the system by the length of time you want the machine to operate without a recharge. See Equation 3-3.

$$H_d = I_d \times T_d \qquad \text{Equation 3-3}$$

where H_d = the minimum ampere-hour rating of the battery, I_d = total average current drain (from Equation 3-2), and T_d = maximum desired running time in hours.

For example, suppose your system draws an average of 10A, and you want the machine to run at least three hours before requiring a recharge. Equation 3-3 shows you need a battery having an ampere-hour rating of at least 30 A-h.

You can bet some overly enthusiastic robot experimenters want their machines to run maybe eight hours at a time without stopping for a recharge. Lots of luck to them. Anyone trying to do that with a 10A robot system will need a battery having an 80 A-h rating. It could be done, of course, by paralleling two 40 A-h auto batteries, each weighing about 45 pounds. The cost isn't bad, probably less than $100. But look at the *weight* of the machine.

Being overly optimistic about the continuous running time of your machine can get you into a lot of trouble. Long running times mean big batteries; big batteries mean a lot of weight; a lot of weight means bigger motors; bigger motors mean more current drain; and more current drain means bigger batteries. The unending, irrational loop is established.

Do whatever you want in terms of maximum, continuous operating time, but keep the figure at two hours or less. I don't like this situation any better than anyone else, but it is one of those nasty compromises we all must live with until something better comes along.

So far, this discussion has brought us to the point of calculating the minimum ampere-hour rating of the battery. Get a catalog that lists lead-acid batteries by voltage and ampere-hour, or one of the ampere-hour equivalents described earlier. Try a *Sears, Roebuck and Co. catalog* if you want to see a lot of different batteries listed together.

For instance, a Harley-Davidson 12V, 9 A-h battery weighs only eight pounds and costs less than $30. Such a battery would be nice for a system calling for a minimum ampere-hour rating on the order of 7 or 8 A-h. Or to allow a bit of extra power for emergency situations, go for a 12 A-h Honda battery that weights just 10 pounds and costs about $2 more than the 9 A-h version. You can get all sorts of valuable specifications, including physical dimensions, from a good battery catalog.

SELECTING BATTERY CHARGER SYSTEMS

Batteries, whether they are lead-acid, nickel-cadmium or gel types, must be recharged properly if they are to serve a useful lifetime. In a practical sense, proper recharging is a matter of making sure the battery is recharged at a rate no faster than the manufacturer specifies for the battery you have at hand.

Calculating Maximum Recharge Current

Some battery charts list the nominal recharge rate in terms of a current level. If that is the case in your situation, you won't have to make any further calculations. More often than not, however, you have to figure the recharge current rate yourself.

Strictly speaking, a lead-acid battery must not be routinely recharged at a rate that exceeds its 10-hour discharge rate for a long period of time. As mentioned earlier in this chapter, some batteries are rated according to their 10-hour discharge rate. That makes it easy to figure the maximum, continuous recharge current; the two figures are identical. The same notion applies to nicads and gel batteries as well.

Suppose you have a battery with a specified 10-hour discharge rate of 2.5A (that would be a 25 A-h battery, by the way). That battery should not spend a great deal of time being recharged at a rate greater than 2.5A. If your battery charger happens to have an adjustable *trickle current* feature, it should be set for that 2.5A level.

In most instances, however, the battery specifications include only the ampere-hour figure; the recharge current and 10-hour discharge ratings are not shown. But as mentioned earlier, most ampere-hour ratings are based on a 10-hour discharge rate, and coming up with that 10-hour rate is a simple matter of dividing the ampere-hour rating by 10. Once you get the 10-hour discharge rate, you are also looking at the nominal recharge current.

A battery cited earlier in this chapter had an ampere-hour rating of 9 A-h. Dividing by 10, it turns out that its 10-hour discharge rate is 0.9A. The nominal recharge current is that same figure. So that particular battery should not be recharged at a current much greater than 0.9A for an extended period of time. Doing so will ruin it after a while. In the case of nicads, excessive recharge current can cause an explosion.

Battery Recharge Characteristics

Figure 3-1 shows an idealized recharge curve for a typical battery. Actual times and currents are not shown for the simple reason that they vary so much according to the battery specifications and the type of recharger being used.

One important feature of the curve is the *initial surge current*. When the charger is first applied to the battery, there will always be an initial, short-term surge of recharge current. The peak amount of current depends on the characteristics of the charger and how dead the battery is at the time. That initial surge current can exceed the nominal recharge rate of the battery by a factor of 4 or 5, but if the battery is a good one, that surge current will drop off rather rapidly—within five or 10 seconds in most cases.

60

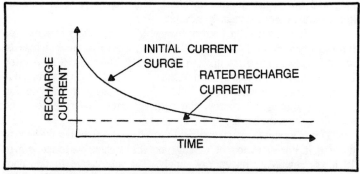

Fig. 3-1. Charge curve for a basic battery and battery-charger system.

The curve shows that the recharge current settles down to a fairly steady level and, in fact, remains at that level for an extended period of time. That level should not exceed the 10-hour discharge rate of the battery.

Assuming the battery is a good one, it will tend to control its own recharge rate. One symptom of a worn-out battery is that it draws the initial surge current for a longer period of time, perhaps indefinitely. So if you don't see that initial spike of current, then a fairly rapid decline to the nominal recharge rate, you can bet the time for buying a new battery is at hand.

In practice, a battery will never show zero recharge current. Unless you have a battery charger with an automatic low-current trip circuit, you will always see some recharge current flowing; even after the battery is fully charged. The main reason is that the charger usually outputs a voltage that is slightly higher than the voltage rating of the battery. The two sources are never balanced, and current always flows from the higher to the lower voltage source, or from the charger to the battery.

So you cannot count on seeing zero recharge current and using that situation as a guide to ending a recharge cycle. Rather, you should consider a recharge current between one-half and three-quarters the nominal, long-term recharge current level. A battery that settles down to a 1A charging rate can be considered fully charged when it draws between 0.5A and 0.75A from the charger.

Actually, this is just a rule-of-thumb principle. The actual full-charge condition is one where the density of the electrolyte reaches a certain level. Few battery specifications list that density figure, and even fewer experimenters care to mess around with a special hydrometer every time they want to recharge a battery.

How long does it take to recharge a good battery? That's a good question, because its answer leads to the overall *operating duty cycle* of the machine—the ratio of running time to the sum of running time and battery recharging time. For most experimenters, the ideal situation is one where the operating duty cycle is as large as possible, where the running time is long and the recharge time is short. Unfortunately, it is virtually impossible to reach the ideal duty cycles.

The time spent at the battery charger is always longer than the free running time. You can count on it taking at least five times longer to properly recharge a battery than to discharge it. The figure is more often 10 times longer. So if your system discharges the battery in two hours, it is going to take between 10 and 20 hours to bring it back to full charge—assuming, of course, the recharge rate doesn't exceed the specifications of the battery.

This is a rather sorry situation. I don't like it, either, but we have to live with it. For every hour of running time, the machine has to spend up to 10 hours rebuilding its energy supply. The alternative to this poor operating duty cycle is to have more than one battery available, with a couple on the battery charger all the time.

A Few Notes About Battery Chargers

Battery chargers are normally specified according to the voltage rating of the batteries they can recharge and a maximum charge current. A commercial charger rated at 12V, 6A, for example, is suitable for charging 12V batteries that can withstand 6A initial surge currents. Figure 3-2 shows the diagram for a typical battery charger. You can clearly see that it doesn't function as an ordinary power supply, and it should not be used as such.

The center-tapped transformer, T1, and diodes D1 and D2 make up a full-wave rectifier circuit. The voltage waveform appearing at the anode of SCR1 is thus a full-wave rectified waveform having a peak value of about 18V (this happens to be a 12V charger circuit). There is no attempt to filter the waveform; in fact, this charger would not work properly if the waveform were filtered.

Assuming for the moment that SCR1 is firing through the full 180 degrees of each half-cycle, you should be able to see that the battery being recharged is subjected to a rather large amount of recharge current. The circuit is blasting it with a voltage that is well above the 12V battery rating, and you can bet that it is recharging quite rapidly.

The recharge current in this case is limited mainly by the current-handling capability of the power transformer. The difference between the voltage at the anode of SCR1 and the battery voltage also plays a role in the amount of recharge current, but it is mainly the power transformer that limits the initial surge current.

If the charger is rated at 4A, the power transformer cannot be expected to deliver an initial current in excess of that amount. A battery having a 10-hour discharge rate of 3 or 4 amperes will have to set at that charger for a very long time. To get reasonable recharge times, the current rating of the charger should be four or five times greater than the nominal recharge current of the battery. That 4A charger would work quite

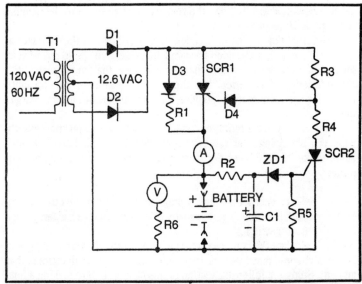

Fig. 3-2. A typical battery charger circuit that features automatic current regulation.

nicely for a battery having a 10-hour discharge figure on the order of 1A (a 10 A-h battery).

Because the circuit in Fig. 3-2 slams the battery with a rather high voltage at first, there has to be some provisions for backing off the charge rate to match the standard charging curve. This is accomplished by firing SCR1 later in each half-cycle of the applied waveform.

SCR1 is gated on through the circuit consisting of D4 and R3. Whenever SCR2 fires, however, it overrides that gating circuit, preventing it from triggering SCR1. So understanding how SCR2 fires is important to understanding how SCR1 does.

As long as the battery voltage is below the voltage rating of zener diode ZD1, SCR2 cannot fire. And as long as SCR2 cannot fire, SCR1 is allowed to fire through most of each half-cycle of the rectified waveform. That means super-maximum recharge current for the battery.

As the battery voltage rises, ZD1 begins to conduct, and SCR2 begins firing at the 90-degree point. SCR1 can still fire earlier than that, still applying a lot of power to the battery. But as the battery voltage continues to rise, SCR2 fires earlier on each half-cycle, eventually firing before SCR1 does. And when that happens, the main power surges through SCR1 cease.

The only recharge current is the one allowed to flow through D3 and R1. That would be the *trickle-charge* current of the charger.

Making R2 a variable resistor provides an adjustable recharge rate. If your battery charger happens to have this sort of adjustment, it should be set to a power where it allows the recharge rate to settle to the nominal level within 10 seconds or so.

The circuit in Fig. 3-2 is intended merely to illustrate the character of battery charger circuits. In no way am I suggesting you build one of your own. Not only is a battery charger a device that must match the characteristics of batteries, but the power transformer can be quite expensive—usually more expensive that a commercial battery charger unit.

Naturally, you can build your own battery charger if you know exactly what you are doing. For the benefit of everyone else, here is a brief summary of items to consider when buying one for your robot/parabot project.

☐ Select a charger having a voltage rating that matches the battery system in your machine. This is usually 6V or 12V. Some chargers have a 6V/12V selector switch.

☐ Calculate the nominal recharge current rate for your battery, and select a charger that has a current rating about four or five times that amount. Study the following section of this chapter before making a final choice, however.

☐ Select a charger that has both a voltmeter and ammeter included in it.

☐ Charge-rate and trickle-charge adjustments can be important in instances where the charger tends to overcharge the battery. They are not absolutely necessary, however.

Power-On Charging and Charging More Than One Battery

The previous discussions concerning batteries and battery charging assume the ideal situation: one where a single battery is charged while it is disconnected from any sort of load. When it comes to charging the battery system in a parabot/robot machine, however, the situation is often far from the ideal one. Rather, it is often necessary to maintain battery power to critical electronic circuits during the charging cycle, and it is sometimes necessary to charge more than one battery at a time. Both situations upset the idea charging scheme.

Power-on charging is especially important in cases where the machine has some violable memory—memory circuits that will lose their contents if the power source is removed for even a few microseconds. What's more, a number of machines suggested in this book have circuits that monitor the charging current, automatically disconnecting the machine from the battery charger when the charging cycle is done. That sort of circuit must have supply voltage during the charging interval, and that power generally comes from the very battery that is being charged.

Figure 3-3A illustrates the power-on charging scheme. Under normal running conditions, the motors and voltage regulator for the electronic circuits operate directly from the battery. The moment the battery charger

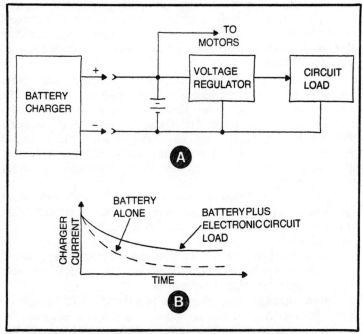

Fig. 3-3. Power-on recharging. (A) Block diagram of a typical power-on recharging system. (B) Charge curve showing the effect of the circuit load.

is connected to the circuit, its current divides between that required for charging the battery and that required for the motors and the electronic circuits.

The motors should not be switched on during the charging cycle. There are a couple of reasons for this. First, the motors probably make the machine move around the floor, and you don't want that happening while the machine is supposed to be standing still at the battery charger, or nest, assembly. Second, the battery charger might not be able to deliver the amount of current necessary for starting the motors; at least the motors will tax the capability of the battery charger.

So through any discussion of power-on battery charging, it is assumed the motors are *shut down until the very end of the cycle*, or in other words, until the battery is fully charged and can supply the start-up current necessary for moving the machine away from the charger.

Figure 3-3B shows how this power-on charging scheme upsets the usual battery charging curve. The dashed lines show the usual charging curve, and the solid lines indicate the power-on charging situation. Notice that the initial surge current is the same in both cases. As mentioned earlier, that initial surge is mainly controlled by the current capacity of the charger, itself. So whether the current demand is large or small, the initial surge current is much the same.

Unfortunately, there is no simple relationship between the charging status of the battery, the current from the battery charger and the source of current for the electronic circuitry. What you will find in actual practice depends so much on the characteristics of your charger, battery and electronic circuit.

If, for example, your battery charger tends to output a voltage that is somewhat greater than the battery voltage—say about 14V for a 12V battery—the charger will supply the lion's share of the current to the electronic circuits, regardless of the charger status of the battery. But if you have an electronically regulated battery charger, the amount of current it supplies to the electronic circuits will taper off with the charging of the battery, allowing the battery, itself, to provide an increasing share of the circuit current.

The only way to determine the situation for yourself is by trying it and observing the results firsthand. To be sure, the amount of current from the charger will be larger than that required for charging the battery when it is removed from the circuit.

You should still observe the dramatic reduction in charger current that always follows the initial surge. If you don't, the charger doesn't have adequate current capability for the system.

When selecting a battery charger for your system, then, you ought to give some special consideration to power-on charging. Suppose, for instance, your battery has a 10-hour discharge current of 1.2A (a 12 A-h battery). The maintaining current for the electronic circuits is 1A, so the worst-case, long-term charging current is on the order of 1.2A + 1A, or 2.2A.

If you want to charge the battery in the shortest possible time, its current rating ought to be at least four or five times the worst-case current level: between, say, 8A and 10A. But that is a rather large and expensive battery charger. It is possible to trade off some of that size and cost by letting the system recharge for somewhat longer periods of time. You can do this by selecting a charger current rating that ignores the circuit current, figuring the capacity to be four or five times the nominal battery-charging current, alone. That would be a charger with a 5A or 6A rating. The curve in Fig. 3-3B, incidentally, is drawn on that basis; otherwise, the initial current surge would be greater for the combined battery-and-load curve.

The only way you can get into trouble by "undersizing" the battery charger this way is when the maintaining current for the electronic circuits is substantially greater than the nominal charge current for the battery. And that won't be the case if you have properly matched the battery to its load. You can get into the same trouble if you try to run the motors before the battery is recharged, too; we have already assumed you won't be doing that.

So if you've selected the right battery for the load and keep the motors turned off during the charging cycle, the procedure for selecting the

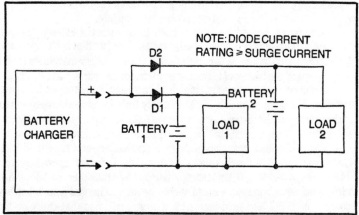

Fig. 3-4. Power-on recharging more than one battery system from a single charger.

battery charger isn't any different than the one outlined at the conclusion of the previous section of this chapter. You will simply notice that the charger current remains above the nominal battery-charging level for a longer period of time, through the entire charging cycle in some cases.

The matter of charging more than one battery from one charger at the same time poses no serious difficulties if the batteries are identical. As shown in Fig. 3-4, the only special procedure is to isolate the two batteries from one another with rectifier diodes.

During normal operation, the diodes prevent the battery having the better charge from discharging some of its current into the battery with the lower charge. Those two batteries cannot exchange current through those diodes.

And yet the diodes are forward biased with respect to the current from the battery charger. If the batteries have identical voltage ratings and similar A-h ratings, they will control their own charging rates. The battery requiring the greater amount of charge will automatically draw the greater share of current from the charger.

The current curve for the charging cycle still has the same general shape as before, showing an initial surge current followed by a rapid decrease and a gradual tapering to a nominal charging level. While the batteries might have identical, or at least nearly identical, specifications, they rarely have the same charge condition when initially connected to the charger. One battery will nearly always require more charge than the other. That situation makes it difficult to draw a typical charging current curve.

There is some risk, however, of overcharging the battery that requires the lesser amount of charge. That is especially troublesome if the battery charge happens to be one of the newer, thyristor-controlled versions. These charges apply rather high voltages at short duty cycles, and a battery that is nearly at full charge, but connected in parallel with one that is not, will suffer some plate damage after a while.

The safest procedure in this case is to select a battery charger suitable for charging just one of the batteries. The fact that it is inadequate for charging two batteries in parallel gives you the safety margin you need for protecting the better-charged battery from deep-charge plate damage.

It will certainly take much longer to recharge the batteries this way, but that's the price to be paid here. The only truly suitable alternative is to make provisions for charging the batteries from separate battery chargers. As you might imagine, that is a rather expensive alternative.

The problem of charging more than one battery at a time is further compounded when the batteries have vastly different specifications. Maybe one is a 12V, 1.0 A-h lead-acid battery, and the other is a 12V, 2 A-h nickel-cadmium battery. A nicad should never be charged from a battery charger built for lead-acid batteries, and you will find that chargers for nicads don't have the current capacity to recharge a lead-acid battery within any reasonable length of time. The safest solution here is to use separate battery chargers, allowing the system to remain connected to the chargers until both batteries are fully recharged.

The Weakest Link In Your System

I sometimes tend to take a rather gloomy attitude toward this matter of self/contained power systems for parabot/robot projects. I regard it as one of the most frustrating technical problems in the whole business.

With microprocessors giving us the capacity for machine control and intelligence thought impossible just a few years ago, we are still stuck with relatively primitive power sources. The matter of amking finely graded manipulator controls and high-resolution sensory systems is one of mere time, effort and patience. But this problem of self-contained power sources is more a matter of principle than procedure. It's like trying to generate power for a major city by making a bunch of mice run around in a cage connected to the generators.

Of course, there are some farfetched solutions that are suitable for conversation with other experimenters at the local pub on a Friday night: solar cells, fuel cells, atomic power pack and the like. Such notions do not get robots working today, however.

Sometimes I wonder if we would be no worse off doing away with the batteries and battery chargers, replacing them with a big, mechanical wind-up system. Yessir, just wind up that big old mainspring, and let it run a DC generator that supplies all the system's electrical power! Too bad the thing couldn't wind up itself every once in a while, though.

Or maybe we could use an internal combustion engine. A 5-hp lawnmower motor could generate a lot of electrical energy, and all we would have to do is work out a program that would cause the machine to drink up a gallon of gasoline every now and then. In principle, that's more suitable than the battery system proposed in the more rational parts of this book.

POWER DISTRIBUTION SYSTEMS

It is unfortunate that so few electronics schools and, indeed, undergraduate engineering colleges teach little in the way of power distribution systems. Unless a technician or engineer picks up the know-how by direct experience in industry, it is likely he or she will not have a real appreciation for the special techniques and problems inherent in distributing and controlling the flow of electrical power through moderately complex systems.

Robot experimenters do not have to deal with very large amounts of current—not normally, anyway. Maybe there is a total of 5 or 6 amperes for the motors under normal running conditions, and special short-term current surges rising as high as 12 or 15 amperes. That really isn't very much current, but it's more than most classroom discussions and lab experiments every consider.

Then there is a matter of distributing power to two or more major subsystems. One might assume that getting power to several major subsystems is a simple matter of running some wire between the primary power source to the assemblies that make up the various subsystems. You have to be careful there, because power source isolation might be necessary; and basic electronics texts and courses rarely consider that subject to any great extent.

Finally, there is the subject of circuit protection. That spells *fuses*, and it means one ought to know how to select the right kind of fuse for the job at hand. Now, that might seem to be a very straightforward task until you look at the bewildering array of fuses listed in a fuse catalog. The size of such a list is a clue that selecting the right kind of fuse is not always a trivial task.

So for our purposes here, the matter of power distribution through a parabot/robot system breaks down into three general categories:

☐ Wiring the subsystems to the main power source in such a way that the scheme can handle a moderately high nominal current as well as surge currents.

☐ Isolating the subsystems with respect to the main power sources.

☐ Fusing the system to protect the circuitry and wiring.

Power Distribution Wiring and Associated Hardware

There are two general rules that must always be applied to power distribution systems that handle more than 1 or 2 amperes of nominal operating current. First, the wire itself must have an adequate current rating. For the sake of economy and convenience, many experimenters

Table 3-1. Recommended Wire Sizes as a Function of the Peak Currents They Will Carry.

Peak Current Rating (Amperes)	Minimum Wire Size (AWG)
3	18
6	16
15	14
20	12
25	10

simply use the same hookup wire throughout the system, giving no special thought to the fact it might be inadequate for certain parts of the system. Second, solder joints must be avoided in circuits carrying more than 1 or 2 amperes. Aside from the fact that it is difficult to modify a circuit that has soldered connections, those connections tend to suffer under high-current surges, eventually breaking down and causing problems.

You might want to add a third important consideration, although most trained technicians and engineers are well aware of it: Keep the power distribution wiring as short as possible. The resistance and stray reactance in long sections of wiring can round off the high-current spikes necessary for starting larger electromechanical devices, especially the main drive motors. Then there is also the matter of electrical noise pickup and radiation along needlessly long wires.

Table 3-1 shows the recommended wire sizes for peak currents between 3A and 25A. You can see from that table that the usual sort of 18-gage hookup wire is adequate for current surges up to 3A. The motors in moderately large parabot/robot systems might draw surge currents on the order of 15A, however, and that means using 14-gage wire; the same gage used in wiring new homes.

You might not be able to find wire having AWG (American wire gage) ratings below 18 at your favorite electronics parts supplier. If so, you will have to shop at the electrical department in any major department or homebuilders store. You might have to buy three-conductor or four-conductor cable and tear it apart to get to the individual lengths of wire. One nice thing about that is the wires are color-coded for you.

Not all the wiring in your power distribution system has to be selected at the maximum size. It would almost be absurd to run 14-gage wire to a circuit board that draws less than 1A from the main power source. In that case, 18-gage wire will work quite nicely.

So the wiring in your power distribution system will probably be composed of several different wire gages—two different gages, anyway. Perhaps that is intuitively obvious to you. But be careful: There is a common, serious mistake lurking in the background. When wiring high-current circuits, make certain the size of the *return* wiring is adequate.

70

Through training and habit, most of us take common ground connections for granted. By convention, this ground connection represents the negative polarity of the power supply. And as mentioned in a later section, connect the power system ground to the major metallic sections of the mainframe.

If you take this common ground connection for granted, it is easy to forget about the current rating of that ground system. A careless experimenter might be conscious that the positive wiring to the motors calls for using a larger-gage wire and then forget to return to the negative terminal of the battery with wiring of the same gage.

Figure 3-5 illustrates this particular point. Figure 3-5A is a schematic diagram of a rather simple power distribution system. The details about the fuses and voltage regulator are not important right now. The important feature is that the two motor circuits require heavier wiring than the voltage regulator does.

Figure 3-5B shows a wiring diagram for the same circuit. According to the wire key accompanying Fig. 3-5B, the circuit is wired with two different wire sizes. Sections of wiring carrying label *A* must be 14-gage wire—the heavier stuff that can handle currents up to 15A. The wiring sections labeled *B* represent common 18-gage hookup wire, which can handle up to 3A.

Notice that all sections of wiring carrying motor current, including the return wiring to the negative side of the battery and jumpers between connections on terminal blocks, are sized at 14-gage. If you let a section of 18-gage wire sneak into any of those places, you are risking some serious problems and possibly a fire.

Incidentally, it is assumed here that the total current to the motors will never exceed 7.5A. That conclusion is dictated by the 15A rating of the MOT COMM terminal—the connection carrying the return current for *both* motors. If each of the motors pulled up to 15A from the battery, that return line and several other sections of the system wiring would have to be boosted to a 30A rating. And as you might suspect from the sequence of numbers on the wiring table (Table 3-1), you should use 8-gage wiring in that case. So consider the current paths for the various subsystems carefully, making certain they are *all* adequate for the current demand that will be placed upon them.

The wiring diagram in Fig. 3-5B also implies the use of solderless connections throughout the power distribution system. This means using crimp-on terminal hardware and barrier strips or terminal blocks. No solder connections appear anywhere in this part of the system. Even the two discrete components, diode D1 and capacitor C1, are connected in a solderless fashion to the terminal blocks.

If you have already selected the wiring, you won't have any trouble specifying the crimp-on terminal hardware. That kind of hardware is specified according to the gage of the wiring with which it is to be used. So in the case of the 14-gage wiring specified in the drawing, you will need spade-tongue or ring-tongue solderless connectors rated at 14-gage. Terminals specified for 16 to 22 gage wiring can be used for the 18-gage sections.

So the current specifications dictate the wire gage, and the wire gage dictates the size of the solderless terminals. Finally, the size and number of solderless terminals dictate the specifications for the barrier strips or terminal blocks.

Plan the specifications and layout of the terminal blocks in advance on paper. Then I suggest you oversize the number of terminals available, leaving yourself some room for future modifications and expansion.

Route the power distribution wiring neatly, but don't be overly generous with the lengths. Keep the wires as short as possible. Use wire fasteners to keep the bundles of wire from flopping around and use color-coded wire so you can later trace the wiring easily. Wherever it is necessary to run power distribution wiring through sheet metal, always protect the wiring from the sharp edges of the hole by using a plastic, clip-in feed-thru grommet.

Power-Supply Isolation

Power-supply isolation isn't a serious problem if you are working with a simple system; say, one having a couple of motors and a single circuit board for the electronic components. Figure 3-5 illustrates such a system. The power supply isolation in this case amounts to little more than inserting a voltage regulator and despiking capacitor ahead of the electronics circuit board.

But suppose you are going to be using more than one circuit board and each board handles up to 1A of supply current. The temptation is usually to locate the voltage regulators somewhere close to the battery source and then run relatively long DC supply wires to the circuit boards. This can get you into trouble, especially if one of the boards happens to have a microprocessor device and some memory chips on it.

Figure 3-6 shows the proper way to distribute DC power to a system having more than one main circuit board. The raw, unregulated power from the battery system goes to the boards, using a few long lengths of wire as possible. Then each board has its own spike-filtering capacitor and voltage regulator.

This technique minimizes the chance of setting up nasty ground loops (a problem often encountered by having an excessive number of widely separated ground connections) and power-supply noise interference between circuit boards. Consider yourself quite fortunate if you have never had to deal with either of these problems; they are terribly difficult to track down. Use the distribution scheme suggested in Fig. 3-6, and you might never have to deal with them.

And if your system is going to include a high-performance microprocessor, you won't be overly cautious to string a couple of ferrite beads

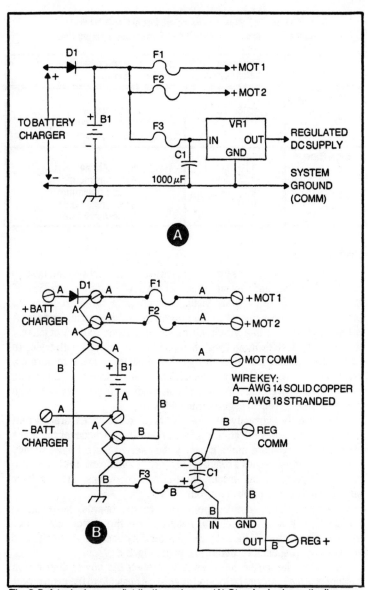

Fig. 3-5. A typical power distribution scheme. (A) Standard schematic diagram. (B) Wiring diagram for the same system.

onto the power-supply wiring running between the individual circuit boards. Don't worry about the technical specifications for those beads. Just select a size that fits nicely onto the wiring you are using. Such beads do wonders when it comes to filtering the rf energy that often sneaks into a power-supply system for microprocessors.

**Table 3-2. Blowing Characteristics of Small Glass
Fuses, Shown as Percent of Rated Current and Clearing Time.**

Type	Percent of Rated Current	Approximate Blowing Time
Fast Acting	100-110	Will not blow
	135	1 hr. maximum
	200	5 sec. maximum
Slow Blow	110	Will not blow
	135	1 hr. maximum
	200	25 sec. maximum
	300	8 sec. maximum
	500	3 sec. maximum

Fuse Protection

There are two reasons for using fuses in an electrical system: first, to protect the wiring, and second, to protect the devices and circuits. That is the order of priorities. It is virtually impossible to select fuses for protecting every component in the system from short-circuit damage. A little 1/4W resistor in a 5V circuit needs only about 100 mA of current through it to destroy it, and you shouldn't bother yourself with trying to fuse the system at that level. If that little resistor is going to burn up, you're going to have to let it go. What is important is protecting the wiring to the board, preventing that short-circuit condition from drawing enough power-source current to begin burning the wiring.

Fortunately, virtually all voltage regulator devices are current limiting; that is, they effectively turn off themselves whenever there is a short circuit anywhere in the circuitry they serve. Since I have already recommended using such a regulator for each circuit board, the boards are already protected from serious short-circuit conditions; the regulator simply shuts down the power for you.

You still have to fuse the wiring to the boards, however, in anticipation of a short circuit anywhere along the power distribution system. The motors, especially the heavier duty ones, ought to be fused individually, too. Such a scheme is shown in Fig. 3-5.

Because the job of the fuses is to protect the power distribution system, the fuses should be located as close as possible to the main power source. Putting them close to the components they protect is generally defeating the whole purpose.

Table 3-2 summarizes the blowing characteristics of some common types of fuses. Assuming you are going to use the small, cylindrical glass fuses, they are specified in two ways: as fast acting or slow-blow, and according to a certain current rating. The current ratings are not shown in Table 3-2; you can find such a list in just about any electronics parts catalog. Rather, the table shows how long the fuse will carry some percentage of its rate current before it blows, or clears.

As the name implies, fast acting fuses are relatively sensitive to overcurrent conditions. According to the table, a fast acting fuse will carry twice its rated current (200 percent) for no more than five seconds before it clears. Of course, it will clear sooner than that if the current is more than 200 percent of its rating.

Slow-blow fuses, on the other hand, are a lot more forgiving in terms of short-term surge currents. Such a fuse will carry twice its rated current up to 25 seconds at a time—five times as long as a comparable fast acting fuse will.

I tend to avoid fast acting fuses. Selected properly, a slow-blow fuse can provide the same, long-term protection, and yet tolerate momentary current surges caused by some of my own clumsiness when tinkering

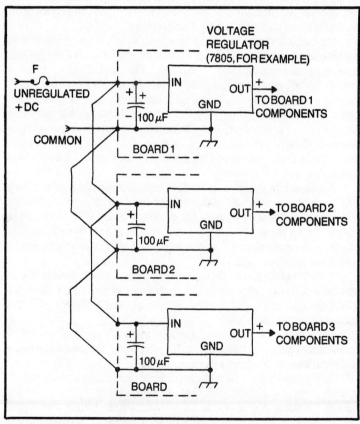

Fig. 3-6. The proper technique for distributing power to two or more electronic circuit boards. Note that each circuit board has its own power supply voltage regulator.

around with the circuitry. The real justification for using slow-blow fuses, however, is to allow large filter capacitors to charge rapidly when power is first applied to them. They are also most suitable for fusing DC motors that routinely draw surge currents during start-up and short-term stall conditions.

To get some general idea about selecting fuses, suppose you are going to pick a fuse for the power distribution/isolation scheme illustrated in Fig. 3-6. Further suppose that each of the three circuit boards has a nominal current rating of 1A. That means the fuse must be able to hold 3A of current for an indefinitely long period of time.

Recall that the regulators will automatically limit the current to the circuitry that they service. And if you are using the popular 7800-series regulators, they have a maximum rating of about 1.2A. That means the worst-case current will be on the order of 3.6A to the circuit. The despiking capacitors will pull an initial surge current when the power if first applied, but that isn't likely to amount to more than 200 percent of the nominal current for more than one second or so.

Putting this altogether, the fuse has to handle 3A for an indefinite period of time and about twice that amount for, say, 10 seconds or so. That short-term surge current makes a fast acting fuse a questionable choice: It can handle 200 percent of the rated current for just five seconds. That's calling it too close. The silly thing might blow every two or three times you apply power to the system. A slow-blow version, however, will handle the turn-on surge for something on the order of 25 seconds. That is longer than we really want, but it's a tolerable situation.

What is the bottom line of the matter? I would select a 3A slow blow fuse in this case.

Now consider the fuses, F1 and F2, used with the motors specified in Fig. 3-5A. Suppose each of the motors has a nominal running current on the order of 4A and a start-up and stall current of about 12A. They will be running at 4A most of the time, but in a normal sort of parabot/robot machine, there will be alot of short start-up operations and some frequent, long-term stall conditions.

Each of those fuses ought to be able to carry the 12A peak current for up to 10 or 20 seconds at a time. That covers the inevitable stall situations, giving the operator or machine ample time to clear it. In this particular case, a 6A slow blow-fuse would work out rather nicely.

Some machine designs should be able to handle motor stall current for an indefinite period of time—for an hour, at least. My favorite systems work that way, and that changes the picture a bit.

Assuming the same current specifications for the motors (4A nominal running current and 12A stall current), I can pick a slow-blow fuse that will handle 135 percent of its rating for an hour. From algebra, the current rating is about 8.9A. Because there is no such thing as a 8.9A fuse, I'll have to settle for a 10A version. If I go for an 8A version, the fuse might blow before the machine has a chance to clear the stall condition.

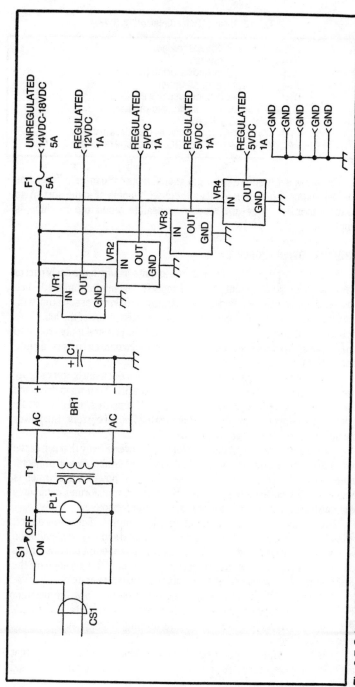

Fig. 3-7. Schematic diagram for a useful auxiliary power supply.

77

Table 3-3. Parts List for Circuit of Fig. 3-7.

CS1—standard 120VAC cordset
S1—SPST toggle switch
PL1—120VAC neon indicator lamp assembly
T1—12V, 5A power transformer
BR1—6A full-wave bridge rectifier assembly
C1—1000uF, 25VDC electrolytic capacitor
F1—5A fast action fuse
VR1—7812 12VDC, 1A voltage regulator assembly
VR2, VR3, VR4—7805 5VDC, 1A voltage regulator

By way of a few additional suggestions, mount the fuses in fuse clips or fuse holders and clearly identify them with labels of some sort. Also situation them so they are readily accessible when the machine is completed.

AUXILIARY POWER SUPPLIES

An auxiliary power supply can come in quite handy in the early part of a parabot/robot development project. Properly designed and applied, such a power supply can set aside the charging problems associated with batteries until the very last part of the construction phase of the job. Or to put it in simple terms: You should have a good DC power supply around to help you test, troubleshoot and calibrate the electronic circuits going into your machine.

This is not to say that you can always replace the main battery supply with a plug-in type power supply, even temporarily. Your main drive and steering motors, for instance, might draw 6 to 12 amperes, and it's a hassle to build a DC power supply that can handle that kind of current; the power transformer is quite large and expensive.

Figure 3-7 is the circuit diagram for a DC power supply that can serve as an auxiliary power supply for most kinds of parabot/robot projects. Table 3-3 is the respective parts list. The 12V power transformer is specified at 5A because that is about the biggest transformer commonly available from electronic parts stores and mail-order houses.

The circuit features a number of different outputs. There is a direct, unregulated DC output that will show between 14V and 18V, depending on the amount of current loading. The maximum current output in this case is on the order of 5A. And because that unregulated output does not have the benefit of a built-in current limiter, it should be fused separately.

The regulator devices for the regulated outputs are the common 78XX-series voltage regulators. In each case, they are rated at about 1.2A. These devices include a built-in current limiting feature, so there is no need to fuse their outputs.

A 12V regulated output is handy for checking circuits that have that particular voltage rating, such as CMOS devices, audio amplifiers and other kinds of linear circuits.

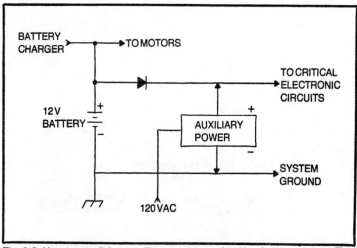

Fig. 3-8. How to parallel an auxillary power supply with a battery system. The rectifier diode isolates the auxillary power supply from the battery and motors.

There are three separate 5V regulated outputs, each capable of driving a load with as much as 1A. These are useful for working with TTL logic devices. You can, of course, include some ammeters and voltmeters, but they are not necessary if you have a decent multimeter available.

There is a second reason for having this sort of auxiliary power supply available. Suppose something goes wrong with your main battery supply or battery charger. You have spent many hours getting programs stored into the RAM (program memory) of the machine and don't want to lose it because of a power failure somewhere along the way. When you see such a crisis developing, you can feed power from the auxiliary power supply to the appropriate circuits, thereby saving a major disaster in terms of lost time and effort.

Figure 3-8 shows some special provisions for paralleling the auxiliary power supply with the usual battery supply in your machine. The diode isolates the main motor system and battery from the auxiliary power source, thereby directing the auxiliary power only to the vital electronic circuits. The auxiliary power supply is thus never taxed by current flow to the motors and battery.

The auxiliary power supply should be built into a separate housing and used as a separate unit through the initial development phase of your project. When the machine is up and running, you then have the option of simply keeping the auxiliary power supply in a convenient place in your workshop or fixing it permanently into the machine, itself.

In the latter case, its output can be fixed to the critical electronic circuits, as shown in Fig. 3-8. That way, getting emergency power to the system is a simple matter of plugging the corset into a nearby 120 VAC power outlet.

Selecting and
Working with DC Motors

It is difficult for most parabot/robot experimenters to imagine a useful and interesting project that doesn't include some sort of mechanical motion. And the need to get some mechanical motion means selecting and using DC motors.

The necessary motors might be fairly large or rather small, depending on the amount of mechanical loading that will be applied to them. One of the important requirements for selecting a motor is to get one that is large enough for the job at hand, but not excessively large. Overly large motors add needless bulk to the project and, in many cases, waste valuable electrical power.

Another part of motor selection is to come up with a motor system that runs at the desired speed. Motors that aren't somehow geared down to achieve lower operating speeds—and, incidentally, higher operating torque—are rarely useful in parabot/robot projects. That speed requirement generally means coming up with a gear motor or a gear train of some sort.

A *gear motor* is a motor assembly that has speed-reducing gears built into its housing. The gearing is viewed as an integral part of the motor system. In some cases, however, you will have to work out your own gearing arrangement, coupling a gear train to the high-speed shaft of the motor.

Ideally, an experimenter should be able to calculate the speed and torque requirements for the motors and then search a motor catalog for a model that comes close to doing the job. In practice, though, the situation isn't always an ideal one; buying new motors according to specifications is usually an extremely expensive affair. It would not be out of line to expect to pay something on the order of at least $50 for a new motor. Considering that most machines require more than one motor, that can be a discouraging price tag.

Experimenters thus find it necessary to use surplus motors and gear motors. It is possible to find some beautiful motors and motor assemblies on the surplus market for $15 or less.

Because of this particular cost situation, all the discussions in this chapter assume you will be specifying and using surplus motors; new ones—good ones to be sure—but economically priced ones.

Using surplus motors presents some problems, but they are solvable. Much of the material in this chapter deals with solving those particular kinds of problems.

GENERAL MOTOR SPECIFICATIONS AND OPERATING CHARACTERISTICS

The subject of DC motor characteristics and specifications can be a highly detailed and complicated one. In practice, however, it is possible to ignore some of the finer details, and work with just a few of the more important ones. Some of the comments and suggestions in this discussion will disturb the conscience of an engineering purist, but it turns out that the notions do indeed work out in the long run.

Motor Current, Voltage and Torque

DC motors are obviously electrically operated devices. That means they have some important voltage specifications. Then, too, DC motors offer some reaction against current flow created by the applied voltage. So there is a matter of voltage, current and reaction against that current. That means a motor dissipates some electrical power.

Some of the power consumed by a DC motor is wasted in the form of heat energy. But it is hoped that a good share of that power will be transformed into useful mechanical energy; that's the point of the whole thing. The efficiency of a DC motor can be reckoned as the ratio of the amount of mechanical power to the total power it consumes from the power source.

If you select any DC motor and measure its winding resistance with an ohmmeter, you are going to notice a certain amount of DC resistance. The resistance can be anywhere between 0.5 ohms and 100 ohms, depending on the design of the motor.

Here is an important point: The greatest amount of current that a motor can draw from the voltage source is equal to the voltage divided by the winding resistance. That is a simple expression of Ohm's law. In this respect, the motor current is directly proportional to the amount of applied voltage.

As a motor responds to its applied voltage and current, it begins to turn. And as it turns, its own generator effect produces a voltage that tends to oppose the externally applied voltage. That's convenient, because the motor current is then less than that found by working with the winding resistance and applied voltage.

81

The self-generated voltage, or counter emf, from a DC motor is often proportional to its running speed, but it never quite reaches the level of the applied voltage. There is always some difference between the applied voltage and counter emf; a difference necessary for overcoming the inherent inefficiency of the motor.

Assuming the supply voltage remains constant, you can always figure that a motor will draw its maximum amount of current when it is not turning. The faster it turns, then, the smaller amount of current it draws from the voltage source.

As mentioned earlier in this section, the power consumed by a DC motor goes into the heat of inefficiency and useful torque. If the motor has no load applied to its shaft, its required torque is practically nonexistent, and the only power consumed is that burned up as the heat of inefficiency. The motor thus draws its least amount of running current when it is running at full speed with no mechanical load applied to the shaft. So here are the two extremes as far as motor current is concerned: maximum current when the shaft is not turning, and minimum current when the motor is running a full speed with no mechanical load on the shaft.

A motor without a mechanical load is useless, however. You are going to have to load the motor by making it turn some wheels or operate a manipulator device. That means the motor will operate somewhere between its peak, nonturning current level and some lower level determined by the amount of mechanical loading.

In the context of parabot/robot machines, a motor draws its peak current under two conditions: when electrical power is first applied to it and whenever the mechanical loading becomes so great that it stops, or stalls, the motor. The first situation is normally a temporary one, lasting only the fraction of a second required to bring the motor up to speed. The stall situation, however, can persist indefinitely, although you will seldom want the stall situation to persist very long.

The start-up and stall conditions draw about the same amount of current from the power source, so the two conditions are frequently considered as one through most of the discussions in this book. You will thus see the expression, *start-up and stall current,* used rather often here.

How can you determine the start-up and stall current for any DC motor you happen to have at hand? The simplest procedure is to measure the DC winding resistance with an ohmmeter and divide the result into the supply voltage level. That does it quite nicely. That gives you the start-up and stall current level of the motor.

Of course, you can also determine that important current parameter by running the motor from its voltage source and stalling the shaft yourself—maybe clamping the shaft with a wrench and holding it still. The current to the motor windings, measured with an ammeter, is the start-up and stall current. You might find this procedure rather difficult when trying to stall a husky gearmotor that is operating from a freshly charged battery.

82

Knowing the start-up and stall current for a motor is paramount to designing the power source and power distribution system for it. This is not to say the voltage rating of the motor isn't important. DC motors tend to run at a speed proportional to the amount of voltage applied to them. Putting it simply, the higher and applied voltage is, the faster the motor runs.

While surplus motor catalogs rarely specify some of the most important parameters, the voltage rating seems to be important enough to include in all cases. Generally, motor voltage ratings fall into increments that are typical of most power-supply voltages: 3V, 6V 9V, 12V and so on. Finding a motor that suits your power-supply requirements is thus a straightforward task.

Most DC motors can be operated at voltage levels that exceed their specified rating. A large number of so-called "hobby" motors, for example, are listed as 6V motors, but they can be run at 12V without causing any serious problems. The only potential problem in this case is overheating the motor whenever it is installed. Because there is a linear relationship between current and voltage when a motor is stalled, it figures that a motor operating at twice the rated voltage will draw about twice the normal start-up and stall current. Don't forget, however, that there is a square relationship between voltage and power dissipation. Double the voltage, and the start-up and stall power dissipation quadruples. Whether or not the motor can safely get rid of that much more heat depends on how much the engineers have overdesigned it. Many motors can take it but some cannot.

There is a positive side to the matter. Increasing the applied voltage also increases the amount of power available for conversion to speed and torque. That fact lines up with the intuitive notion and first-hand experience that a motor runs faster and develops more torque as that supply voltage increases.

After putting together some of these facts about motors, you can come up with some workable generalizations. First, the running speed is proportional to the applied voltage and inversely proportional to the mechanical loading on the shaft. Second, the current demand of the motor is inversely proportional to running speed and proportional to the mechanical loading. It is difficult to portray a DC motor in terms any simpler than these.

Definitions and Units of Measure

Motor catalogs, especially those offering reasonably priced surplus motors, specify only a few important parameters. Sometimes there is enough information for calculating other important parameters, sometimes you have to make some educated guesses, and other times you simply have to cross your fingers and hope the motor will work.

Motor voltage rating is one specification you can count on finding in any listing of surplus motors. If you have the background necessary for understanding this chapter so far, you do not need a formal definition for motor voltage rating.

The catalogs sometimes list *motor current rating*. That's fine, but you have to be careful about it. Such listings seldom specify the test conditions that show that current level. Is that the stall current? Is it the no-load, full-voltage current? Is it a current level measured under some sort of unspecified mechanical load? Usually, you have no way of answering those questions, so the specification is not very helpful in the strictest sense. The motor current specification however, does give you a general notion about how much current the motor will draw from your power supply. At least you know whether it's a ballpark figure of 1A or 10A.

Winding resistance, whenever it is specified, is a useful parameter. As described earlier in this chapter, you can apply the DC winding resistance and voltage rating to Ohm's law, thus coming up with a number for the start-up and stall current of the motor. Knowing the winding resistance and voltage specifications also lets you calculate the maximum input power of the motor: $P = E^2/R$ in watts. That input power figure can be helpful for making other educated guesses, as described later.

So far as motor voltage, current and resistance specifications are concerned, the *voltage rating is the most critical*. Suppliers recognize this fact and clearly specify the value for you. The current rating isn't very helpful because the supplier rarely specifies the test conditions. The winding resistance, on the other hand, is a specific figure that can lead to some unambiguous and useful information.

Motor catalogs nearly always specify a rate of rotation in rpm. But you have to be careful about that figure because, like the motor current specification, the catalog rarely cites the test conditions. Was the rpm measured at full load? Or was it measured at no load?

The rpm specification can, however, infer some useful information for making educated guesses about the motor. Of course, the rpm rating provides a ballpark figure for the speed of rotation. There are a lot of parabot/robot applications, for example, where a 40,000 rpm motor would be wholly inappropriate, whether it is fully loaded or not.

A catalog listing of motor rpm can also tell you whether or not the device is a gear motor. A gear motor is a basic motor that is fixed to a built-on or built-in gear assembly, usually an assembly that steps down the basic rate of rotation of the motor. DC motors, themselves, rarely run at speeds less than about 400 rpm. The speed is usually on the order of 1000 rpm or more. So if you see an rpm specification well below 400 rpm, you can bet it is a gear motor, even though the catalog might not list it as such.

It is important to know whether or not a particular motor assembly is a gear motor because gearing arrangements drastically reduce the efficiency of the system. Whereas you might need the equivalent of 20 watts of mechanical power to perate a part of your parabot/robot machine, the inefficiency of a DC gear motor scheme might make it necessary to dump 50 watts into the motor. If that same motor is not geared, you might need only 30 watts of input power.

There's more to be said about motors and gearing arrangements later in this chapter. The important point now is that a catalog rpm listing infers something about gearing and system efficiency.

For the sake of making some calculations a bit further down the line, it is often helpful to be able to translate rpm (revolutions per minute) into rps (revolutions per second).

$$rps = rpm/60 \qquad \textbf{Equation 4-1}$$

where rps = motor turning speed in revolutions per second, and rpm = motor turning speed in revolutions per minute.

Another catalog specification that is very useful is the torque of the motor. *Torque* is a measure of the ability of the motor to do useful work. In a rather loose sense of the term, it is an indication of its twisting force, or power. The higher the torque rating, the more powerful the motor must be.

Torque ratings are usually specified in units of lb-ft (pound-feet), oz-in (ounce-inches) or, occasionally, lb-in (pound-inches). You might be familiar with similar measures of work expressed in units turned around the other way: ft-lb, in-oz and in-lb. Some catalogs express motor torque that way, too. It's all the same. Physicists like to express torque with the force-distance order of things to separate it from linear, straight-line, work. The difference between a ft-lb and lb-ft is really trivial—don't worry about it, unless you happen to be a stickler for doing things the proper way.

Figure 4-1 illustrates the most general definition of torque: a force (pounds or ounces) multiplied by a unit of distance (feet or inches) between the force and axis of rotation. In the context of motors, torque is a measure of the amount of force the motor can apply to a mechanical load. But like current and rpm specifications, the catalogs seldom spell out the conditions for getting the torque figure. It is safe to assume, however, that the torque figure is derived by measuring the force of the motor while it is stalled at its specified voltage.

Because motor torque is specified in several different units of measure, it is necessary to convert from one set of units to another. Here

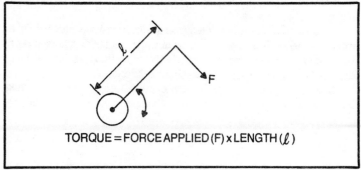

TORQUE = FORCE APPLIED (F) x LENGTH (ℓ)

Fig. 4-1. A simple definition of torque.

are some equations that will help you with that little task.

$$lb\text{-}ft = (oz\text{-}in)/192$$
$$oz\text{-}in = 192(lb\text{-}ft)$$
$$lb\text{-}ft = (lb\text{-}in)/12$$
$$lb\text{-}in = 12(lb\text{-}ft)$$
$$oz\text{-}in = 16(lb\text{-}in)$$
$$lb\text{-}in = (oz\text{-}in)/16$$

where lb-ft = torque rating is pound-feet, oz-in = torque rating in ounce-inches, and lb-in = torque rating in pound-inches.

Some motor specifications show a horsepower rating in lieu of a torque. Horsepower (hp) and torque are two different things; the former being a measure of power, or rate of doing work, and the latter being a measure of work, alone. Motor power is really a measure of torque per unit of time.

Basically, mechanical power is defined as the amount of force required to move a given mass a certain distance in a particular period of time. For instance, you are working at a rate of 1 ft-lb/sec when you lift a 1-pound weight 1 foot in 1 second. That's mechanical power.

A horsepower is figured as 550 ft-lb/sec.

$$P_h = Fd/550t \qquad \textbf{Equation 4-2}$$

where P_h = mechanical power (hp), F = applied force (lb), d = distance moved (ft), and t = time of motion (sec).

As an example, suppose you lift 100 lb to a height of 5 ft in 2 sec. To do that weightlifting job, you have to exert

$$\frac{(100)(5)}{550(2)} = 0.45\text{hp},$$

or just a bit less than ½ horsepower. To do that same job in just one second requires nearly 1 hp.

So the higher the horsepower rating of a motor, the more work it can do in a specified period of time. Or taking a different approach to the same situation: The higher the horsepower rating, the less time it takes to do a specified amount of work.

Because a DC motor is an electrical device, it is often more helpful to work with mechanical power in terms of watts, rather than horsepower. Watts and horsepower express exactly the same thing—a rate of doing work. Only their constants of conversion are different.

One horsepower is equal to 740W of mechanical power. Converting from horsepower to watts is thus a simple matter of multiplying horsepower by 746. See Equations 4-3 and 4-4.

$$P_w = 746P_h \qquad \textbf{Equation 4-3}$$

$$P_w = 1.36Fd/t \qquad \textbf{Equation 4-4}$$

where P_w = mechanical power (watts), P_h = mechanical power (hp), F = applied force (lb), d = distance moved (ft), and t = time of motion (t). Lifting a 100-lb weight to a height of 5 ft in 2 sec thus calls for developing 1.36(100)(5)/2, or 340W of mechanical power.

Here is where the matter of expressing mechanical power in terms of watts comes in handy. In the example just cited, the job required 340W of mechanical power, and that means a motor for the job must have a power rating of at least 340W, not taking into account any inevitable element of inefficiency.

If you want an electrical motor that at least comes close to lifting 100 lb to 5 ft in 2 sec, it must be rated at least 340W. There is a straightforward relationship between the mechanical power a motor system generates, expressed in watts, and the electrical power rating of the motor, also in watts. The only thing that separates them is the inefficiency of the system.

Suppose, at least for the sake of an example, the mechanical system in these examples is about 45 percent efficient. That sounds terrible, perhaps, but it can be a realistic figure in many instances. At any rate, if you have to output 340W of mechanical power and the system is 45 percent efficient, it figures the input power to the system is 340/.45, or about 756W.

Are you beginning to see what all this means in terms of preliminary designs for your parabot or robot system? By looking at the sort of mechanical work you want the machine to perform, you can begin getting a good idea of how large the motor system will have to be.

If the notion of using a 756W motor doesn't give you a perspective on the job, assume the motor will run from a 12V source and calculate the current. Current equals power divided by voltage, so in this case, the current is 756/12, or 63A. Now that is a sobering figure. Maybe you'll want to change your mind about having the machine able to lift 100 lb 5 ft in 2 sec.

Playing with these notions by way of a realistic example, suppose you want to move a 20 lb robot at the rate of 0.5 ft in 1 sec. The mechanical power at the output end of the system can be found from Equations 4-3 and 4-4: P_w = 1.36(20)(.5)/1 = 13.6W. If the gearing arrangement for the system is a relatively simple one, the efficiency might be as high as 60 percent. Dividing the 13.6W required by 0.6, the electrical power rating of the motor comes out to about 22.7W. And if it is a 12V motor, the current demand on the battery will be 22.7/12 = 1.89A. Indeed, that is a realistic figure which falls into line with a lot of practical experience with systems of this particular type.

Perhaps this is the place to formalize the matter of adjusting the mechanical output power figure to get the electrical power rating of the motor.

$$P_m = P_w/e \qquad \text{Equation 4-5}$$

where P_M = electrical power rating of the motor (watts), P_w = mechanical power required for doing a job (watts), and e = mechanical efficiency (a unitless number between 0 and 1).

To complete this discussion, it is necessary to put together some meaningful relationships between motor torque, rpm and power. Look at Equation 4-6. It shows mechanical power as a function of torque and motor rpm.

$$\frac{P = 2\pi LM_r}{60} \qquad \textbf{Equation 4-6}$$

where P = mechanical power (ft-lb/sec), π = constant 3.14..., L = torque (lb-ft), and M_r = motor speed (rpm).

Most people have trouble relating to power expressed in terms of ft-lb/sec. That P term in the equation, however, represents the Fd/t expression that appears in a couple of earlier equations. From Equation 4-2, for instance, you can solve for horsepower:

$$P_h = \frac{2\pi LM_r}{(60)(550)}$$

All those constants make the equation messy, and you can tidy it up by working them into a single constant. A more satisfactory version of this horsepower equation is:

$$P_h = 1.9 LM_r \times 10^{-4}$$

There you have motor horsepower as figured from the torque and rpm ratings.

As pointed out earlier in this section, expressing power in watts often leads more directly to some meaningful results. Applying the procedures just described to Equation 4-4:

$$P_w = \frac{(1.36)2\pi LM_r}{60}$$

Doing away with the constants:

$$P_w = 0.14 LM_r$$

This is the mechanical power in watts of a motor developing torque L at M_r rpm.

Equations 4-7 and 4-8 summarize these results for future reference purposes.

$$P_h = 1.9\, LM_r \times 10^{-4} \qquad \textbf{Equation 4-7}$$
$$P_w = 0.14 LM_r \qquad \textbf{Equation 4-8}$$

where P_h = mechanical power (hp), P_w = mechanical power (watts), L = motor torque (lb-ft), and M_r = motor speed (rpm).

Trying on these equations for size, suppose you have a motor that lists an rpm of 120 at a torque of 0.5 lb-ft. The horsepower rating, according to Equations 4-7 and 4-8, is $1.9(0.5)(120) \times 10^{-4} = 0.011$ hp. The power rating—the mechanical power in watts—is $0.14(0.5)(120) = 8.4$ W. That's a pretty small motor. Estimating an efficiency of 0.6, its input electrical power (Equation 4-5) is $8.4/.6 = 14$ W. Dividing by 12V to find its 12V current level, the little motor pulls about 1.2A.

As another example, suppose you have the option of buying a different motor rated at 120 rpm with a torque of 100 oz-in. Is that motor more powerful than the one just described? To find out, first convert the oz-in torque rating to lb-ft by the appropriate conversion given earlier: 100 oz-in/192 is 0.52 lb-ft. Because the first motor has a torque of 0.5 lb-ft, the two are much the same. As far as output power is concerned, it makes no difference which one you choose.

ESTIMATING MECHANICAL REQUIREMENTS FOR MOTORS

The first step in selecting a motor is to determine as reasonably as possible the mechanical specifications of the system relevant to the motor: rpm, torque and power. Only after doing that can you expect to get realistic figures for selecting the motor and determining the electrical specifications for the power supply that operates it.

This section considers only a few basic kinds of mechanical systems. The information, combined with that presented in earlier sections of this chapter, is sufficient for dealing with just about any sort of parabot/robot motor system you would ever care to use. Where your own system differs from these basic examples, you will have to generalize the information and procedures yourself.

Motors for Main Drive Wheels

Most experimenters at least want their machine to roll around the floor under its own power. That situation calls for having at least one drive wheel operated from one motor.

An analysis of how much mechanical power is required at that drive wheel begins with the equation, $P=Fd/t$, where P is the required power in ft-lb/sec, F is the amount of weight to be moved, and t is the time interval required for moving the weight distance d.

It turns out that the ratio of d/t can represent the speed of the motion. A distance divided by a time interval always cranks out a figure for speed. The basic equation just cited can thus be represented in this useful form:

$$P=Fs \qquad \text{Equation 4-9}$$

where $P=$ mechanical power required (ft-lb/sec), $F=$ weight of the machine (lb), and $s=$ linear, or straight-line, speed (ft/sec).

So if you are thinking about having a parabot or robot machine that weighs about 15 lb and moves at a speed of 2 ft/sec, the mechanical power developed at the drive wheel must be at least $(15)(2)=30$ ft-lb/sec.

Mechanical power, expressed in ft-lb/sec, doesn't mean much to most people, so you can convert to horsepower by dividing by 550 (30/550=0.05 hp), or to mechanical watts by multiplying by 1.36 (1.36 x 30 = 40.8W). Working these conversion units into Equation 4-9 yields this result:

$$P_h = Fs/550 \qquad \textbf{Equation 4-10}$$

$$P_w = 1.36Fs \qquad \textbf{Equation 4-11}$$

where Ph = required horsepower, Pw = required power in watts, F = weight of the machine (lb), and s = speed of the machine (ft/sec).

Equations 4-9 through 4-11 give you a pretty good idea of the magnitude of the job. You reckon the weight of your machine, perhaps taking an educated guess, and you determine for yourself how fast you want it to move across the floor. The equations then turn out the power required for the job. The equations reflect some rather obvious principles: The heavier the machine is, the more power you need to move it; the faster you want it to travel, the more power you need. The power available at the drive wheel, in other words, is directly proportional to the product of machine weight and speed.

That's rather simple, but here's the tricky part: The motor must develop a certain rpm in order to get the straight-line speed you want, and the equations do not deal directly with motor rpm. You have to know the necessary rpm in order to specify the motor assembly; knowing just the mechanical output power is not enough.

The distance a wheel moves with each revolution is equal to its circumference. From elementary geometry, the circumference of a circle (wheel) is given by $2\pi r$, where r is the radius of the circle (wheel). Thus:

$$d = 2\pi r$$

where d=the linear distance moved with each revolution of the wheel, and r=the radius of the wheel.

For example, a wheel having a radius of 2 in will move $2\pi(2)$=12.6 in with each revolution.

But your motor is going to turn more than once, and you can take this into account in the distance formula by multiplying by the total number of complete revolutions:

$$d = 2\pi nr$$

where n= total number of complete revolutions.

Who wants to sit around counting complete revolutions of the wheel, though? That n term has to be replaced with something more practical. It figures that the total number of complete revolutions is equal to the turning rate multiplied by the time of observation: n=turning rate(revolutions per second) x time (seconds).

90

The turning rate of a motor is normally specified in rpm, rather than rps (revolutions per second). Using a figure for rpm and dividing by 60 gives the turning rate in rps. The equation for the number of complete revolutions thus becomes:

$$n = M_3 t/60$$

where n = total number of complete revolutions, M_r = rotational rate of the wheel (rpm), and t = elapsed time (sec).

Substituting this expression for n into the equation for distance:

$$d = \frac{2\pi M_r tr}{60}$$

where d = distance moved (in), M_r = wheel speed (rpm), t = elapsed time (sec), and r = wheel radius (in).

This is still an awkward situation, however, because it calls for observing absolute distance and elapsed time. You probably don't want to sit around measuring those quantities any more than you want to count rotations of the wheel. To get out of the situation, try dividing both sides of that last equation by time t. That gives you a d/t expression on the left-hand side of the equation.

And what is the meaning of d/t? That is the basic expression for linear speed, or how fast the machine moves in response to the turning wheel. The equation now looks like this:

$$s = \frac{2\pi M_r r}{60}$$

where s = straight-line speed (in/sec), M_r = wheel speed (rpm), and r = radius of the wheel (in).

When selecting a motor for a parabot/robot project, it is the motor rpm you are trying to find. Applying some algebra, the last equation can be reworked to solve for M_r:

$$M_r = \frac{60s}{2\pi r}$$

From this equation, you can figure the wheel rpm after specifying the speed (in/sec) and wheel radius (in). Normally, though, we like to express speed in ft/sec and the size of the wheel in terms of its diameter. Multiplying a speed in ft/sec by 12 gives the speed in in/sec, and dividing a diameter by 2 gives the radius of the wheel.

Substituting these conversions:

$$M_r = \frac{60s(12)}{2\pi(D/2)}$$

where s = the linear speed in ft/sec, and D = the wheel diameter in inches. After cleaning up the equation by working out the constants, we come up with a valuable design equation:

$$M_r = \frac{720s}{\pi D} \qquad \textbf{Equation 4-12}$$

where M_r = turning rate of the wheel (rpm), s = linear speed of the machine (ft/sec), and D = diameter of the wheel (in).

For example, suppose you want your machine to travel at a peak speed of 1.5 ft/sec, and you have some wheels with a diameter of 4 in. What rpm is required? Plugging the given numbers into Equation 4-12:

$$M_r = \frac{720(1.5)}{\pi(4)} = 86 \text{ rpm}.$$

The equation shows what might be intuitively obvious: The required rpm for the wheel is directly proportional to the speed you want the machine to move and inversely proportional to the diameter of the wheel. With a given wheel diameter, getting a faster speed from the machine is a matter of turning a higher rpm. But given a certain rpm, you can make the machine move faster by increasing the diameter of the wheels. It's possible to massage that equation in a lot of different ways, supporting a number of notions that might be of practical significance to you.

Before leaving this set of ideas, imagine yourself in a common situation. You already have a motor at hand as well as a wheel it will drive. What you want to know is how fast that assembly will drive a parabot/robot machine. Reworking Equation 4-12 to solve for speed, s:

$$s = \pi D M_r / 720$$

If the motor is rated at 120 rpm and the wheel is 6 inches in diameter, the speed is $\pi(6)(120)/720 \approx 3$ ft/sec. That machine will zip right along.

To this point in the discussion, you can calculate the amount of mechanical power the wheel/motor system must develop (Equations 4-10 and 4-11) and the required rpm (Equation 4-12). The equation for the required rpm is already worked out in commonly specified terms. Most small motor systems, however, are specified by torque, rather than power. The power equations are still important for later use, but things have to be worked out in terms of torque as well. What is the torque rating of the motor/wheel assembly?

Equation 4-6 solves for mechanical power (ft-lb/sec) in terms of torque (lb-ft) and rotational rate (rpm). That equation can be reworked to solve for torque, L:

$$L = \frac{60P}{2\pi M_r}$$

And from Equation 4-9, P=Fs. Substituting into the torque equation yields:

$$L = \frac{60Fs}{2\pi M_r}$$

If you have already calculated the necessary rpm from Equation 4-12, you can fit that figure into this equation, combining it with the machine weight (F) and linear speed (s) to come up with the required amount of output torque.

Using the M_r equation in this way makes it necessary for you to specify some speed and diameter parameters a couple of different times. By making some suitable substitutions and cancellations of like terms, it is possible to come up with an even simpler equation for required torque: L= FD/24, where F and D are the weight of the machine and wheel diameter, respectively. It is left to you to figure out how this simple equation evolves. Do you still wonder why one has to study so much algebra in electronics school?

Here is a summary of the useful torque equations:

$$L = \frac{60Fs}{2\pi M_r} \qquad \textbf{Equation 4-13}$$

$$L = FD/24 \qquad \textbf{Equation 4-14}$$

where L= required torque (lb-ft), F= weight of the machine (lb), s= linear speed of the machine (ft/sec), M_r= rotational speed of the wheel (rpm), and D= diameter of the wheel (in).

The equation, L=FD/24, suggests some important practical ideas that might not be obvious to you. Suppose, for instance, you have a motor that has a certain torque rating. How do you go about getting that motor to move a greater amount of weight than you originally proposed? The equation states you can carry more weight by simply reducing the diameter of the wheel. Sound strange? Maybe so, but it's hard to argue with what the equation states.

There is a trade-off, though. If you reduce the diameter of the wheel, the speed of the machine will drop by a proportional amount (see Equation 4-12).

Indeed, equations can tell you things you might not have realized before. Again, that is the reason why it is so important to study all the math you can. And that's why this chapter goes into math into such great detail.

In a practical sense, there are two ways to couple mechanical power from a motor assembly to a drive wheel. The simpler procedure is to buy a gear motor assembly and connect its output shaft directly to the axle of the wheel. The procedure assumes the output rpm of the gear motor and torque fall into line with your calculations for those parameters. The gear motor torque and rpm are transferred directly to the drive wheel.

The second procedure involves a gear or pully arrangement you insert, yourself, between the output shaft of the motor assembly and the drive wheel. Maybe you want a chain drive mechanism between the motor and wheel, or maybe you can come up with a simple gearing assembly of your own. Small machines can be operated by a friction-gear arrangement or a set of pulleys that are mechanically ganged with a tough rubber band.

No matter what the exact details are, the principles are the same in this second case. There is usually some change in the torque and rpm ratings you must take into account.

Figure 4-2 illustrates a basic gear or pulley arrangement. The shaft of the motor or gear motor is connected directly to wheel 1. The radius of wheel 1 is specified as r_1. The wheel having radius r_2 is fastened directly to the shaft of the drive wheel. Wheels 1 and 2 are mechanically ganged by a chain, belt or direct gear connections.

Ideally, all of the power generated by the motor assembly at wheel 1 is transmitted to wheel 2 and the drive wheel. Some losses will occur along the way, making the power available at the drive wheel somewhat less than that developed by the motor. This particular loss of power will be disregarded for the time being.

The power at wheel 1 and wheel 2 is the same. This leads to a simple mathematical expression, $P_1 = P_2$, where P_1 is the power available from the motor, and P_2 is the power transmitted to the drive wheel.

Substituting elements from Equation 4-6, we get the relation:

$$\frac{2\pi L_1 M_{r1}}{60} = \frac{2\pi L_2 M_{r2}}{60}$$

And cancelling like terms:

$$L_1 M_{r1} = L_2 M_{r2}$$

where L_1 = torque at wheel 1 (lb-ft), L_2 = torque at wheel 2 (lb-ft), M_{r1} = rotational speed at wheel 1 (rpm), and M_{r2} = rotational speed at wheel 2 (rpm).

Through a similar algebraic process involving Equation 4-12, it is possible to show:

$$\frac{M_{r1}}{M_{r2}} = \frac{r_2}{r_1}$$

Working these two results together yields yet another important equation:

$$L_1 M_{r1} = L_2 M_{r2} \qquad \textbf{Equation 4-15}$$

Summarizing for the sake of future reference:

$$M_{r1} r_1 = M_{r2} r_2 \qquad \textbf{Equation 4-16}$$

$$L_2 r_1 = L_1 r_2 \qquad \textbf{Equation 4-17}$$

where L_1 = torque developed at wheel 1, L_2 = torque developed at wheel 2, M_{r1} = rotational rate of wheel 1, M_{r2} = rotational rate of wheel 2, r_1 = radius of wheel 1, and r_2 = radius of wheel 2.

94

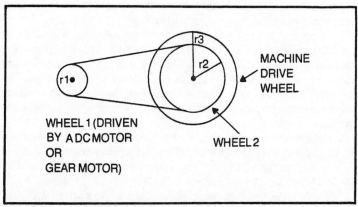

Fig. 4-2. A typical gearing or pulley arrangements.

Equations 4-15 through 4-17 tell you some things you might recall from a course in basic physics and mechanics. If, for example, the radius of wheel 2 is greater than that of wheel 1 (a common situation in projects of this kind), wheel 2 will turn more slowly than wheel 1. See that from Equation 4-16: $M_{r2} = M_{r1}(r_1/r_2)$. Supposing r_1 is 0.5 inches, r_2 is 2 inches, and M_{r1} is 200 rpm, the equation states that $M_{r2} = (200)(0.5/2) = 50$ rpm. There is a reduction in rpm in this case, which isn't always a bad thing; maybe your motor is running too fast anyhow.

What is lost in rpm is gained in torque, as suggested by rearranging Equation 4-15: $L_2 = L_1 M_{r1}/M_{r2}$. If wheel 1 generates a torque of 2 lb-ft, and the other specifications are the same as in the previous example, $L_2 = (2)(200/50) = 8$ lb-ft. This represents a 4:1 increase in torque. With a radius increase of 1:4, the rpm decreases by 4:1 and the torque increases by 1:4.

The Practical Side of Drive Motor Specification

While all these principles and equations can help you pin down specifications for the drive system of your project, they are all too cumbersome when it comes to making preliminary estimates in the real world. The following examples and procedures are intended to help you rough out some ideas about the drive motor scheme in a real sense.

Here is a rather common sort of situation. You have a robot machine that weighs about 15 lb. You want to move it across the floor at about 1.5 ft/sec. The drive wheels you have in your shop have a diameter of 4 in., and there is a gearing mechanism connected to the wheel that has a diameter of 3 in. The gearing mechanism represents wheel 2 in Fig. 4-2. On the other hand, you have a 120 rpm gear motor with a torque rating of 0.75.

The first question is this: Will that particular motor be able to mold the robot as you have specified? If so, what gear diameter do you need at the motor (wheel 1 in Fig. 4-2)? If the motor will not handle the job, what are the minimum specifications for one that will.

The first question is one of those go/no-go questions. Will the motor work or not? There is a quick and easy way to answer that. First, figure the mechanical power required to move the robot as you have specified and then figure the mechanical power that can be developed by the motor. The motor will work if its mechanical power exceeds the required power of the robot (remembering to allow a bit of extra power to cover losses in the system).

From Equation 4-9, the required mechanical power is 1.5 ft/sec x 15 lb, or 22.5 ft-lb/sec. Next, the equations leading up to Equations 4-13 and 4-14 show that the mechanical power developed by a motor is found by $P=2\pi LM_r/60$; or in this case, $P=2\pi(.75)(120)/60= 12.6$ ft-lb/sec.

There is no way that motor will do the job. No matter which gear ratios you might use, a motor generating just 12.6 ft-lb/sec of mechanical power cannot possibly drive a system calling for 22.5 ft-lb/sec. Forget about using that motor for driving applications. You might be able to use it somewhere else, but certainly not here.

So what are the minimum motor specifications? Well, it will be a motor that generates a bit more than 22.5 ft-lb/sec. But that works out to be an infinite number of combinations of rpm and torque. Those two figures, multiplied together have to come out to be greater than 22.5. Maybe this is where the money you invested in this book begins paying off for you!

Look at the graph in Fig. 4-3. It shows the torque-rpm product of a motor as a function of required mechanical power. To see how the graph works, consider that 22.5 ft-lb/sec is the power required for moving the robot in this example. Find that figure on the horizontal axis of the graph, which is a point about halfway between 20 and 25.

Then run a vertical line up to the line on the graph, and a horizontal line over the vertical axis. That last line intersects the vertical axis at about 225. That means the motor you select must have a torque and rpm rating such that, when the two figures are multiplied together, the number is at least 225.

A lot of numbers still fit the requirement, but at least you are getting things narrowed down a bit. You can, for example, use a 450 rpm motor that has a torque rating of just 0.5 lb-ft, a 225 rpm motor having a torque rating of 1 lb-ft, or a 112.5 rpm motor with a torque of 2 lb-ft. You should be able to come up with a motor that will meet, or preferably exceed, those figures.

Any motor satisfying the graph in Fig. 4-3 will, in principle, do the job for you; at least it will have the power capability. The next step is to figure out how to arrange the gearing to get the desired rpm at the wheel. You are exceedingly lucky if you find a motor that has the right rpm at the outset.

In this example, the wheel-driving gear has a diameter of 3 inches. From other information and from Equation 4-12, the necessary rpm for the

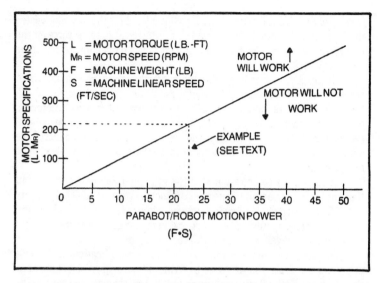

Fig. 4-3. Graph for determining motor torque and rpm as a function of mechanical output power.

drive wheel comes out to be $M_r = \dfrac{720(1.5)}{(4)\pi} = 86$ rpm.

The gear motor selected for the job probably doesn't have a speed rating of 86 rpm, so you have to select a radius (or diameter) for a wheel/gear to be fixed directly to the gear motor shaft. The size of that wheel or gear is determined by applying Equation 4-16.

By that eqation, $r_1 = M_{r2}r_2/M_{r1}$. If the speed of the gear motor shaft happens to be 240 rpm, $r_1 = (86)(1.5)/240 = 0.54$ in. You thus bring everything up to specifications by finding a primary gear (wheel 1 in Fig. 4-2) that has a radius of 0.54 inches. It doesn't have to be exactly that figure; you can probably tolerate some slop in the running speed of the machine.

About the only real practical problem with this gear selection process is caused when a motor with a very high rpm is selected. In that case, your calculations might show a gear radius that is smaller than the radius of the motor shaft, making it necessary to either find a motor with a lower rpm or devise a multiple-wheel, speed-reducing gear train of your own.

Motors for Vise-Action Manipulators

Of all the possible mechanical configurations for manipulators, I have found those based on a vise action to be the most versatile in the long run. Even where a gear or pulley-type might seem more feasible at first, I tend to stay with the vise configuration.

Figure 4-4 illustrates the basic vise-action principle. As the motor shaft rotates, it turns a threaded rod fixed to that shaft. The fixed frame portion of the vise grip has a tapped, or threaded, hole that accommodates the threaded rod. So as the threaded rod turns, it screws through the fixed-frame hole, moving the motor assembly and moveable portion of the vise grip with it.

Mechanically speaking, the main advantage of this manipulator scheme lies in the mechanical advantage of the screw mechanism. As will be shown in this section, it is possible for relatively low-power motors to exert a surprising amount of force at the grips. But that's just one side of the matter. The real advantage of the system is that it takes an incredible amount of external force at the grips to make them move apart, even when motor is turned off.

Thus, a force can be applied at the grips by running the motor, but then the force can be maintained after the motor is turned off. That is a very useful feature in most parabot/robot manipulator systems. In the long run, it is an efficient system in an electrical sense. Most alternate schemes require motor power to maintain a force.

The scheme shown in Fig. 4-4 is just one application of the vise principle. Certainly you will want to do more than grip an object between the two grips. Instead of squeezing an object, for instance, you can use the moveable frame to push or pull a lever that operates a jointed section of a manipulator. Most alternate schemes can only pull, making it necessary to use two motors and a clutch assembly to achieve motion and force in two different directions.

The main drawback of the vise scheme is its mechanical inefficiency. Some friction will always be at the point where the threaded rod screws through the tapped hole in the fixed-frame assembly. In smaller systems, this friction might reduce the mechanical efficiency to 50 percent or less. Efficiency thus becomes an important factor in the process of selecting a motor for this sort of task.

Many of the equations for selecting drive motor and for motors in general apply here. The main differences are the equations for calculating the amount of work to be done.

With drive motors, force was defined in regards to the weight of the machine. The task of the drive motor is to move the entire machine against gravity and its own frictional forces and inertia. In the case of this vise-type manipulator, force is defined in terms of the amount of pressure, in pounds, that the moveable grip will apply. So the work that the system has to do is defined by the equation, $W=Fd$, where W is the amount of work done in ft-lb, F is the force applied in lb, and d is the distance the moveable grip travels under that amount of applied force (in ft).

Mechanical power is the time-rate of doing work:

$$P=Fd/t$$

where P=power applied (ft-lb/sec), F=force applied (lb), d=distance moved (ft), and t=elapsed time (sec).

And because the ratio, d/t, defines speed, it is possible to come up

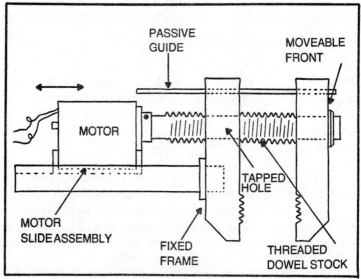

Fig. 4-4. A simple vise-action manipulator assembly.

with a useful alternative to the power equation:

$$P=Fs$$

where P=power applied (ft-lb/sec), F=force applied (lb), and s=speed of motion (ft/sec).

Work should be expressed in ft-lb/sec, but the scale of manipulator projects makes it more desirable to specify the speed in in/sec. So dividing the last equation by 12 provides a key equation for this job:

$$P=Fs/12 \qquad \textbf{Equation 4-18}$$

where P=required mechanical power (ft-lb/sec), F=desired amount of force (lb), and s=desired speed of motion (in/sec).

Suppose you want to be able to close the grips—or in a more general sense, move the moveable grip—at a rate of 0.5 in/sec and apply a force of 100 lb. The amount of mechanical power the scheme must develop by Equation 4-16, is (100)(0.5)/12=4.2 ft-lb/sec. You can apply Equations 4-10 and 4-11 to get the power expressed in horsepower ($P_h = 4.2/550) = 7.6 \times 10^{-3}$hp) or watts ($P_w = 1.36(4.2) = 5.7$ w).

If those power figures seem rather small compared to those found for drive motors, there is a good reason for it: The larger forces required for manipulators are more than balanced by very low operating speeds. The slower you move a manipulator section, the more force it can develop for a given expenditure of mechanical power.

The next step is to come up with some figures for the amount of torque that must be applied by the screw mechanism to develop the necessary power. Equation 4-13 $L = \dfrac{60Fs}{\pi 2M_r}$ is helpful, but it isn't altogether suitable because the speed, s, is expressed in feet per second. Inches per second is more appropriate for most manipulators. Dividing the equation by 12 remedies the matter, and the equation looks like this:

$$L = \frac{2.5Fs}{\pi M_r}$$

where L=required torque (lb-ft), F=desired force (lb), s=desired operating speed (in/sec), and M_r = rotational rate of screw (rpm).

The only trouble with this equation is that it cannot be solved with the available information. You can specify the force and speed, but you cannot solve for the torque until you know the rpm—and you cannot simply guess at the rpm. How can rpm be worked into this scheme?

The screw mechanism, itself, will have a certain *pitch*, which is the distance between adjacent threads. Each revolution of the screw will move the assembly by a distance equal to that pitch distance. Those facts can be assembled into a simple, but very useful, equation:

$$M_r = 60sb \qquad \textbf{Equation 4-19}$$

where M_r = rotational rate of the screw (rpm), s=desired speed of motion (in/sec), and b=thread density of the screw (threads/in.)

Play with that equation for a moment, and you'll find that it makes sense. For instance, a higher thread density will cause a higher rates of linear motion for a given rpm.

Now, the rpm equation can be substituted for the M_r term in the torque equation:

$$L = \frac{F}{24\pi b} \qquad \textbf{Equation 4-20}$$

where L=required torque (lb-ft), F=desired force (lb), and b=thread density of the screw (threads/in).

That equation proves that using a higher thread density will require a lower torque to produce a given amount of force. The screw arrangement does, indeed, give this system a real mechanical advantage. It trades off operating speed for force, but no matter how you look at it, the scheme provides a lot of force for a surprisingly small amount of input power.

To test the equations in a real-life situation, suppose you want to operate the vise manipulator at a rate of 2 in/sec and apply a force of 200 lb. If the screw has a thread density of 10 turns per inch, Equation 4-19 shows that the motor rpm should be $60(2)(10) = 1200$ rpm. And Equation 4-20 shows that the motor should develop a torque of at least $\dfrac{200}{24\pi(10)} =$ 0.27 lb-ft.

The torque figure is low, however, because it does not take into account the frictional forces of the screw mechanism. With the motor disconnected from the assembly and no forces applied to the grips, you might find the torque required to turn the screw is on the order of 0.25 lb-ft. The figure depends on so many different physical conditions: the pitch of the thread, how well the thread matches the tapped hole in the fixed grip, the amount of surface area within the tapped hole, the amount and quality of lubricant, etc.

Whichever way you might go about determining the frictional torque, it can be added to the calculated torque in Equation 4-20 to get a total torque figure. That one should be used for selecting the motor torque rating. Once you have figured motor rpm (Equation 4-19) and torque (Equation 4-20), use the procedures specified for drive motors to come up with a motor that can do the job for you.

Basic Motor Circuits
And Direction Controls

It is difficult to overestimate the importance of DC motors in parabot/ robot systems. The primary modes of expression of a motor almost inevitably take the form of some kind of motor reaction. Therefore, an experimenter must have at least a nodding acquaintance with motor control circuits.

This chapter deals with basic motor control circuits as they apply specifically to manipulator and main direction/drive schemes. The same information can be generalized to include other kinds of motor-controlled systems as well. Chapter 6 then deals with motor speed controls, and Chapter 7 takes up the subject of closing the loop to create more sophisticated sorts of servo systems.

SIMPLE ON/OFF MOTOR CONTROLS

The simplest and most obvious way to turn a motor on and off is by connecting a SPST (single-pole—single-throw) toggle switch or a normally open pushbutton switch in series with the motor and its power source. Both circuits require some mechanical interaction, usually from a human operator. Manually setting the toggle switch to its ON position or depressing the pushbutton completes the electrical circuit, applies battery power to the motor, and causes the motor to run. The complementary sort of manual action, setting the toggle switch to OFF or releasing the pushbutton, interrupts the circuit and allows the motor to stop running. See the diagrams in Fig. 5-1.

By getting some measure of automatic control into a simple motor switching scheme, it is possible to use more than one switching mechanism in the series-circuit arrangement. In Fig. 5-2, two kinds of switches are connected in series. S1, in that instance, is a SPST toggle switch while S2 is normally closed microswitch.

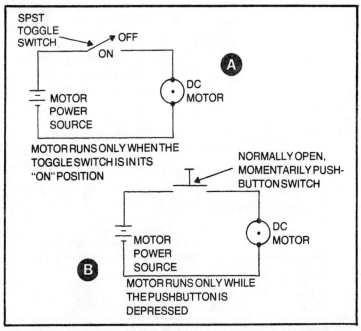

Fig. 5-1. Simple on/off motot controls. (A) A toggle switch control. (B) A pushbutton control.

Assuming the microswitch is in its closed position, the operator can apply power to the motor by setting the toggle switch to its ON position. The motor then runs until the operator returns the toggle switch to its OFF position, *or* some sort of mechanical condition arises whereby the microswitch is forced open.

Such a two-switch scheme can be useful for both machine drive circuits and manipulator controls. In the former case, the microswitch can serve as an obstacle-contact sensor. The operator gets the machine running across the floor by closing S1, but when the machine runs into an obstacle that activates the microswitch, S2, the machine stops running until the contact situation is removed.

Used in a manipulator system, the microswitch in Fig. 5-2 can serve as a limit switch that immediately removes power from the motor as the manipulator reaches an extreme limit of its motion in one particular direction. Using microswitches in this fashion, as limit switches for manipulators, is not an altogether trivial matter. Even systems operating from sophisticated electronic controls should have normally closed microswitches to remove power from the motor in case the control scheme fails somehow.

Figure 5-3 illustrates how it is possible to switch a pair of motors off and on with a single toggle switch, but still allow the motors to respond

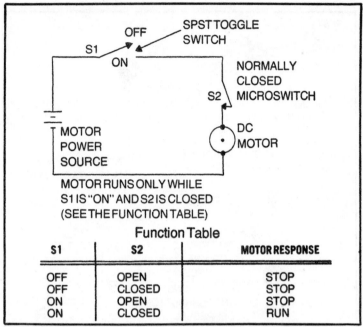

Fig. 5-2. A toggle switch on/off control with microswitch limiting. The microswitch, S2, overrides any setting of the ON/OFF switch whenever the manipulator mechanism reaches an extreme limit of travel.

individually to micro-microswitches of their own. Assuming for the moment that both microswitches, S2 and S3, are in their normally closed position, the operator can maintain complete off/on control over both motors, simultaneously, by working toggle switch S1. Any mechanical situation that energizes one of the microswitches, however, will override the toggle switch and cause the corresponding motor to stop running.

The exact meaning of the situation depends on what the running motors are supposed to be doing and how the microswitches are arranged. For example, the system can be a crude form of contact-sensing and control scheme for robot drive motors. Both motors run until the machine encounters an obstacle that opens one of the microswitches. It then responds by turning off that particular motor. But with the other motor still running, the machine will turn from its original path, perhaps steering itself away from the obstacle. Of course, the entire circuit will shut down whenever both microswitches are energized.

The success of such a simple, microswitch-controlled drive scheme depends on a carefully planned layout of the microswitches. In fact, it is possible to work out a complete robot blunder control scheme based on arrays of contact-sensing microswitches. Combinations of normally open and normally closed microswitches connected to each motor can be set up to handle just about any conceivable sort of obstacle-contacting situation.

While such an idea might appeal to experimenters who have no understanding of semiconductor logic circuits, the notion of using arrays of microswitches for control purposes is generally impractical in this day of low-cost and simple logic IC circuits. And for that reason, there is no point in pursuing the notion any further. Bear in mind, however, that microswitches do indeed make fine backup mechanisms for overriding more sophisticated electronic motor controls that can go awry.

Any switch connected in series with a DC motor must have a voltage and current rating that matches or exceeds the corresponding specifications for the motor it controls. This is not normally a problem, considering that most switches are rated at 120 VAC, 10A. Most motors for parabot/robot systems call for operating voltages less than 24 VDC, so using a switch rated at 120 VAC or 240 VAC works quite nicely. Some microswitches, however, have current ratings on the order of just 1 or 2 amperes. Of course a larger switch is in order if the motor normally runs with a power source drain of 8 or 10 amperes.

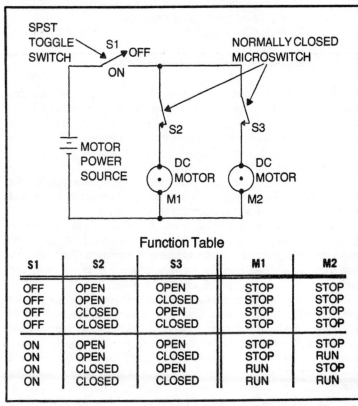

Function Table

S1	S2	S3	M1	M2
OFF	OPEN	OPEN	STOP	STOP
OFF	OPEN	CLOSED	STOP	STOP
OFF	CLOSED	OPEN	STOP	STOP
OFF	CLOSED	CLOSED	STOP	STOP
ON	OPEN	OPEN	STOP	STOP
ON	OPEN	CLOSED	STOP	RUN
ON	CLOSED	OPEN	RUN	STOP
ON	CLOSED	CLOSED	RUN	RUN

Fig. 5-3. A simple ON/OFF control for two different motors. The toggle switch control is common to both motors, but the limit switches each control one motor.

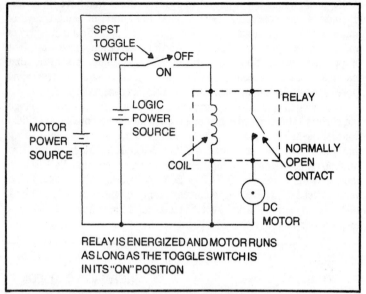

Fig. 5-4. Turning a motor on and off via a relay interface.

In any case, it is better to use switches having specifications that exceed the circuit requirements than those that just barely make it or fall short. If you cannot determine the voltage and current ratings of a switch by consulting a catalog or by reading the information impressed on the side of the switch, itself, it would be wise to select another switch having specifications you can determine.

The motor control circuit shown in Fig 5-4 introduces a relay into the on/off scheme. Using a relay in this manner permits electrical isolation of the motor circuit from the logic circuitry that controls it.

The relay contact, a normally open one in this case, is connected in series with the motor and its power source. The relay contact carries full motor current whenever the relay assembly is energized. The relay coil, however, is operated from an entirely different power source, one assigned to the logic or control circuitry of the system.

One of the main advantages of relay interfacing between logic circuits and motors is current isolation. Whereas the motor might be rated at 8 or 10 amperes, the relay coil might draw only 50 milliamperes from the logic power supply. In short, that means you can control the operation of a high-current motor from a relatively low-current logic control circuit. A relay, in other words, can provide a great deal of current gain. For example, 50 mA to 10 A amounts to a current gain of $10/50 \times 10^{-3} = 200$.

The current-isolating feature means you need to use heavier wire only for the motor circuit. The control circuit can use ordinary electronic hookup wire (22 or 24 gage) or even finer stranded wire.

And by using two separate power supplies as suggested in Fig. 5-4, fluctuations in the motor-supply circuit have no affect on the more critical logic power supply. Electrical noise generated in a supply line feeding a running motor can play havoc with control logic, especially microprocessor systems. Isolating the two power sources and interfacing them with a relay virtually eliminates that difficult problem.

Using a relay to control a motor also allows voltage isolation between the motor circuit and the logic that controls it. If the logic system is built around 5V TTL devices, there is no problem controlling a 12V motor. As long as the voltage rating of the coil matches the logic-circuit supply voltage, the motor can be operated at any other voltage level it requires.

The advantages of using relays to control DC motor circuits far outweigh any disadvantages, most of which are imaginary disadvantages anyway. The two most common complaints about relays is that they are slow and have short lifetimes. Both of these objections arise when comparing relays with solid-state relay devices, such as SCRs and triacs. The objections, in the context of motor controls for parabots and robots, are empty ones. Indeed, it might take something on the order of 50 ms to close a relay contact, but in a mechanical system having mechanical response times of 500 ms or more, relays are quite adequate.

Relay contacts do wear out, but relay manufacturers normally guarantee something on the order of a quarter-million operations. And that is generally a very conservative figure. If a relay is properly treated, it will last far longer than a robot experimenter will care to use it.

The only real drawback of mechanical relays, compared to semiconductor power-switching circuits, is their relatively high cost. The higher cost of relays should be traded off against the inability of SCRs and triacs to function very well in DC circuits. It is possible to set up such semiconductors for DC power control, but the cost and complexity of the necessary auxiliary circuitry far exceeds that of a simple relay. So if you are intent upon ruling out relays as interfacing and isolation devices between control and motor circuits, it must be on grounds other than operating speed, reliability and even cost.

The circuit shown in Fig. 5-4 is mainly for illustrative purposes. Little is to be gained by using this circuit, having a toggle switch control the logic power applied to the relay coil. It does the same job as the simpler and less expensive circuit shown in Fig. 5-1A.

Figure 5-5 shows a more compelling application of the relay interfacing circuit. In this case, the relay is energized and deenergized by means of a logic level applied to the base of transistor Q1. It is current flowing from the collector circuit of the transistor that energizes the relay.

The transistor is necessary whenever the logic circuitry is capable of driving only 10 mA or 5 mA current loads, such as TTL or CMOS circuitry. Most relay coils are rated at 20 mA to 100 mA—far too much current for TTL and CMOS IC devices. The transistor thus boosts the current level available from those sources, making it possible to energize the relay.

The relay in this circuit, as mentioned earlier, is energized whenever transistor Q1 is conducting. Turn off the transistor, and the relay is deenergized. Using a normally open relay contact in series with the motor, it follows that the conduction of the transistor is in step with the running of the motor. When Q1 is switched to its ON state, or fully saturated, it energizes the relay and turns on the motor. Turning off the transistor deenergizes the relay and turns off the motor.

A convenient way to turn on this transistor—and the motor as well—is by connecting the INPUT connection of the circuit to the positive side of the LOGIC SUPPLY voltage. That forward biases the transistor, and if the values of the resistors have been properly chosen, the transistor will go into saturation. The transistor conducts to energize the relay and turn on the motor circuit.

The transistor and motor can be turned off by either connecting the INPUT to ground or by making no connection at all to the INPUT terminal. In either case, the transistor is not biased on; it cannot conduct, it cannot energize the relay, and it cannot make the motor run.

The motor cannot run whenever there is no connection to the INPUT connection. This is a fail-safe feature that should be built into all motor control schemes. A broken connection to the INPUT, for example, will cause the machine to stop running, as opposed to the much less desirable alternative of causing the machine to run away when something goes wrong at the INPUT.

Using a transistor driver circuit for a relay coil retains all the isolation characteristics described for the switch-controlled circuit shown in Fig. 5-4. The current rating of the transistor, for instance, does not have to be anywhere near that of the motor circuit. Also, the logic supply voltage can be quite different from that required for the motor circuit. Just make sure the voltage rating of the relay coil is compatible with the logic supply voltage.

Take note, however, that the positive voltage applied to the INPUT connection should be adequate for saturating the transistor. The logic supply voltage applied to the relay coil and collector of the transistor can be different from that applied to the INPUT connection, but you must take steps to make certain that INPUT voltage level can switch on the transistor.

How should you go about selecting the transistor and resistor values for a circuit such as the one shown in Fig. 5-5? It isn't difficult at all if you are willing to abide by a couple of rule-of-thumb design procedures.

As far as the transistor is concerned, it should have a collector current rating and emitter-to-collector voltage rating that exceeds those specifications for the relay coil. So if you are using a relay coil rated at 6 VDC at 25 mA, Q1 must be able to switch 6V (there are few transistors that cannot hold off 6V—most are rated at 30V or more) and carry a collector current in excess of 25 mA. It is rather easy to look through a list of transistor ratings and come up with a host of candidates for the job.

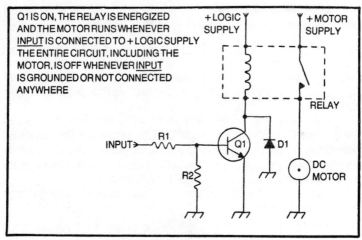

Fig. 5-5. Controlling a motor relay by means of a transistor.

Here is the first rule-of-thumb procedure: Assume any transistor you select will have a current gain of 10 or more. With the exception of a few very-high-power transistors (which you don't want to use in this application anyway), just about any transistor you pick will have an h_{fe}, or beta, specification far in excess of 10. You want a current gain of at least 10; anything over that is just icing on the cake. That is all you need for picking the transistor.

The only purpose of resistor R1 in Fig. 5-5 is to limit transistor base current whenever the INPUT terminal is connected to the positive side of the logic supply voltage. R1 just goes along for the ride whenever INPUT is connected to ground or nowhere at all.

For the sake of working out an example, suppose the positive voltage to the INPUT connection if 6V. Further assuming the emitter-base junction of the transistor will drop about 0.5V whenever it is forward biased means R1 will have to drop 6V-0.5V, or 5.5V.

Returning to the rule-of-thumb assumption that the transistor will have a minimum gain of 10, you can figure that the turn-on base current is one-tenth that of the required collector current. And because the collector current must be 25 mA to drive the relay coil, it follows that the base current should be 2.5 mA, or one-tenth the collector current.

Enough information now exists for using Ohm's law in determining the value of R1. That resistor, whenever INPUT is connected to +6 VDC, has to drop 5.5V at about 2.5 mA. Working Ohm's law with those values, the calculated value of R1 comes out to 2.2K. If you are one who tends to be a bit conservative when it comes to designing circuits of this kind (as I am), you might want to opt for a value that is just a bit smaller, such as a 2K resistor. Doing so, you allow a bit of extra drive current that will cover for some "errors" in the opposite direction elsewhere in the circuit.

In any event, the general equation for selecting the value of that series-connected base resistor is:

$$R_i = (V_{cc} - 0.5)/I_b \qquad \textbf{Equation 5-1}$$

where R_i = the value of the series base resistor, V_{cc} = the positive voltage level applied to INPUT, and I_b = the base current required for the transistor.

The value for the series base resistor rarely happens to come out to be a standard resistor value, such as 2.2K in the previous example. Whenever the calculation turns up an oddball resistance value, just pick the next *lower* standard value. It is better to drive the base circuit more than necessary than to drive it with too little current.

Here is a more general version of the equation for selecting the value of that input resistor:

$$R_i = (V_i - V_{eb})/0.1I_{rc}) \qquad \textbf{Equation 5-2}$$

where R_i = the value of the series base resistor, V_i = the positive voltage level applied at the INPUT terminal, V_{eb} = the emitter-base voltage of the transistor (usually around 0.5V for small transistors), and I_{rc} = the current rating of the relay coil.

The purpose of resistor R2 in Fig. 5-5 is to ensure the transistor is switched off whenever the INPUT is either connected to ground or not committed to any connection at all. The rule-of-thumb procedure for selecting that resistor value is to pick one that is about 10 times the value of the series input resistor. So if solving Equation 5-1 or Equation 5-2 turns up a value of 2.2K for R1, you should use a 10 x 2.2K, or 22K resistor for R2.

The diode connected across the transistor is not absolutely necessary for proper operation of the circuit. Any circuit having a semiconductor device driving an inductive load, such as a relay coil or motor winding, should include this reverse-connected diode.

The purpose of the diode is to clip the high-voltage spikes that occur whenever the current through the inductive load is suddenly switched off by the semiconductor. Without that diode in place, the semiconductor will have to hold off a voltage equal to about twice the supply voltage for several milliseconds each time it switches off the current flow. Such voltage spikes can eventually break down the transistor, so try to include the diode wherever a semiconductor drives a coil of wire. It can be any small rectifier diode.

Consider the following specifications when selecting a motor-control relay: the *voltage and current rating of the contacts*, and the *voltage and current rating of the coil*. Because the relay contacts serve the function of a switch connected in the motor circuit, consider the contact ratings in the same light you would the ratings of a switch. The voltage rating and current specification must equal or exceed that of the motor.

Quite often, you will end up having to use a relay that has contact configurations far more complicated than a simple on/off motor circuit calls for; the circuit in Fig. 5-5, for instance, requires only a normally open, SPST relay contact. You might find that a DPST (double-pole—single-throw) or even a DPDT (double-pole—double-throw) relay is easier to locate and, in some instance, less expensive than the simpler counterpart.

That is no real problem, though. It is possible to wire any of the more complicated contact configurations to operate in a normally open, SPST mode. In fact, if you are an avid robot experimenter, you can work with DPDT relays exclusively. As your projects grow in complexity and sophistication, you can save the time and expense of getting new relays for each project.

As far as relay coil ratings are concerned, the *voltage rating* is the more critical one. You can cover for just about any current rating by working out a transistor driver circuit that will supply the necessary current. Relay coil voltages most useful for parabot/robot projects are those of 5 VDC, 6 VDC, 9 VDC and 12 VDC. Just be sure to pick one that matches the available voltages as closely as possible.

ON/OFF AND DIRECTION CONTROLS

The usefulness of parabot/robot motor controls multiplies an immeasurable amount when direction-reversing features are added to a basic on/off control system. Being able to change the direction of the motor and stop it at will is often all that is necessary for many kinds of motor-driven systems. Rodney, my most sophisticated and meaningful working robot system, works quite well with a two-motor drive scheme built around this on/off two-direction, single-speed idea.

Figure 5-6 shows the most reliable sort of motor direction control. It is possible to design simpler motor-reversing schemes that use a single relay or a couple of semiconductor devices, but those alternatives either do not have a built-in mechanism for stopping the motor or they have the potential for creating short circuits across the motor power source.

Note that there are two SPDT relays in the circuit. When neither relay is energized, the motor cannot run because both of its terminals are connected to the common side of the motor power source. Without any difference in electrical potential across the motor, it cannot possibly run. By the same token, energizing both relays prevents the motor from running. The only way the motor can run is by energizing one relay or the other. In a manner of speaking, the circuit works like an exclusive-OR logic system. It is energized only when the two inputs have opposite on/off states.

Whenever switch S100 is OFF and S101 is ON, for example, relay RL100 is deenergized and RL101 is energized. Consequently, the positive side of the motor is at power-source common and the negative side is at the

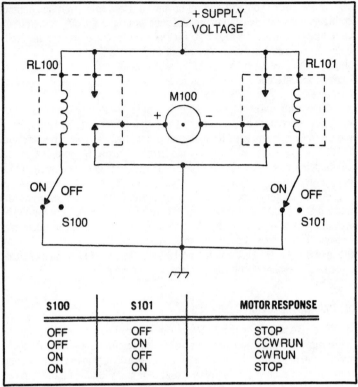

S100	S101	MOTOR RESPONSE
OFF	OFF	STOP
OFF	ON	CCW RUN
ON	OFF	CW RUN
ON	ON	STOP

Fig. 5-6. A basic two-direction motor control circuit.

positive power-source potential. The motor is thus connected across the power source in a manner that makes it run at full speed in a counterclockwise direction.

Reversing the settings of the switches (turning ON S100 and turning OFF S101) reverses the direction of the motor. In that condition, RL100 is energized, RL101 is deenergized, and the terminals of the motor are connected to the motor power source in a way that makes it run in a clockwise direction. There is no way to short circuit the power source by manipulating the control switches in this circuit, and the motor can always be stopped by setting the switches to identical ON/OFF conditions.

The circuit shown in Fig. 5-7 demonstrates how the toggle switches can be replaced with relay-driver transistors. This is a two-direction version of the simple on/off circuit already described for Fig. 5-5 in the previous section of this chapter. The reason for using the driver transistors is to boost the current level from any low-power logic circuitry that might be used for determining the operating status of the motor.

The inputs to the circuit are labeled MRF and MRR, and Table 5-1 shows how the motor responds as all four possible combinations of logic

Table 5-1. Logic Table for Circuit of Fig. 5-7.

MRF	MRR	MOTOR RESPONSE
0	0	STOP
0	1	CCW RUN
1	0	CW RUN
1	1	STOP

0 = CONNECTION TO COMMON ($\perp$)
1 = CONNECTION TO +CONTROL SUPPLY

levels are applied to these two inputs. Whenever a logic 0, or ground potential, is applied to both MRF and MRR, neither relay is energized and the motor cannot run. And incidentally, the system takes on that same status if no connections at all are made to the inputs. That is the fail-safe feature of the circuit. Applying a logic-1 level to both inputs also causes the motor to stop running.

The only way to get the motor running is by applying opposite, or complementary, logic levels to MRF and MRR. According to Table 5-1,

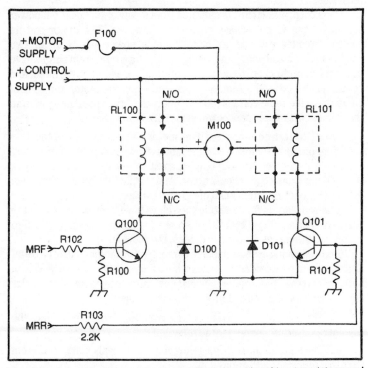

Fig. 5-7. A two-direction motor control featuring relay-drive transistors and separate supply voltages for the motor and its control circuitry.

113

applying a logic 0 to MRF and a logic 1 to MRR causes the motor to run in a counterclockwise direction. The logic-0 level at MRF turns off Q100, blocking any current flow to RL100. RL100 is thus deenergized and the positive side of the motor is connected to ground. The logic-1 level at input MRR forward-biases Q101 to turn it on. That allows current to flow through the coil of RL101 to energize it. The negative side of the motor is thus connected to the positive side of the motor power source. With full power-source potential applied to the motor, it runs at full speed. The polarity of the applied voltage dictates that it runs in a counterclockwise direction.

Reversing the logic levels applied to the two inputs, setting MRF to 1 and MRR to 0, establishes the opposite set of operating states throughout the circuit. RL100 is energized and RL101 is deenergized, the positive side of the motor is connected to the positive side of the motor supply and the negative side is grounded, and the motor runs in a clockwise direction.

The values for the input and base resistors for Q100 and Q101 can be calculated according to Equation 5-1 or Equation 5-2. The values depend on the amount of current required for driving the relay coils and the 1-and-0 logic voltage levels from whatever sort of circuit drives the two input terminals.

This particular circuit is operated from a pair of independent power supplies. That is not always necessary, however. The motor and its control circuitry can operate from a single power supply under two conditions: when the motor is so small that its turn-on current surges have no appreciable effect on the power-supply voltage, and when the motor and control circuit have the same voltage, and when the motor and control circuit have the same voltage rating. Otherwise, try to plan the system so that it uses two independent power sources, as shown here.

Fuse F100 can be eliminated if the motor is a relatively small one. I used here on the assumption that the motor supply is capable of providing current in excess of 10A—enough current to set the circuit afire if a short-circuit catastrophe should occur. And that is a likely condition when the motor is operated from a hefty lead-acid battery.

The procedures for selecting the relays have been described in the previous section of this chapter. Also see the discussion for Fig. 5-5 to appreciate the purpose of the two diodes, D100 and D101.

Before looking at some specific parabot/robot motor control circuits, it is important to consider instances where the relay coil should be driven by a pair of transistors, rather than just one. Such an instance arises under any one or a combination of the following conditions:

☐ The rated output current of logic circuits is less than 10 times the current required for energizing the relay coil.

☐ The logic circuitry driving the relay has a supply voltage different from that required for the relay coil.

☐ The logic circuitry generates active-low logic levels; that is, the relay is to be energized whenever the logic circuit outputs a logic 0, and the relay is to be deenergized whenever the logic generates a logic 1.

☐ There is a reasonable chance the experimenter will want to use the same relay driver circuit in an entirely different robot/parabot project.

The circuit shown in Fig. 5-8 satisfies any one of these conditions, or any combination of them. In a sense, it is a universal relay driver circuit. First notice that the circuit can be operated from three independent supply voltages. There is a +LOGIC SUPPLY for transistor Q1. This is the voltage supply used for operating any previous logic circuitry, usually TTL, CMOS or NMOS IC devices.

The +CONTROL SUPPLY provides collector voltage for Q2 through the relay coil. That voltage level matches the voltage rating of the relay coil

Finally, there is a +MOTOR SUPPLY. As you might suppose, this voltage level matches that of the voltage specification of the motor.

All three of these voltages can be different. The preceding logic ICs, for example, might operate from +5 VDC, while the relay coil is rated at 6 VDC and the motor runs from a +12 VDC source. In instances where the power-supply levels for the logic circuitry and relay coil are indentical, the +LOGIC SUPPLY and +CONTROL SUPPLY connections on the drawing can be connected together as one.

In other instances, all three of these supply connections can be strapped to the same power source. Of course, this setup is not desirable

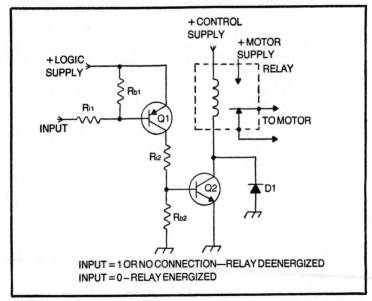

INPUT = 1 OR NO CONNECTION—RELAY DEENERGIZED
INPUT = 0 – RELAY ENERGIZED

Fig. 5-8. A transistor interface for driving a motor-control relay from a low-power logic source.

whenever the motor is a relatively large one that will stir up a lot of spikes on the power-supply line—spikes that can mess up the preceding logic circuits.

This high level of flexibility with regard to operating voltages is made possible by the isolation character of the two-transistor circuit configuration. Transistor Q1 isolates the logic circuitry from any +CONTROL SUPPLY voltage applied to the relay coil and Q2. The relay, itself, isolates the +CONTROL SUPPLY from the +MOTOR SUPPLY.

The circuit can be interfaced with very low current IC circuits, including NMOS IC devices. As shown shortly, the two transistors provide a total current gain on the order of 100. Operating a 50 mA relay coil calls for an IC current drain of only 1 mA, which is a figure well within the specifications for sensitive IC devices.

Finally, you must realize that the circuit inverts the correspondence between the logic level applied at INPUT and the action of the motor. It still has a desirable fail-safe feature, inasmuch as the relay is deenergized whenever there is no connection at all to the INPUT terminal. A logic-1, positive level applied to INPUT also deenergizes the relay, while a logic-0 level at INPUT energizes it. This logic-inversion feature turns out to be an advantage in more cases than not.

Now for the task of selecting the values of the resistors. The best place to start is at R_{i2}. This resistor is in a series current path for the emitter-collector circuit of Q1 and the emitter-base circuit of Q2. (The two transistors are always in conducting or nonconducting states at the same time.)

Running through that particular series current path, assume the forward voltage drop across the emitter-collector circuit of Q1 is about 0.7V when it is conducting. That is a typical saturation voltage for small transistors.

Then assume the emitter-base voltage for Q2, under the same circuit condition, is about 0.5V That, too, is a typical forward-biasing voltage for small transistors.

Add these two junction voltages together, and you can account for 0.7 + 0.5 = 1.2V in that series circuit. The voltage remaining to be dropped across R_{i2} is thus the +LOGIC SUPPLY voltage, minus 1.2V. If the +LOGIC SUPPLY voltage happens to be 12V (as in the case of CMOS logic circuitry), the voltage across R_{i2} will be 12 − 1.2, or 10.8V, whenever the two transistors are conducting. And this figure fixes the voltage for solving Ohm's law. The next trick is to come up with a current figure for R_{i2}.

The amount of current through R_{i2} depends on the current rating of the relay coil. Recall from the first section of this chapter that every transistor in this design procedure is assumed to have a gain of 10 or more. So the base-drive current for Q2 should be one-tenth the rated coil current, which means the current through R_{i2} should be set at 0.1 times the coil current for the relay.

116

If the coil happens to have a current rating of 50 mA and, from the example cited a bit earlier in this discussion, R_{i2} has to drop 10.8V, Ohm's law dictates the resistance value of R_{i2}: $R_{i2} = 10.8/5$ mA $= 2.16$K. Rounding down to the nearest standard resistor value, select R_{i2} to have a value of 2K. Gathering the assumptions into one equation, the value of R_{i2} comes out of Eqution 5-3A.

$$R_{i2} = (V_{LS} - 1.2)/(0.1I_{rc}) \qquad \textbf{Equation 5-3A}$$

where R_{i2} = the value of resistor R_{i2}, V_{LS} = the voltage rating of the + LOGIC SUPPLY, and I_{rc} = the current rating of the relay coil.

Equation 5-3B is a more general version of the same calculation.

$$R_{i2} = \frac{V_{ls} - (V_{ec} + V_{eb}5}{.1I_{rc}} \qquad \textbf{Equation 5-3B}$$

where R_{i2} = the value of resistor R_{i2}, V_{LS} = the voltage rating of the +LOGIC SUPPLY, V_{ec} = the emitter-collector saturation voltage of Q1, V_{eb}= the emitter-base junction voltage of Q2, and I_{c} = the current rating of the relay coil.

The purpose of R_{b2} in Fig. 5-8 is simply to ensure turn-off of Q2 whenever it is not being switched on by collector current from Q1. Its value is generally reckoned as 10 times the series base resistor, R_{i2}. So after using Equation 5-3A or Equation 5-3B to determine the value of R_{i2}, figure R_{b2} equal to 10 times that value.

$$R_{b2} = 10R_{i2} \qquad \textbf{Equation 5-4}$$

The next step is to calculate the value of R_{i1}, the input resistor for Q1. The purpose of this resistor is to limit the current from the preceding logic circuits whenever the logic level is 0, which is the condition necessary for forward-biasing Q1 and turning it on.

Assuming a current gain of 10 for transistor Q1, it follows that the base current through R_{i1} should be one-tenth that of the collector current for the same transistor. And since the collector current for Q1 is one-tenth the collector current for Q2, it figures that the current through R_{i1} is ultimately equal to one-hundredth the relay coil current. So if the relay coil is rated at 50 mA as in a previous example in this discussion, figure the current through R_{i1} to be 50mA/100 or 0.5 mA.

It is a bit trickier to find the voltage to be dropped across that resistor, but it must be done in order to calculate its value. The base circuit of Q1, you see, is operated from a logic-0 level from the preceding logic circuit. That implies you must know what the logic-0 voltage is—and it's rarely zero. Finding the logic-0 level of the IC device that drives the base of Q1 calls for looking up the figure in a data book. The figure is 0.4V maximum for TTL circuits, but it changes with the supply voltage level for CMOS and NMOS circuits. Use the appropriate logic data book to find a term generally expressed as V_{OL} *maximum,* which is the maximum expected voltage from a digital IC generating its logic-0 level.

For the sake of a specific example, suppose the base of Q1 is to be driven from the output of a TTL circuit. The V_{OL} maximum is listed at 0.4V. Now there is enough information to get things going smoothly.

The voltage at INPUT is thus figured as the +LOGIC SUPPLY level, minus the V_{OL} rating of the device, minus the forward junction potential of the emitter-base circuit of Q1. Assuming the logic circuits are operating from +5V, the V_{OL} of the preceding circuit is 0.4V and the emitter-base junction voltage of Q1 is 0.5V, it turns out that R_1 has to drop 5-0.4-0.5, or 4.1V.

The current, figured earlier in this example as 0.5mA, can be worked into Ohm's law to yield a value of $R_{il} = 4.1V/0.5mA$, or 8.2K. It so happens that 8.2K is a standard resistor value for 5 percent-tolerance resistors.

Equation 5-5 puts this all together:

$$R_{il} = \frac{V_{LS} - (V_{eb} + V_{oL}t}{0.01I_{rc}} \qquad \textbf{Equation 5-5}$$

where R_{il} = the value of resistor R_{il}, V_{LS} = the voltage rating of the +LOGIC SUPPLY, V_{eb} = the emitter-base junction voltage of Q1, V_{OL} = the maximum logic-0 output voltage of the IC device that drives the base of Q1, I_{rc} = the current rating of the relay coil.

The final resistor in the circuit, R_{bl}, makes certain the base voltage of Q1 is pulled up to the +LOGIC SUPPLY level whenever the preceding logic circuit is generating a logic-1 level. The value of that resistor should not exceed 10 times the value of R_{il}.

$$R_{bl} = 10R_{il} \qquad \textbf{Equation 5-6}$$

TWO-DIRECTION MOTOR CONTROLS FOR MANIPULATORS

Parabot/robot manipulators can be mechanically complicated, but their motor controls can be quite simple. This section draws upon the presentations in the previous one to illustrate how it is possible to assemble some relatively simple, yet useful, motor schemes for manipulators.

A Basic Two-Direction Manipulator Control

Figure 5-9 shows a simplified motor control scheme for a two-direction manipulator. The feature that sets it apart from a system drive motor circuit is the need for sensing and responding to the limits of mechanical motion.

This circuit looks very much like a basic two-direction, relay-operated motor control. The main difference is the use of a pair of limit

118

switches, S103 and S104. These limit switches make it desirable to include a RESET pushbutton, S102, as well.

Notice that the limit switches are of the normally closed variety, and that the pushbutton is normally open. For the time being, assume those switches are all in their normal states.

That being the case, the motor is controlled by a pair of normally open pushbutton switches labeled REV and FWD. Depressing the FWD switch applies CONTROL SUPPLY power to relay RL100, thus energizing it and applying the positive side of MOTOR SUPPLY to the positive terminal of the motor. The motor, under those conditions, runs in its forward direction.

Depressing the REV switch energizes RL101 and connects the positive side of MOTOR SUPPLY to the negative terminal of the motor. The motor responds by running the reverse direction. Depressing both FWD and REV has no effect on the motor, because both of its terminals are connected to the positive side of MOTOR SUPPLY. So as fas as the human operator is concerned, depressing the FWD pushbutton makes the motor run in a forward direction and depressing the REV button makes the motor run in reverse. It is a simple idea, but fully adequate for a pushbutton-operated, two-direction motor control.

The purpose of the two limit switches is to stop the motor automatically whenever the manipulator reaches its extreme limits of travel. These switches are physically located at the extreme-limit points in the mechanical system, and are intended to prevent stall current from flowing through the motor for extended periods of time.

Suppose, for example, the operator is running the motor in its forward direction. The manipulator is responding in an appropriate fashion but eventually reaches the mechanical limits of its motion in that direction. If the FWD LIMIT switch, S103, is properly located, it opens the moment the manipulator reaches that limit. And when that switch opens, it interrupts the MOTOR SUPPLY connection to the motor. Even if the operator is still depressing the FWD pushbutton, the motor stops running.

This is a fundamental sort of parabot mechanism, so it is up to the human operator to take corrective action—action aimed at getting the manipulator out of this extreme-limit, stall condition. Simply releasing the FWD pushbutton will not clear the problem; in fact, neither will depressing the REV button. Once the manipulator reaches one of its extreme limits and the corresponding limit switch is opened, the only way to clear the stall condition is by shorting out the limit switches by depressing the RESET button, S102.

Depressing the RESET button overrides the limit switches and allows the operator to apply power to the motor by depressing one of the direction-control pushbuttons at the same time. So if the manipulator is stalled and then stopped at its forward limit, the situation can be cleared only by depressing the RESET and REV pushbuttons at the same time. In a similar fashion, clearing a limit-stop condition in the reverse direction is a matter of depressing the RESET and FWD pushbuttons at the same time.

Technically speaking, the reason for using this limit-switch and RESET-button scheme is to prevent excessive stall current from flowing through the motor whenever the manipulator reaches an extreme limit of travel. In a practical sense, it covers a situation where a human operator is busy working a lot of other pushbuttons at the same time and cannot give full attention to the position of the manipulator.

The circuit shown in Fig. 5-9 can be incorporated, as shown, in a manipulator system. Indeed, the reset procedure is cumbersome, but it serves its intended purpose. The next step in this discussion of manipulator motor controls is to show a control that performs the same general task in a more convenient way.

A "Smart" Two-Direction Parabot Manipulator Motor Control

The circuit shown in Fig. 5-10 represents a marked improvement over the basic version just described. This version has some "smarts" that make it easier to use in an operator-controlled, parabot manipulator scheme.

As in the previous example, this circuit includes two normally closed microswitches and a normally open RESET pushbutton in the power supply path of the motor. In this case, however, these switches disable the motor only under extreme limit conditions brought about by a failure of the basic control scheme. If all goes well, these switches will never be used.

Under normal operating conditions, the operator sets the direction of the motor with the FWD and REV pushbuttons, S102 and S103. Here, these control switches are isolated from the motor relay circuits by means of some logic circuitry.

Before seeing how these FWD and REV switches do their intended task, note two additional, normally open microswitches, OLR and OLF. These two microswitches are physically arranged at the desired operating limits of the manipulator mechanism, OLR, for instance, is fixed in a position where it is closed as the manipulator reaches an operating limit in the reverse direction—a limit that is most often just short of a mechanical stall limit in that direction. Microswitch OLF is set up in a position that defines the useful operating limit in the forward direction of motion.

What is the functional difference between microswitches OLR and OLF, and EXLR and EXLF? Switch OLR defines the desired operating limit of travel in the reverse direction. It is fixed at a point in the manipulator where you, the system engineer, want to define a reasonable limit of travel. That operational limit of travel should be short of an extreme condition that risks binding up the mechanical system and stalling the motor. But in the event that something fails along the line, there is an extreme-limit sensor for reverse motion: the EXLR microswitch. Perhaps the ORL and EXLR microswitches are fixed only a fraction of an inch apart, but they are always arranged so that ORL is energized before EXLR is. That way, the OLR switch will pick up a limit of reverse travel before EXLR does. And, as mentioned earlier, EXLR will never be energized if OLR and its logic circuitry are doing the job. OLF and EXLF do the same

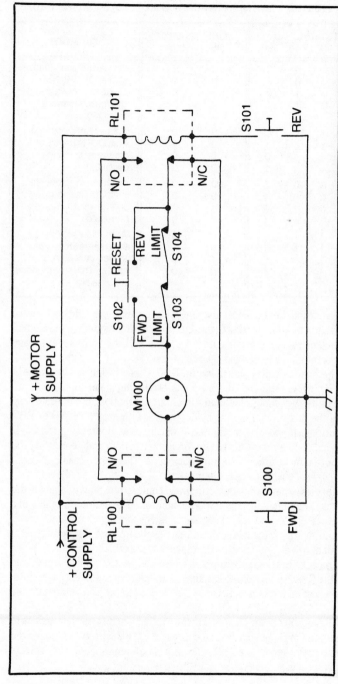

Fig. 5-9. A two-direction, pushbutton-operated motor control for manipulators. The motor stops automatically whenever either of the two limit switches is energized by the mechanical system. Depressing the RESET button lets the operator get things moving again.

Table 5-2. Logic Table for the Smart Manipulator Control Circuit of Fig. 5-10.

INPUT STATUS OLR	OLF	FWD	REV	MOTOR RESPONSE FWD	REV	COMMENTS
0	0	0	0	0	0	Normal running mode;
0	0	0	1	0	1	neither OL switch is
0	0	1	0	1	0	energized; forward and
0	0	1	1	0	0	reverse motions are permitted.
0	1	0	0	0	0	Mechanism at forward
0	1	0	1	0	1	limit; OLF is energized;
0	1	1	0	0	0	only reverse motion is
0	1	1	1	0	0	permitted.
1	0	0	0	0	0	Mechanism at reverse
1	0	0	1	0	0	limit; OLR is energized
1	0	1	0	1	0	only forward motion is
1	0	1	1	0	0	permitted.
1	1	0	0	0	0	Mechanism at both limits
1	1	0	1	0	1	at the same time — an
1	1	1	0	1	0	unlikely condition; for-
1	1	1	1	0	0	ward and reverse motions are possible.

do the same job, but the sense the operating and extreme limits of forward manipulator motion. In short, the OL microswitches define the normal operating limits of the manipulator, while the EXL switches cover for any failure of the OL switches and logic circuitry.

The logic circuitry shown in Fig. 5-10 responds only to the status of the FWD, REV, OLR and OLF switches. Assuming the manipulator is not in one of its two impossible operating limits of motion, the operator can control the direction of motion by depressing either of the FWD or REV pushbuttons. If you haven't figured it out by now, depressing the FWD button makes the motor turn in its forward direction, and depressing the REV button makes the motor turn backward. The motor does not turn if neither *or both* pushbuttons are depressed.

Here is where the logic circuitry comes into play. Suppose the operator is holding down the FWD pushbutton. The motor is running in a forward direction and the manipulator is responding in an appropriate fashion. Eventually the manipulator reaches the forward-direction operating limit of travel, as sensed by the closure of the OLF microswitch. When that happens, the logic circuitry disables the motor control relays and stops the forward motion. Continuing to hold down the FWD pushbutton has no effect on the motor, because the logic circuitry is keeping it turned off.

All that's necessary for setting the mechanism into motion again is to release the FWD pushbutton and depress the REV button. Assuming the extreme-limit sensor, EXLF, has not been energized, the system immediately responds by moving in the reverse direction.

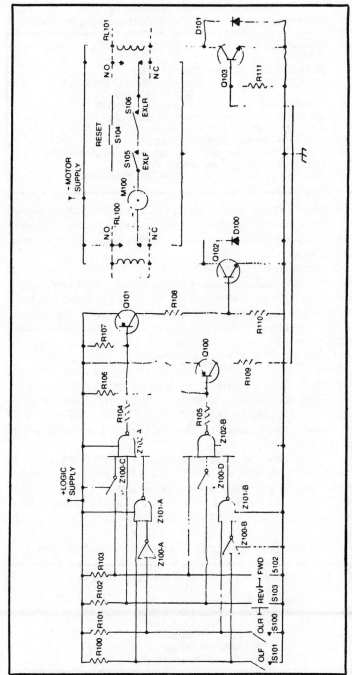

Fig. 5-10. A "smart" two-direction parabot manipulator motor control.

123

Table 5-3. Parts Lists for the Circuit of Fig. 5-10.
Components Common to Both the 5V and 12V Logic Supply Versions.

```
Components Common to 5V and 12V LOGIC SUPPLY Versions:

M100—12VDC motor, 3A stall current or less
RL100,RL101—SPDT (or DPDT) relay; 120VAC, 3A contacts;
                                        12VDC, 50mA coil
D100,D101—any small rectifier diode such as 1N458
Q100,Q101—any small PNP transistor such as 2N3906
Q102,Q103—any small NPN transistor such as 2N3904
S100,S101—normally open microswitch
S102,S103,S104—normally open, momentary pushbutton switch
S105,S106—normally closed microswitch
R100,R101,R102,R103,R104,R105—22K, ¼W resistor
```

The logic circuitry works in conjunction with the OL switches to avoid motor-stall conditions altogether. The manipulator automatically stops at a predefined operating limit that is always short of an extreme, motor-stalling condition.

It's a nice parabot system that lets the operator control the motion of the manipulator with full confidence it will not stall or latch up while he is giving attention to some other control. Things simply come to a stop and turn off until the operator gets around to doing something else with it.

This is also the first example of a working circuit that interfaces some logic ICs through a set of driver transistors with a two-direction motor control. The transistor function as described in Fig. 5-8.

Relay R100, for example, is energized only when transistors Q101 and Q102 are both switched on. The only way to turn on those transistors is by pulling the base of Q101 down to logic 0. The same is true for energizing RL101 by means of transistors Q100 and Q103. A logic-1 level, or no potential at all, applied to the base of Q101 or Q100 deenergizes their respective relays.

The logic level appearing at the bases of these two transistors is dictated by the outputs of some three-input NAND gates. Table 5-2 is a complete logic table for the "smart" part of the circuit.

According to Table 5-2, the system is in its normal operating mode whenever neither of the OL microswitches is energized. The operator is

Table 5-4. Additional Components for the 12V Version of Fig. 5-10 Circuit.

```
Additional Parts for CMOS, 12 VDC LOGIC SUPPLY:

Z100—4009 hex inverter
Z101—4011 quad 2-input NAND gate
Z102—4023 triple 2-input NAND gate
R106,R107—220K, ¼W resistor
R108,R109—1.8K, ¼W resistor
R110,R111—22K, ¼W resistor
```

free to determine the direction of the motor by working the FWD and REV pushbuttons. In this case, energizing the REV pushbutton causes the motor to turn in its reverse direction, and energizing the FWD button causes the motor to turn in its forward direction. Depressing neither or both pushbuttons prevents the motor from running at all.

The second group of four conditions specifies that the OLF microswitch is energized, which means the mechanism is at its operational limit in the forward direction. The only motor response permitted under this set of circumstances is that of moving in reverse—the opposite direction.

The third group of conditions assumes the OLR microswitch is energized. The mechanism has thus reached the operating limit in the reverse direction, and the only permissible motion is in the opposite direction, or forward.

The final group of conditions might seem a bit absurd, because it specifies the system having reached both the forward and reverse limits at the same time. This should never happen, but in the remote chance it should happen, the operator is able to move the motor in either direction.

Specifications for the components in Fig. 5-10 depend mainly on the supply voltage levels and the current and voltage ratings of the motor and drive relays. Tables 5-3 through 5-5 summarize the values for the components, assuming the system uses 12 VDC for the MOTOR SUPPLY and either 5 VDC or 12 VDC for the LOGIC SUPPLY.

Table 5-3 is a parts list for components that are the same, whether the LOGIC SUPPLY is 5 VDC or 12 VDC. Table 5-4 lists the remaining components where the LOGIC SUPPLY voltage is 12 VDC—a situation calling for the application of CMOS logic circuits. Table 5-5 lists the same additional parts for 5 VDC operation.

Naturally there is a chance you will want to use power-supply configurations and components other than those specified in these examples. In that case, you can work out the values for the components, using the information described in earlier sections of this chapter.

A Two-Direction Manipulator Motor Control for Processor Systems

The circuit shown in Fig. 5-11 performs much the same set of tasks as the other two-direction controls already described in this section. There

Table 5-5. Additional Components for the 5V Version of Fig. 5-10 Circuit.

Additional Parts for TTL, 5 VDC LOGIC SUPPLY OF FIG. 5-10:

Z100—74LS04 hex inverter
Z101—74LS00 quad 2-input NAND gate
Z102—74LS10 triple 3-input NAND gate
R106,R107—220K, ¼W resistor
R108,R109—680-Ohm, ¼W resistor
R110,R111—6.8K, ¼W resistor

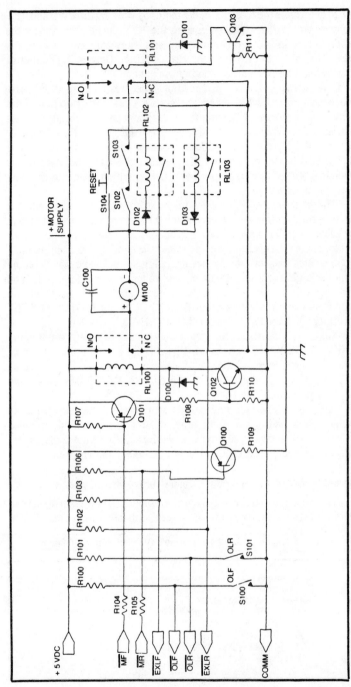

Fig. 5-11. Circuit for a two-direction manipulator control that can be interfaced with a bus-oriented processor system.

Table 5-6. Input Logic Function Table for Circuit of Fig. 5-11.

$\overline{MR}$	$\overline{MF}$	Motor Response
0	0	STOP
0	1	RUN REVERSE
1	0	RUN FORWARD
1	1	STOP

are no operator switches, however, because it is designed for use with a bus-oriented I/O port. In other words, it has the potential for being a two-direction motor control in a true robot system.

The two main operating input connections to the circuit are labeled $\overline{MR}$ and $\overline{MF}$, and they influence the motor described in Table 5-6. Whenever these two signal inputs have the same logic level, the motor is stopped. Setting $\overline{MF}$ to logic 1 while keeping $\overline{MR}$ at logic 0 makes the motor run in reverse, because that particular input situation energizes relay RL101, but leaves RL100 deenergized. Reversing that input situation, setting $\overline{MR}$ to 1 and $\overline{MF}$ to logic 0, causes the motor to run in its reverse direction, because RL101 is deenergized and RL100 is energized.

The four remaining signal connections are limit-switch outputs. As described in Table 5-7, OLF and OLR pick up normal manipulator operating limits for forward and reverse directions of motion, respectively. They are active-low signals going to logic 0 whenever they are energized by the manipulator system.

Signals EXLF and EXLR mark the manipulator's extreme limits of motion—those occuring only when something drastic goes wrong with the manipulator or its motor control system. If, for example, the system fails to respond to normal operating limit OLF, it is backed up by a zero-going signal from EXLF.

While the normal operating limits of motion are picked up by microswitches labeled OLF and OLR, the extreme limits are sensed by microswitches S102 and S103. For your own purposes, you can label the extreme-limit switches EXLF and EXLR, but it turns out that it doesn't matter which is which, just as long as they are physically arranged to be activated if the manipulator systems should ever move beyond one of the normal operating limits.

Table 5-7. Definition of Limit-Switch Outputs of Circuit of Fig. 5-11.

$\overline{OLF}$—Normally at logic l; goes to 0 as the manipulator reaches its operating limit in the forward direction

$\overline{EXLF}$—Normally at logic l; goes to 0 if the manipulator ever exceeds the forward operational limit and goes to the extreme forward position

$\overline{OLR}$—Normally at logic l; goes to 0 as the manipulator reaches its operating limit in the reverse direction

$\overline{EXLR}$—Normally at logic l; goes to 0 if the manipulator ever exceeds the reverse operational limit and goes to the extreme reverse position

**Table 5-8. Parts List for the Circuit
of Fig. 5-11. Components Common to Most Versions.**

```
D100, D101, D102, D103—any small rectifier diode such as 1N458
C100—0.1 uF capacitor
Q100, Q101—any small PNP transistor such as 2N3906
Q102, Q103—any small NPN transistor such as 2N3904
S100, S101—normally open microswitch
S102, S103—normally closed microswitch
S104—normally open, momentary pushbutton switch
R100, R101, R102, R103—22K, ¼W resistor
```

In the event S102 and S103 is energized, power is immediately removed from the motor. Those switches, you see, are connected in series with the main current path of the motor. The fault condition will stop all motor action until the operator depresses the RESET switch. You won't have any trouble justifying human intervention in this case, if you bear in mind that the fault condition occurs only under a peculiar set of circumstances that threaten the well-being of the entire system. With a properly designed manipulator, the situation will rarely, if ever, occur.

Relays RL102 and RL103 are energized only when one of the extreme-limit switches is energized. The purpose of these two relays is to inform the bus system which extreme limit has been reached. The diodes connected in series with the coils for these relays are responsible for determining which of the two will be energized. For example, the polarity of diode D102 is such that it will let RL102 be energized only when the system reaches an extreme limit in the forward direction. D103, on the other hand, allows RL103 to be energized only when the system reaches an extreme limit in the reverse direction.

And the point of all this is to generate the EXLF and EXLR signals that tell the bus system that an extreme limit has been reached and which limit it happens to be. Because the system cannot take its own corrective action (the operator's RESET pushbutton, S104, is the only way to get things running again), connections EXLF and EXLR might only operate a set of LEDs that signal which extreme limit has been reached.

Perhaps it seems that the circuit shown in Fig 5-11 places undue emphasis on the extreme-limit fault conditions. That is a matter of opinion. Experienced robot experimenters are often haunted by the inexorable working of Murphy's law: If anything can go wrong, it will. The extra dollars spent on the extreme-limit circuitry in Fig. 5-11 will return a lot of benefits in terms of saving battery power and giving the operator confidence that the system will respond in the best possible fashion to an extreme fault condition. Even if the processor system feeding this circuit fails, the circuit will still shut down if the manipulator happens to override the normal operating limit switches.

Tables 5-8 through 5-10 suggest some component values for the circuit under two different sets of design conditions. The primary

**Table 5-9. Additional Components for Operating
High-Power, 12 VDC Motors in Circuit of Fig. 5-11.**

M100—12VDC motor with stall current up to 10A
RL100, RL101—SPDT (or DPDT) relay; 120VAC, 10A contacts; 12VDC, 75mA coil
RL102,RL103—SPST (or SPDT) relay; 120VAC, 1A contacts; 12VDC, 10mA coil
R108,R109—470-Ohm, ¼W resistor
R110,R111—4.7K, ¼W resistor
R104,R105—5.1K, ¼W resistor
R106,R107—47K, ¼W resistor

assumption is that the processor system's I/O connections to the signal lines runs at the standard +5 VDC level.

Table 5-8 specifies the parts that are common to either motor-circuit operating voltage. Table 5-9 specifies additional parts, assuming you are using a 12 VDC motor that draws about 10A under its stall condition. Table 5-10 specifies the additional parts required if the motor is rated at 6 VDC and has a current drain of 1A or less.

Capacitor C100, connected across the motor terminals, merely helps reduce electrical noise from the brush contacts. This is especially important in processor-operated systems.

TWO-DIRECTION MOTOR CONTROLS FOR MAIN DRIVE SYSTEMS

Like the motor controls for manipulators, those for the main drive motors can be rather simple. If you are planning to use a two-motor drive/steer system, the only point of complication is that the scheme uses a pair of identical motor controls.

A Basic Two-Direction, Two-Motor Drive System

The circuit shown in Fig. 5-12 is an expanded version of the single-motor control already described in connection with Fig. 5-6.

**Table 5-10. Additional Components for Operating
a Low-Power, 6 VDC System in Circuit of Fig. 5-11.**

M100—6VDC motor with stall current up to 1A
RL100,RL101
RL102,RL103—SPDT (or DPDT) relay; 120VAC, 1A contacts; 6VDC, 12mA coil
R104,R105—33K, ¼W resistor
R106,R107—330K, ¼W resistor
R108,R109—2.7K, ¼W resistor
R110,R111—27K, ¼W resistor

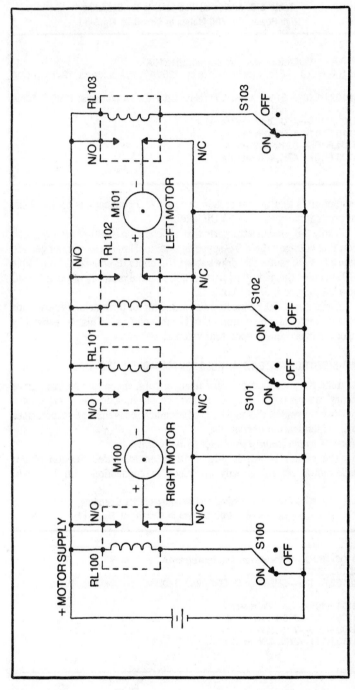

Fig. 5-12. A simple, switch-operated drive/steering motor control.

Table 5-11. Function Table for the Basic Two-Direction, Two-Motor Drive/Steer Control Circuit of Fig. 5-12.

S103	S102	S101	S100	LEFT MOTOR	RIGHT MOTOR	PARABOT RESPONSE
OFF	OFF	OFF	OFF	STOP	STOP	STOP
OFF	OFF	OFF	ON	STOP	FORWARD	FWD, LEFT TURN
OFF	OFF	ON	OFF	STOP	REVERSE	REV, LEFT TURN
OFF	OFF	ON	ON	STOP	STOP	STOP
OFF	ON	OFF	OFF	FORWARD	STOP	FWD, RIGHT TURN
OFF	ON	OFF	ON	FORWARD	FORWARD	FORWARD
OFF	ON	ON	OFF	FORWARD	REVERSE	CW SPIN
OFF	ON	ON	ON	FORWARD	STOP	FWD, RIGHT TURN
ON	OFF	OFF	OFF	REVERSE	STOP	REV, RIGHT TURN
ON	OFF	OFF	ON	REVERSE	FORWARD	CCW SPIN
ON	OFF	ON	OFF	REVERSE	REVERSE	REVERSE
ON	OFF	ON	ON	REVERSE	STOP	REV, RIGHT TURN
ON	ON	OFF	OFF	STOP	STOP	STOP
ON	ON	OFF	ON	STOP	FORWARD	FWD, LEFT TURN
ON	ON	ON	OFF	STOP	REVERSE	REV, LEFT TURN
ON	ON	ON	ON	STOP	STOP	STOP

Everything is doubled to give the operator manual control over a pair of identical motors.

The RIGHT MOTOR is controlled by switches S100 and S101, with S100 setting the forward motion and S101 setting the reverse motion of that motor. In a similar fashion, the LEFT MOTOR is controlled by S102 and S103, with those two switches controlling the forward and reverse directions, respectively.

Table 5-11 illustrates the operation of this circuit. For instructive purposes, you ought to run through this table carefully, comparing each line with the function of the circuit, itself.

In passing, you should note that there are four overall STOP responses, two of each of the motions having a turn associated with them, and just one each for CW SPIN, CCW SPIN, FORWARD and REVERSE. The table deals with 16 different combinations of switch settings, but there are only nine different kinds of motion; seven of the possible motions are redundant. Computer enthusiasts will recognize an opportunity for doing some code compression (reducing the number of possible input codes to eliminate unnecessary redundancy), but it turns out to be impractical at this point. Some of the circuits presented in Chapter 6 will take advantage of the opportunity to insert additional functions in place of the redundant ones that appear here.

The only critical design requirements for this circuit are that the current ratings of the relay contacts match or exceed the stall current of the motors, and that the voltage rating of the relay coils match that of the motors. Of course, it is possible to get around the second requirement by using separate power supplies for the motors and relay coils, modifying the circuit in a fashion similar to that shown in Fig. 5-7 by separating the wiring between the positive sides of the motor relay contacts and relay coils.

For beginners, the circuit can be a lot of fun. It is possible, you see, to get full control over the motion of a parabot by means of the four switches. Everything but the switches can be fixed to a suitable mainframe, and the switches can be mounted to a little project box that is connected to the circuit by means of a long umbilical cord. The operator can then walk along with the parabot, controlling its motion at will.

Interfacing for Logic-Level Control

The next step toward more sophisticated controls calls for some semiconductor interfacing between the electronic control scheme and the motor control circuit. This is illustrated in a general way in Fig 5-13.

The circuit is really nothing more than a doubled-up version of the two-transistor interface circuits already described in this chapter. Determining the values of the resistors is a matter of applying Equation 5-1 through 5-6.

Table 5-12 summarizes the operation of the circuit. Inputs $\overline{\text{MRF}}$ and $\overline{\text{MRR}}$ control the forward and reverse motion of the RIGHT MOTOR,

Table 5-12. Function Table for the Logic-Level Motor Control Circuit of Fig. 5-13.

$\overline{MLR}$	$\overline{MLF}$	$\overline{MRR}$	$\overline{MRF}$	LEFT MOTOR	RIGHT MOTOR	PARABOT RESPONSE
0	0	0	0	STOP	STOP	STOP
0	0	0	1	STOP	REVERSE	REVERSE, LEFT TURN
0	0	1	0	STOP	FORWARD	FORWARD, LEFT TURN
0	0	1	1	STOP	STOP	STOP
0	1	0	0	REVERSE	REVERSE	REVERSE, RIGHT TURN
0	1	0	1	REVERSE	REVERSE	REVERSE
0	1	1	0	REVERSE	FORWARD	CCW SPIN
0	1	1	1	REVERSE	STOP	REVERSE, RIGHT TURN
1	0	0	0	FORWARD	STOP	FORWARD, RIGHT TURN
1	0	0	1	FORWARD	REVERSE	CW SPIN
1	0	1	0	FORWARD	FORWARD	FORWARD
1	0	1	1	FORWARD	STOP	FORWARD, RIGHT TURN
1	1	0	0	STOP	STOP	STOP
1	1	0	1	STOP	REVERSE	REVERSE, LEFT TURN
1	1	1	0	STOP	FORWARD	FORWARD, LEFT TURN
1	1	1	1	STOP	STOP	STOP

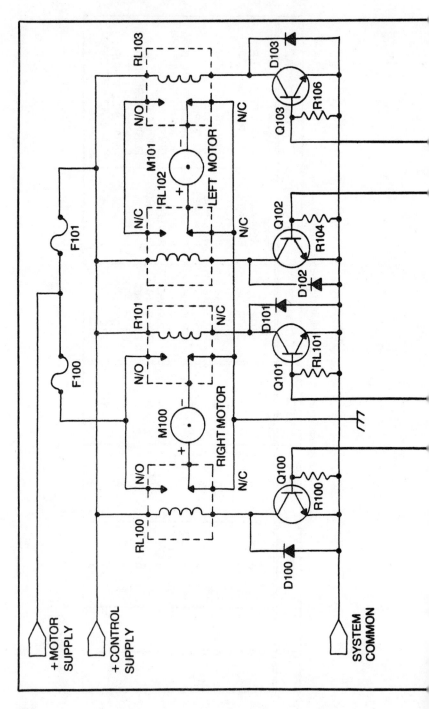

134

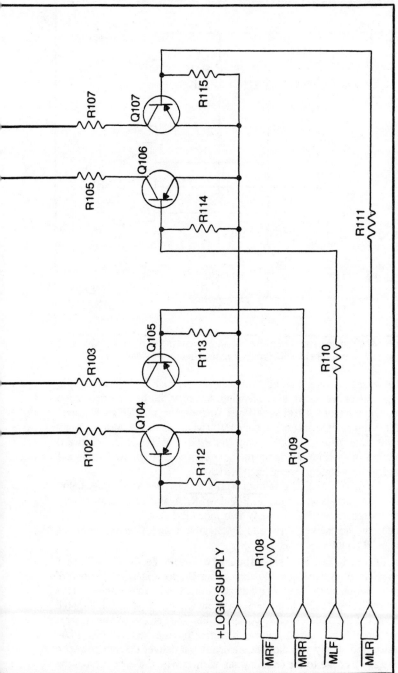

Fig. 5-13. Circuit for a two-direction drive/steering motor control that features transistor interfacing to the motor control relays.

135

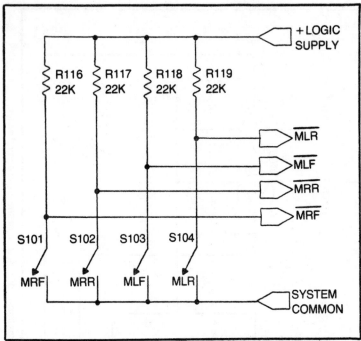

Fig. 5-14. A simple switch input for the control circuit in Fig. 5-13. The values of the resistors assume a LOGIC SUPPLY voltage between 5V and 12V.

M100 while inputs $\overline{\text{MLF}}$ and $\overline{\text{MLR}}$ control the LEFT MOTOR, M101. These inputs are active-low, meaning that they do their particular job whenever they are set to a logic-0 level. Whenever input MRF goes to logic 0, for instance, it turns on Q104 and Q100 which, in turn, energizes relay RL100. Energizing RL100 sets up the RIGHT MOTOR for a forward direction of motion, assuming that the opposing relay, RL101, is *not* energized by a logic-0 level at input $\overline{\text{MRR}}$.

The LEFT MOTOR section works exactly the same way, using active-low signals at inputs $\overline{\text{MLF}}$ and $\overline{\text{MLR}}$. Again, this is not a very complicated circuit; it is simply two indentical motor control circuits which, in themselves, are actually simpler than those required for manipulator motor controls.

Fuses F100 and F101 should be used whenever the MOTOR SUPPLY is capable of delivering more than about 2A for extended periods of time—long enough to set the circuit on fire or, at least, cause some serious damage. These fuses should be slow-blow versions that match the stall currents of the motor as closely as possible. The fuses aren't necessary, however, when using low-power batteries for the MOTOR SUPPLY, because the supply will die long before it can deliver enough power to cause any damage to the circuitry and wiring. With the exception of the

Table 5-13. Parts List for the Circuit of Fig. 5-15.

M100,M101—6VDC permanent-magnet motor with stall current less than 1A
RL100,RL101,RL102,RL103—SPDT relay; 6VDC, 12mA coil; 120VAC, 1A contacts
F100,F101—Not required
PS—6VDC, low-power source, such as four C-cells in series
D100,D101,D102,D103—Any small rectifier diodes, such as 1N458
Q100,Q101,Q102,Q103—Any small NPN transistor, such as 2N3904
Q104,Q105,Q106,Q107—Any small PNP transistor, such as 2N3906
R100,R101,R104,R106—39K, ¼W resistor
R102,R103,R105,R107—3.9K, ¼W resistor
R108,R109,R110,R111—33K, ¼W resistor
R112,R113,R114,R115—330K, ¼W resistor

motors and relays, this entire circuit can be assembled on a single circuit board.

Figure 5-14 shows a simple switch circuit that can be connected to the four input terminals to provide parabot-class motor control. It can also serve as a test circuit for checking the operation of the motor control before making a commitment to running it from a more sophisticated sort of logic system.

The block diagram in Fig. 5-15 shows how to interconnect the motor control and switching circuit to created a simple, low-power parabot.

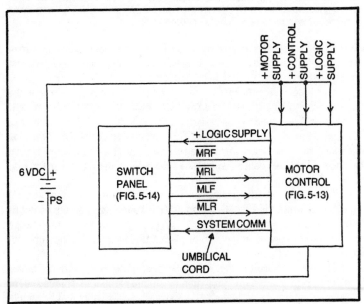

Fig. 5-15. Block diagram for a complete drive/steering motor control. The assumption here is that all three power elements (MOTOR SUPPLY, CONTROL SUPPLY AND LOGIC SUPPLY) operate from the same 6 VDC source. This is suitable for economical, low-power parabot systems.

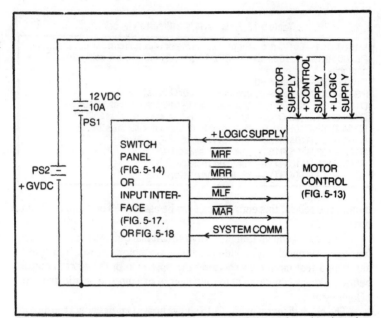

Fig. 5-16. Block diagram and suggested parts specifications for a drive/steering motor control. In this case, the LOGIC SUPPLY is rated at +6V (TTL compatible), while the MOTOR SUPPLY and CONTROL SUPPLY are taken from the same high-currend, 12 VDC source.

Everything is operated from a single 6 VDC source that can be made up of four C-cells connected in series. Table 5-13 is the recommended parts list for Fig. 5-15.

The block diagram in Fig. 5-16 illustrates a situation where the motors are rated at 12 VDC and have stall currents on the order of 12A. Further, it is assumed in this case that the +LOGIC SUPPLY is close to a TTL operating level—say, 6 VDC. This scheme can be used as a working parabot, but it also makes up a fine testing circuit for robot systems that provide TTL-level control signals. Table 5-14 is the parts list for Fig. 5-16.

There are actually countless possible variations of the general scheme illustrated in Figs. 5-15 and 5-16. If your requirements do not fit either of the examples exactly, it is up to you to apply the design procedures discussed in the earlier part of this chapter, coming up with component specifications of your own.

The switch circuit in Fig. 5-14 interfaces quite easily with the motor control circuit, and it's great for running simple parabot machines and for testing the operation of the more complicated motor control. The simple switch circuit suffers from a little problem that can turn out to be a big nuisance if it is used in a more sophisticated, high-performance robot system.

The motor control circuit responds to the instantaneous switch settings and can be a problem whenever you want to change those switch settings. It is difficult to go from one switch combination to another without having to go through a couple of unwanted settings along the way. Of course, you could train yourself to throw several different switches in different directions simultaneously, but that's a nuisance.

The switching circuit in Fig. 5-17 offers an alternative. In this case, the control circuit does not respond to any changes in switch settings until the operator momentarily depresses the LOAD pushbutton, S104. Upon depressing and releasing that button, the switch status is loaded into a four-bit latch-type memory, and the control circuit responds to that pattern of signals from the switches.

Once the latches are loaded, however, any changes in the switches do not affect the signals going to the control circuit. The control circuit behaves as though the switch positions are not being changed at all. This means you can set up the four-bit control command at the switches, having no effect on the operation of the control circuit until you depress the LOAD button. Then the control circuit responds as you prescribed—at least until you set in a different control code and depress the LOAD button again.

The scheme provides smoother control over the motors. They do exactly what you want them to do after you depress the LOAD button. Intermediate switch settings are completely blocked out.

The insert in Fig. 5-17 illustrates the operation of one of the four latch sections in Z100. Technically speaking, resistor R124 holds the enabling inputs at logic 0 most of the time. For a 7475 quad latch circuit, this is the "latch" state. Depressing the LOAD pushbutton, however, raises the enabling inputs to a logic-1 level, and whatever logic pattern is set at the control switches at that moment is fed directly through the latches and to the control outputs. Releasing the pushbutton returns the enabling inputs of Z100 to logic 0, and the circuit "remembers" the switch setting for you.

Figure 5-18 carries the general development of this switch input interface to a higher level. In many instances, it is necessary to be able to control the motors from one of two different data sources. One of those

Table 5-14. Parts List for the Circuit of Fig. 5-16.

```
M100,M101—12VDC permanent-magnet motor with stall current up to 10A
RL100,RL101,RL102,RL103—SPDT (or DPDT) relay; 12VDC, 75mA coil;
   120VAC, 10A contacts
F100, F101—10A slow-blow fuse (or to suit motor stall current)
PS1—12VDC, 12 A-hr lead-acid or gel-cell battery
PS2—6VDC, low-power source, such as four C-cells in series
D100,D101,D102,D103—Any small rectifier diode, such as 1N458
Q100,Q101,Q102,Q103—Any small NPN transistor, such as 2N3904
Q104,Q105,Q106,Q107—Any small PNP transistor, such as 2N3906
R100,R101,R104,R105,R108,R109,R110,R111—6.8K, ¼W resistor
R102,R103,R110,R111—620-Ohm, ¼W resistor
R112,R113,R114,R115—68K, ¼W resistor
```

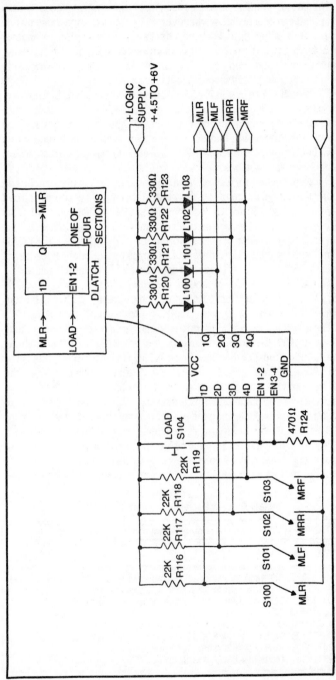

Fig. 5-17. An improved switch-control input interface for the motor control circuit in Fig. 5-13. The LEDs indicate the present motor-control code, but are optional.

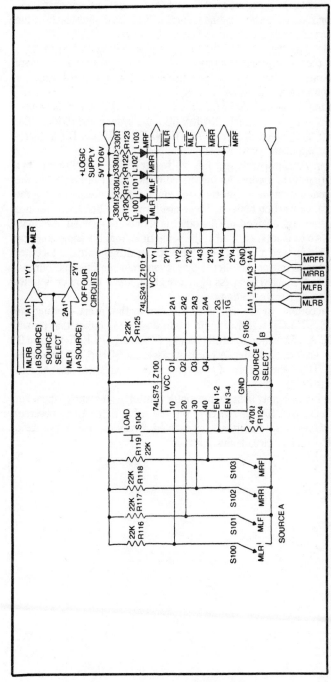

Fig. 5-18. A motor-control input interface that allows one of two difference sources of control codes to reach the motor circuit. Source A is the same switch/latch scheme shown in Fig. 5-17. Source B is any alternate source of motor-control codes, including an output port from a microprocessor system.

141

sources ought to be the four control switches used in all of these examples. The second source can be any other source of TTL-level signals—from the output port of a microprocessor system, for instance.

As long as the SOURCE SELECT toggle switch,S105, is in its A position, the four signals from this circuit and to the motor control board are generated by the switch/latch circuitry. This gives the operator manual control over the motor system.

Setting the SOURCE SELECT switch to its B position, however, disconnects the switch/latch circuitry from the system and picks up the four controlling commands from ALTERNATE SOURCE B (a microprocessor output port or whatever).

The key to the operation of this two-source data selector is a 74LS241 noninverting, octal buffer. As illustrated in the insert of Fig. 5-18, the corresponding control bits for the two sets of data sources are fed to the inputs of noninverting buffers. These are three-state buffers, and since one is active-high enabled and the other is active-low enabled, it follows that the two cannot be enabled at the same time from the same enabling signal. This feature makes it possible to connect their outputs together without running into any conflict of logic levels from two different sources; one is always inabled and passing information, while the other is in its three-state, disabled condition.

So according to the simplified diagram in the insert, setting the SOURCE SELECT signal to logic 1 will enable the buffer that passes SOURCE A information to MLR. Setting SOURCE SELECT to logic 0, however, enables the buffer section that passes the SOURCE B logic level to MLR. The same pattern of thinking applies to all four such sections included in Z101.

The LEDs simply indicate the motor control input status. Whenever the MRL lamp, L100, is switched on, the MLR (left motor, reverse) portion of the control circuit is being enabled. Whenever that lamp is dark, the MLR part of the circuit is switched off.

Motor Speed Controls

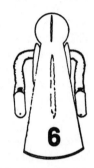

6

While a parabot/robot system need not employ any sort of variable speed control for its motors, such controls are desirable under many circumstances. My first major machine project, Buster, operated quite successfully from a three-speed drive system; my more sophisticated machine-intelligence demonstrator, Rodney, used only two-direction, on/off motor controls.

There is much to be said in favor of simpler on/off motor controls of the sort described in Chapter 5. But then one can rightly argue in favor of smoother manipulator and drive/steer action that comes about only by using motor speed controls.

The speed control circuits described here are all linear in the control sense of the word; that is, there are no provisions for closed-loop, feedback servo control. That is the subject of Chapter 7. Nevertheless, the open-loop speed controls featured in this chapter have their place in parabot/robot designs.

It is a basic fact of motor theory that the speed of a DC motor is proportional to the amount of voltage applied to it, and inversely proportional to the mechanical load applied to the shaft. Assuming a constant mechanical load, for example, it follows that the motor runs faster with an increasing amount of applied voltage.

With the applied voltage fixed at some suitable level, the motor will run slower as more mechanical load is applied to the shaft. And, incidentally, under this fixed-voltage, variable-load condition, it turns out that the current demand of the motor is proportional to the amount of mechanical loading.

In short, a DC motor is a dynamic electrical system, and the complexity of the matter is compounded when attempting to view the

143

operation in the light of parabot/robot speed controls. Fortunately, it is entirely possible to ignore most of the complicated dynamics of DC motors by applying a few rule-of-thumb principles. In most cases, simply knowing the voltage specification and stall current of the motor is adequate for building any sort of motor control system.

HOW NOT TO BUILD A MOTOR SPEED CONTROL

When it comes to controlling the running speed of a DC motor, one of the main rule-of-thumb principles is to avoid circuits such as those illustrated in Fig. 6-1. The only reason for giving attention to these negative examples is that they happen to represent the most obvious way to control the speed of a motor. And, indeed, the idea works. The main problem is that the scheme is terribly inefficient. Control components get very hot and, worse yet, the circuits burn up valuable doses of battery power.

The circuit shown in Fig. 6-1A uses a variable resistor connected in series with the DC motor and its power source. It is a basic voltage divider circuit that lets the operator adjust the amount of voltage reaching the motor winding. The larger the value of the resistance is, the smaller the amount of voltage is that reaches the motor—and the more slowly the motor runs.

At first thought, the scheme might seem just fine, but there is a problem in the very nature of voltage dividers. What happens to the supply voltage that is not dropped across the motor? It is dropped across the series resistance, of course. And with the motor current running through that resistor, it only takes a simple application of the I x E power formula to see that the resistor can dissipate a great deal of power when its voltage and current are relatively large (as in a case where the motor is stalled at a low-speed setting). That resistor power is absolutely wasted as heat energy, and the situation is intolerable as long as there are more efficient alternatives available.

The circuit in Fig. 6-1A is most efficient only when the resistor is set for zero resistance and the motor is running directly from the power source. In that one instance, the resistor is not dissipating any power at all. It might have plenty of current running through its wiper arm, but there is no voltage to create a power-dissipating situation. Bear that little fact in mind. It will be extremely important later.

An additional problem with the voltage divider approach stems from the fact that a motor must draw its stall current, or a goodly portion of it, in order to begin running from a dead stop. If some resistance is placed in series with the motor, there is a good chance that it will limit the start-up current or stall current to an extent that the motor cannot begin running.

The motor, in other words, cannot be started at a low-speed setting. You have to reduce the value of the series resistance far enough to let start-up current flow through the motor. Then when the motor starts running, you have to back off the resistance to slow it down to the desired low speed. Smooth, low-speed starts are very difficult to achieve.

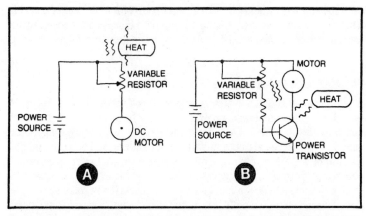

Fig. 6-1. How not to control motor speed. (A) A variable resistor connected in series with the motor. (B) A class-A transitor amplifier in series with the motor.

Using a transistor in series with the motor, as illustrated in Fig. 6-1B, doesn't really change the matter at all. The variable resistor certainly burns up very little power at any setting, because it carries only the base current for the transistor. This is still a voltage divider circuit, however, and it is the transistor that must dissipate the unused power. All you are doing here is trading excessive and wasteful resistor power for equally excessive and wasteful transistor power.

The transistor in that circuit dissipates its minimum power only when it is completely turned off. Full supply voltage is dropped across its emitter-collector terminals, but there is no current flowing. Any voltage times zero current equals zero power dissipation.

The power dissipation of the transistor isn't too bad at the opposite extreme: where it is completely switched on, or saturated. Current flow through the circuit is at its maximum level, but the emitter-collector voltage of the transistor is at its lowest conducting point—say, something on the order of 1V for a power transistor. So if the motor has a stall current of 12A and the transistor is saturated, the power dissipation of the transistor is only about 12W.

However, if you adjust the resistor in Fig. 1-6B so that the transistor drops something such as 6V and the motor current happens to be 6A at the time (not at all an unlikely condition), the power dissipation of the transistor will be 6V x 6A=36W. It could be worse at other speed settings.

The moral of the story, as far as the circuit in Fig. 1-6B is concerned, is to operate the transistor in just two different conditions: fully off or fully saturated. Run the circuit anywhere between those two extremes, and you find the transistor getting extremely hot.

And what good is a transistor-controlled speed circuit that is set up to run at full speed or full stop? That is the topic of the remainder of this chapter. It's the way out of the motor speed control dilemma.

So *never* use either of the circuits in Fig. 6-1 for motor speed control. They are far too inefficient. But bear in mind the discussion concerning the transistorized version and its low power dissipation at two extremes of conduction.

VARIABLE DUTY CYCLE MOTOR SPEED CONTROLS

Figure 6-2 shows a switch-operated, on/off motor control circuit. It is, indeed, quite simple and very much like the basic on/off motor control described in Chapter 5. What does it have to do with controlling motor speed? Suppose you are able to switch the motor on and off very rapidly. The motor will attempt to run at full speed, drawing full start-up current if necessary, as long as the switch is closed. The motor tends to slow to a stop, however, whenever you open the switch. So if you could switch the circuit on and off very rapidly, the motor would seem to run at some speed between full speed and full stop.

Of course if you toggle the switch very slowly, maybe on and off once every second or so, the motor will start and stop in the same sequence. But if you happen to have a superfast hand that can operate the switch at a rate of about 100 Hz, the motor will run rather smoothly because its inertia will tend to keep it spinning through the relatively short OFF intervals. If you keep toggling the switch at a 100-Hz rate, but vary the ratio of on-to-off times, you will notice the motor running at different speeds. The greater the proportion of ON time, the faster the motor seems to run.

The Concept of Operating Duty Cycle

In an on/off switching circuit, the proportion of ON time can be conveniently expressed in terms of duty cycle:

$$D = t_{on}/(t_{on} + t_{off}) \qquad \text{Equation 6-1}$$

where D = duty cycle, and t_{on} = time ON, and t_{off} = time OFF.

Experimenting with Equaiton 6-1, you can see that the duty cycle is equal to 1 whenever there is no OFF time. The duty cycle reduces to zero, however, whenever there is no ON time. Throughout this book, *duty cycle* will most often be specified as a unitless number between 0 and 1, although it is equally correct to multiply that figure by 100 and come up with an expression for *percent duty cycle*.

So here you are, toggling the switch in Fig. 6-1 on and off very rapidly. If you can somehow determine the ON and OFF times, you can calculate the duty cycle. That's nice, but what's the point? Before describing how such a thing can be useful, consider an alternate version of the duty-cycle equation:

$$D = t_{on} f \qquad \text{Equation 6-2}$$

where D = duty cycle, t_{on} = time ON in seconds, and f = frequency in hertz.

146

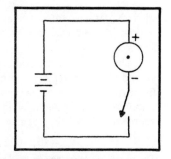

Fig. 6-2. The simplest sort of efficient motor speed control.

This version of the equation comes about by first realizing that t_{on} + t_{off} in the denominator of Equation 6-1 is really the period of the switching waveform. And since the period of any waveform is equal to $1/f$, Equation 6-2 works out quite naturally by applying some basic algebra.

The equation now shows that the duty cycle is proportional to both the ON time and the frequency of the waveform. If you choose a constant frequency, it is possible to vary the duty cycle by changing the ON time. The longer the ON time is, the greater the duty cycle is.

And alternately, fixing the ON time at some point and varying the frequency achieves the same sort of result. The higher the toggling frequency is, the larger the duty cycle.

But all this still does not explain why the duty cycle is important to a motor speed control scheme. Consider this:

$$V_{ma} = DV_s \qquad \text{Equation 6-3}$$

where V_{ma} = the apparent, or average, voltage applied to the motor in volts, D = duty cycle (unitless figure between 0 and 1), and V_s = motor source voltage in volts.

We must acknowledge the fact that the speed of a motor is proportional to the amount of voltage applied to it. But now it is also possible to say that the speed of the motor is proportional to the apparent, or average, amount of voltage applied to it. Taking that into account, you should be able to see that the speed of a motor that is being switched fully ON and fully OFF will take on an apparent running speed that is proportional to the duty cycle of the switching action and the amount of supply voltage.

If, by way of an example, you are running a motor from a 12 VDC source and switching at a duty cycle of 0.8, the apparent voltage to the motor is 12V x 0.8, or 9.6V. If the switching is taking place rapidly enough, it will seem to run as though you are applying a constant 9.6V to it. It is thus possible to vary the apparent, or average, voltage applied to a motor between 0V (duty cycle=0) to full source voltage (duty cycle=1).

A useful variation of Equation 6-3 is one that substitutes Equation 6-2 for variable D:

$$V_{ma} = t_{on}fV_s \qquad \text{Equation 6-4}$$

147

where V_{ma} = apparent, or average, voltage applied to the motor (volts), t_{on} = time ON (seconds), f = switching frequency (hertz), and V = motor source voltage (volts).

The implication is that the running speed of a DC motor being switched on and off rather rapidly is proportional to the ON time and operating frequency as well as the source voltage. Working a specific example, what if the frequency of the toggling is 100 Hz, the ON time is 6 ms and the source voltage is 12V. Plugging those values into Equation 6-4:

$$V_{ma} = (6 \times 10^{-3}\text{sec})(100\,\text{Hz})(12\text{V}) = 7.2\text{V}.$$

The motor, in other words, will run as though it has a 7.2 VDC applied to it, even though the actual applied voltage is switching between 0V and 12V.

You probably aren't very enthusiastic with the prospect of controlling the speed of a DC motor by hand-toggling a switch, on and off at a 100 Hz rate. The way around the problem is to let an electronic circuit do the job for you. Transistors can be switched on and off very rapidly and at any desired duty cycle. Before seeing all the ways this job can be done automatically, however, consider a few details concerning power transistors driving DC motor loads.

Switching Transistors and Motor Loads

The circuit in Fig. 6-3 doesn't really improve on the basic toggle switch idea very much. At any rate, the motor will be deprived of supply voltage whenever the switch in Fig. 6-3 is open. The reason is that the transistor, in that condition, has no forward-biasing base current; it cannot conduct. It is completely turned off by virtue of emitter-base resistor R2.

Closing S1, however, completes a path for current flow between the emitter, base and positive side of the power source. That forward-biases the transistor, turns it on and allows it to apply power to the motor. And if the transistor is turned on hard enough, the collector current will be adequate for running the motor very close to its full-power speed.

Assuming the values of the components have been selected properly, the motor can be switched on and off by means of switch S1. And if the switching action is fast enough, the motor responds as though varying levels of supply voltage are being applied to it. All things considered, this circuit can be used exactly the same way as the one already described in connection with Fig. 6-2.

As mentioned earlier, a transistor operating in series with a DC motor is most efficient at two extremes of conduction: whenever the transistor is switched off and whenever it is switch fully on, or saturated. When the transistor is switched off by opening the toggle switch, S1, its collector current drops to zero. Plenty of voltage is across the emitter-collector circuit—full supply voltage, in fact. The transistor is very efficient in this condition for the simple reason that it isn't dissipating any power: Zero current times any amount of voltage is zero watts of power dissipation.

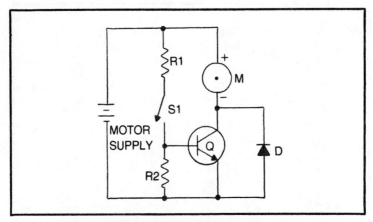

Fig. 6-3. Motor speed control with a switching transistor.

Closing the switch forward-biases the transistor. If this forward-biasing current is sufficiently high, it will saturate the transistor and apply nearly the full source voltage to the motor. A lot of current will be flowing through the collector circuit of the transistor, but the emitter-base voltage will be quite low. By way of another set of rule-of-thumb figures, the emitter-collector saturation voltage for a small transistor is about 0.7V, while the same figure for a power transistor is closer to 1V.

So the efficiency of the transistor switching circuit is excellent whenever the transistor is turned off, and it isn't very bad when the transistor is fully on, or saturated. The only time the efficiency becomes really poor is when the transistor is operated somewhere in between. That in-between condition must be avoided in any sort of workable switching circuit.

The first step in selecting a transistor and values of the resistors in Fig. 6-3 is to find a transistor having a collector current rating that matches or exceeds that of the stall current of the motor. If you find the stall current of the motor is on the order of 10A, a Motorola MJ3055T will do the job because it has a collector current rating of 10A. If you tend to be a bit more conservative, you might prefer a 2N3055 because it has a collector current rating closer to 15A.

If the motor has a lower stall current, you can get away with using smaller switching transistors. Just make sure the maximum continuous collector current rating of the transistor meets or exceeds the stall current of the motor.

Selecting the value of R1 is a matter of applying the one-tenth rule-of-thumb described in Chapter 5. In this instance, it is safe to assume the transistor will have a gain of 10 as far as currents are concerned. So if the collector is supposed to drive the motor with 10A, the base current should be at least 1A when the switch is closed. The value of R1 is thus

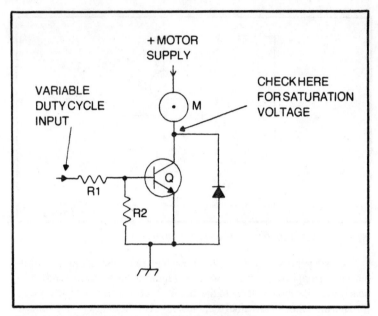

Fig. 6-4. Operating a switching transistor for a DC logic source.

chosen by Equation 5-2. The value of R2 is then selected to have a value about 10 times that of R1.

Again, all of this is described in great detail in connection with a similar sort of circuit shown in Fig. 5-5. The only real difference in this case is that the current and power ratings tend to be much higher.

The real point of this whole discussion, however, is to convince you that a power transistor can be used for controlling the average power applied to a DC motor, provided the transistor is never allowed to operate in its linear region (anywhere between cutoff and saturation). Incidentally, the purpose of the reverse-biased diode in Fig. 6-3 is to clip high-voltage spikes from the motor winding that occur every time the transistor is switched from on to off.

Principles of Pulse-Width and Frequency Modulation

The next step in the development of a workable motor speed control scheme is to replace the mechanical switch in Fig. 6-3 with an electronic version of the same thing. Take a look at the circuit in Fig. 6-4. In principle, it isn't much different from the previous transistor switching circuit. This one, however, can be operated from another electronic system that generates a rectangular waveform that will snap the transistor on and off. The positive portion of that waveform will turn on the transistor and, hopefully, drive it into saturation. The zero-voltage portion of that same waveform ought to turn the transistor off.

150

Recall from an earlier discussion in this chapter that the apparent voltage applied to the motor is proportional to the duty cycle of the waveform applied to the base of this transistor. Getting complete control over the speed of the motor is a matter of adjusting the duty cycle of that applied waveform.

There are two systematic ways to go about varying the duty cycle of a rectangular waveform: *pulse-width modulation* (PWM) or *frequency modulation* (FM). Both have their special advantages and drawbacks in parabot/robot motor speed control systems, so it is necessary to describe both in some detail.

Figure 6-5 illustrates the operation of a PWM scheme. In this case, the operating frequency is fixed at some appropriate clock rate. See

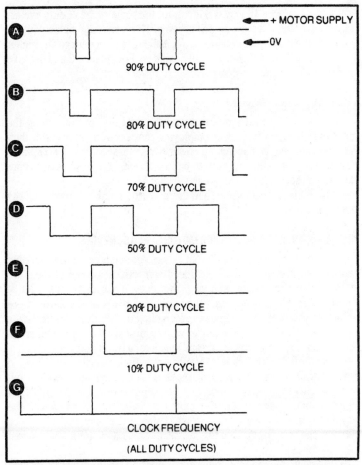

Fig. 6-5. Some pulse-width modulation (PWM) waveforms for variable duty cycle motor speed control.

waveform G and note that the positive-going edge of each of the other waveforms begins as that clock pulse occurs.

Those constant-frequency clock pulses will trigger a circuit that functions as a monostable multivibrator, or one-shot multivibrator. Whenever that one-short circuit is triggered, it generates a positive pulse that lasts for some prescribed period of time. In waveform A, the positive part of the waveform lasts for a relatively long time, providing a duty cycle of 90 percent, or 0.9. In waveform F, however, the positive portion of the triggered signal does not last very long. It is high only about 10 percent of the time between successive triggerings, so it has a duty cycle of 0.1.

Fixed frequency and variable time ON are the essence of PWM motor speed control. And you can back up the claim by applying the idea to Equations 6-2 and 6-4.

The waveforms in Fig. 6-6 illustrate the same range of duty cycles, but they are derived from an FM scheme. Here, the high pulse width is fixed, and the duty cycle is adjusted by changing the frequency. The higher the frequency is, the higher the duty cycle is. Equations 6-2 and 6-4 give the idea mathematical credence.

A Final Word Before Moving On

There are currently no more attrative circuits for controlling the speed of DC motors than those employing constant-voltage, variable duty cycle techniques. The switching power transistors work with acceptable efficiency, the circuits are easy to build and test, and they provide more than enough speed control for parabot/robot systems.

All of this is hard to beat, and there is little point in considering any other approach.

DRIVER/INTERFACE CIRCUITS FOR MOTOR SPEED CONTROLS

There are a number of different ways to go about generating variable duty cycle waveforms for motor speed controls, and some of them are illustrated in the section following this one. While there are a lot of different circuits for generating the variable duty cycle waveform, there isn't much choice when it comes to assembling transistor driver/interface circuits—circuits having low-power waveform sources as their inputs and high-power switching transistors and the motor as their outputs.

Figures 6-7 and 6-8 show four such driver/interface circuits. Those in Fig. 6-7 are listed as noninverting types. That is, the motor receives full MOTOR SUPPLY voltage whenever the input signal is at its HIGH, or logic-1, state. Conversely, the motor is switched off whenever the waveform input is in its LOW, logic-0 state. The only real difference between the two diagrams in Fig. 6-7 is the amount of motor stall current they can handle.

The diagrams in Fig. 6-8 are both inverting types. A logic-1 level at the INPUT switches the motor off, while a logic-0 level turns it on. Again,

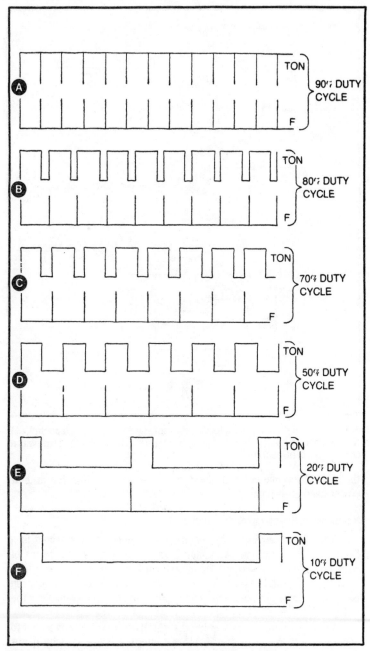

Fig. 6-6. Some frequency modulation (FM) waveforms for variable duty cycle motor speed control.

153

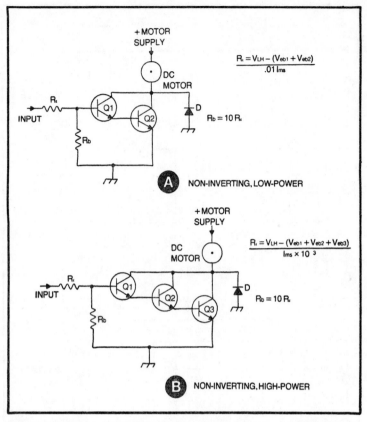

Fig. 6-7. Darlington amplifiers for variable duty cycle motor speed control. (A) Noninverting type with a current gain of at least 100. (B) Noninverting type with a current gain of at least 1000.

the main difference between these two schematics in Fig. 6-8 is their power-handling capability.

Noninverting Driver/Interface Circuits

The circuit in Fig. 6-7A features two transistors connected in a Darlington configuration. Assuming a gain of at least 10 (in terms of current) for both transistors, the overall current gain of the circuit is at least 100. In a practical sense, this means the stall current of the motor can be up to 10 times greater than the maximum current drive available at the INPUT.

Working from the assumption that the INPUT will be connected to some sort of IC device that is capable of driving the base of Q1 with 10 mA of current, it figures that Q2 can saturate with a collector current of about 1A. The figure will most often be much greater than that, but it is a good

154

idea to stay with the conservative current-gain figure of 10 for each transistor.

Transistor Q2 is selected so that its maximum continuous collector current rating matches or exceeds the stall current of the motor. Q1 is then selected so that its collector can drive the base of Q2 with one-tenth the stall current of the motor.

So if it happens that the motor has a stall current of 1A, Q2 must have a collector current rating of at least 1A, and Q1 should have a collector rating of at least 100 mA. You should have no problem locating such transistors at a very reasonable cost.

The equations accompanying the circuit in Fig. 5-7A show how to calculate the values of the resistors. These equations are based on the assumptions already described for a similar situation in Chapter 5. In those equations, V_{LH} is the logic-1, turn-on voltage level of the waveform at the INPUT connection, V_{eb1} is the emitter-base junction voltage for Q1, V_{eb2} is the same parameter for Q2, and I_{ms} is the stall current of the motor.

To work out a specific design example, suppose you are working with a 12 VDC motor and want to work things out so that the circuit can handle stall currents up to 1A. Further suppose that the positive portion of the variable duty cycle waveform at INPUT is 4.8V.

A 2N3724 transistor will work quite well for Q2, because it has a constant collector current rating of 1A. A less expensive and smaller transistor, such as a 2N3904, will do the job for Q1. That takes care of selecting the transistors.

Using the specifications for this example, the value for resistor R_i works out to be:

$$R_i = \frac{(4.8V)-(0.5V-0.5V)}{.01(1A)} = 380 \text{ ohms}$$

The nearest standard resistor value is 360 ohms, so use a 360-ohm resistor for R_i. The value of R_b is 10 times that amount, or a 3.6K resistor. Using these resistor values, the circuit will pull little more than 10 mA from whatever sort of circuit drives the INPUT connection.

The circuit shown in Fig. 6-7B uses a triple Darlington configuration to boost the current gain to at least 1000. The same amount of drive current at the INPUT can thus operate a motor having 10 times the stall current of the circuit in Fig. 6-7A.

It isn't difficult to select the transistors for the triple Darlington circuit if you bear in mind that the final stage, Q3, must be able to handle the stall current of the motor. Q2 should then be able to carry one-tenth that amount, and Q1 should handle one-tenth the collector current for Q2.

For example, suppose the motor has a stall rating of 8A at 12 VDC. An MJE-3055 transistor can handle that so use it for Q3. Transistor Q2 should be able to carry 0.8A, so a 2N3724 will work in that spot. Finally, Q1 has to deal with currents of only 80mA so try a 2N3094.

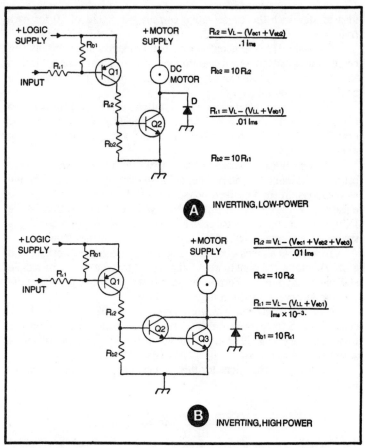

Fig. 6-8. Switching amplifiers for variable duty cycle motor speed control. (A) Inverting type with a current gain of at least 100. (B) Inverting type with a current gain of at least 1000.

The equations for determining the values of the resistors in this case are almost identical to those in Fig. 6-7A. The only differences are the scaling factor for I_{ms} and the addition of a third base voltage, V_{eb3}:

$$R_i = \frac{(4.8V) - (0.5 + 0.5 + 0.5)}{10 \times 10^{-3}} = 430 \text{ ohms}$$

The nearest standard resistor value is exactly 430 ohms, and that takes care of the value of the input resistor. The base resistor, R_b, can be fixed at 10 times that value, or 4.3K.

The example assumes the logic-1 input voltage level is 4.8V. Using any other amount of drive voltage promises to change the values of the resistors. Of course any change in the stall current of the motor, I_{ms}, will also call for changing the resistor values and, perhaps, the current ratings of the transistors.

Inverting Driver/Interface Circuits

The circuits in Fig. 6-8 do the same basic sort of job, but they are intended to cover instances where the variable duty cycle waveform at INPUT is inverted: A logic-0 level at INPUT is supposed to turn on the motor, and vice versa. This inverting configuration is also most suitable when driving the INPUT connection from TTL logic devices, because these devices have a higher output-low drive capability.

You can count on the circuit in Fig. 6-8A having a current and power gain of 100. If it is being operated from a 5V TTL circuit at the INPUT connection, the circuit can readily handle 1A motor stall currents at 12 VDC.

The equations accompanying Fig. 6-8A use some variables that have not been defined yet. V_L is the LOGIC SUPPLY voltage level, V_{ecl} is the emitter-collector saturation voltage of Q1, and V_{LL} is the logic-0 voltage level of the circuit driving the INPUT connection. The remaining variables can be figured out in the light of the discussion of Fig. 6-7.

Suppose you are working with a motor that has a stall current of 1A, V_L is the standard 5.5V supply level for TTL circuits, and V_{LL} is the typical logic-low level of 0.4V for TTL devices. This is all you need to determine the values of all four resistors in Fig. 6-8A.

$$R_{i2} = \frac{5.5V - (0.7V + 0.5V)}{0.1(1A)} = 43\,ohms$$

This works out quite nicely. A 43-ohm resistor is available as a 1/4 W, 5 percent-tolerance resistor. But be careful! What about its power rating?

Table 6-1. Parts List for the Circuit of Fig. 6-8B.

Motor stall current=12A
LOGIC SUPPLY = 5.5V
Logic-0 level at the INPUT=0.4V
All emitter-base junction voltages=0.5V
All emitter-collector saturation voltages=0.7V

Suggested Parts List

Q1—Any PNP transistor with a collector current rating of 120mA or more (2N2906)
Q2—Any NPN transistor with a collector current rating of 1.2A or more (MJE-3055T)
Q3—Any NPN transistor with a collector current rating of 12A or more (2N3055)
R_{i1}—430 Ohms, 1/4 W resistor
R_{b1}—4.3K, 1/4 W resistor
R_{i2}—33 Ohm, 1 W resistor
R_b—330 Ohm, 1/4 W resistor

157

The current through R_{i2} is going to be one-tenth the stall current of the motor, or about 0.1A in this case. And if you apply the power formula, $P = I^2 R$, you will find $P_{i2} = (0.1)^2 \times 430 = 0.43$ watts. A little 1/4W resistor cannot handle that, so a 1/2W resistor must be used in its place.

You might not be able to find a 430-ohm, 1/2W resistor, so you'll have to settle for the next lower standard value of 39 ohms. The use of Darlington circuits in Fig. 6-7 gets around the problem of inserting higher-power resistors in the main current paths. You'll have to watch the power ratings of R_{i2} in these two circuit configurations, however.

The value of R_{b2}, as usual, is taken as 10 times that of its corresponding input resistor: either 430 ohms or 470 ohms. Take your pick—it doesn't make any difference which value you use. A 1/4W version works quite well in any case.

Finding the value of R_1 is the next step.

$$R_{i1} = \frac{5.5V - (0.4V - 0.5V)}{0.01(1A)} = 460 \text{ ohms}$$

Rounding down to the nearest 1/4W, 5 percent value, R_{i1} should be set at 430 ohms. The low current level at this point allows you to use a 1/4W resistor here. Finally, $R_{b2} = 10R_{i1}$, or 4.3K.

If you have been following these discussions and examples carefully, you should have no trouble figuring out the operation of the higher-power, inverting drive circuit shown in Fig. 6-8B. Assuming the following circuit specifications, see if you agree with the comments on the resulting parts list shown in Table 6-1. How much logic-0 drive current should be available from the TTL device driving the INPUT to this circuit? The answer is 12 mA.

Some parabot/robot motor speed control situations call for driving motors with currents greater than 10A or 12A. The general procedure for designing these super power drivers is to tack on an additional power transistor, Darlington fashion, to the circuits shown in Fig. 6-7B or Fig. 6-8B. If it becomes impossible to find transistors for the final driver, you can always resort to connecting power transistors in parallel, remembering to insert a low-value power resistor in series with each of the collector circuits. The resistors are necessary for making certain all the paralleled transistors are able to take on their own saturation voltages. See the 50A examples in Fig. 6-9.

Some Construction Hints

All power transistors specified in the previous circuits should be mounted to aluminum heat sinks. Use generous doses of heat-sink compound to ensure efficient heat transfer from the transistor to the sink assembly. The smaller components, including the low-power and medium-power transistors, can be assembled on any desire sort of circuit board.

158

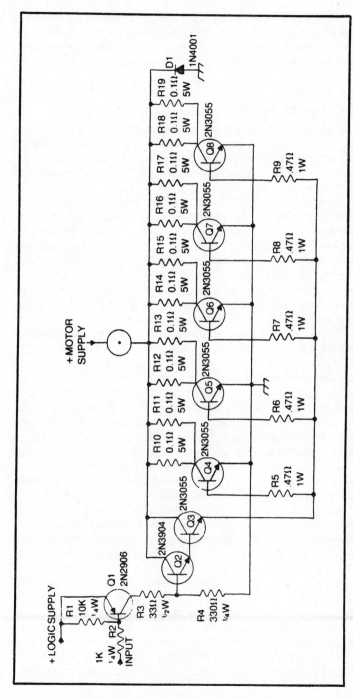

Fig. 6-9. Inverting, super-power switching amplifiers for variable duty cycle motor speed control. This circuit can operate a 50A DC motor.

159

Also try to keep the wiring to the main driver transistors as short as possible. Any stray inductance created by using overly long wires will slow down the switching time of the circuit, leaving the power transistors in their heat-producing linear region longer than necessary.

Some Troubleshooting Hints

The most common problem associated with high-power switching circuits is that the driver transistors get so hot that they soon fail. You will usually notice the motor running at a moderate speed under this fail condition, no matter what sort of control signal you apply at the INPUT connection of the circuit.

Aside from improper heat sinking, the cause of overheating is the failure of the transistor to go into complete saturation when it is supposed to be switched on. Connect an oscilloscope across the final drive transistor (from emitter to collector), drive the INPUT with a waveform having a 50 percent duty cycle, and carefully note the voltage levels of the waveform on the oscilloscope.

If, indeed, the transistor is not going into saturation, you will find the low-level voltage at about +1V or more. The higher this low-level voltage is, the more severe the heating problem will be.

The saturation voltage should be about +0.7V or less. Check the specifications for your particular driver transistor to find out what $V_{ec(sat)}$ should be.

There are several reasons why a power transistor is not being driven into saturation. For one, it is possible it isn't being driven hard enough at the base circuit. To determine whether or not this is the case, increase the base-drive current as follows. If you are using one of the noninverting circuit configurations in Fig. 6-7, increase the base drive by connecting a resistor in parallel with R_i. The value of the resistor can be equal to that of R_i, itself. And if you are using one of the inverting versions in Fig. 6-8, simply connect a jumper across R_{i2}. Do not allow the circuit to operate more than a few seconds at a time under these test conditions. Run the test just long enough to see whether or not the final transistor stage is showing a lower saturation voltage. If that is the case—if the final transistor shows a low-level voltage on the order of 0.7V—you will have to use a high-power version of the same circuit. The circuit requires additional current gain.

Suppose the test just described does not settle down the saturation problem, however, and the transistor still runs terribly hot. The next possibility is that the final driver transistor simply does have the necessary amount of current gain. Despite calculations and assuming every transistor has a current gain of 10 or more, things simply don't work out as expected. The simplest solution to this particular problem is to replace that transistor with one having a higher current-gain or h_{fe}, rating.

When using one of the Darlington configurations, the same trouble could be caused by a low current gain for any one of the transistors. It is most likely the final one, however.

These tests and recommended remedies should cure most situations where a power transistor is overheating because it is not being driven into saturation. Sometimes, however, a transistor will be driven into saturation with no trouble, but fails to turn off completely. This appears on an oscilloscope check as a high-level voltage that is somewhat less than that of the +MOTOR SUPPLY voltage.

If that is the case, and you are using one of the noninverting power circuits shown in Fig. 6-7, connect a jumper across resistor R_b and note the waveform across the final transistor stage. If the high-level voltage is equal to the +MOTOR SUPPLY level, remove the jumper and remedy the trouble by cutting the value of R_b in half.

When working with one of the inverting versions shown in Fig. 6-8, you have to run the jumper test for both R_{b1} and R_{b2}—one at a time. If shorting out either or both of these resistors lets the output power transistor switch completely off, the resistor should be replaced with one having one-half the calculated value.

On the outside chance that shorting out the resistors does not cure the switch-off problem, a leaky transistor appears somewhere in the circuit. Test or replace the transistors with new or better versions.

Finally, if the transistor is going into saturation and the cutting off properly, but overheating persists, look carefully at the waveform where it makes its high-to-low transistion. This is the turn-on switching interval. If it appears sloped and rounded, the power transistor is not being switched on cleanly; it is spending too much time in its linear region.

The cure for this problem is relatively simple, assuming you have already taken the precaution of keeping the wiring as short as possible. Simply connect a 0.47-μF capacitor in parallel with R_i (if you are using one of the circuits shown in Fig. 6-7) or R_{i2} (if you are using a circuit from Fig. 6-8). This serves the purpose of a speed-up capacitor, creating a short-term, high-current path for driving the bases of the following transistors about as hard as they can be driven.

FITTING TOGETHER DIRECTION AND SPEED CONTROL CIRCUITS

While it is quite possible to build a fine parabot/robot system using only a two-direction motor control, there is good reason to question the value of a system that has a lot of speed control, but no direction control. So all of the material in this chapter isn't really very useful unless the circuits are integrated with a two-direction control scheme from Chapter 5.

Figure 6-10 shows a simplified control circuit that includes driver transistors for both direction and speed controls. Transistors Q100 and Q103 control relays RL100 and RL101 to achieve full direction control over the motor. Q101 is the speed-control driver. Note that Q101 is

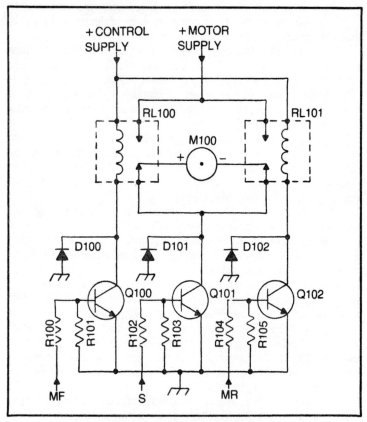

Fig. 6-10. Putting together motor-speed and direction controls.

connected in series with the motor circuit, between the normally closed contacts of the relays and systems ground.

The inputs to the circuit are labeled MF, MR and S. All three are active-high in this case. Applying logic-1 levels to the MF and MR inputs turns on their respective transistors and energizes their relays for motor-direction control. A variable duty cycle waveform applied to input S applies power to the motor whenever it is at logic 1, and turns off that power whenever the waveform is at logic 0. The circuit, as it is, can actually be used in a very small parabot/robot system that employs a motor drawing less than 1A from its power source.

A Two-Direction, Variable-Speed Control for Manipulators

The circuit shown in Fig. 6-11 is the power section of a typical motor control for a manipulator system. It is no accident that the circuit resembles the one shown in Fig. 5-11: This is the same circuit, extended to include the motor speed control transistors, Q101, Q104 and Q105.

162

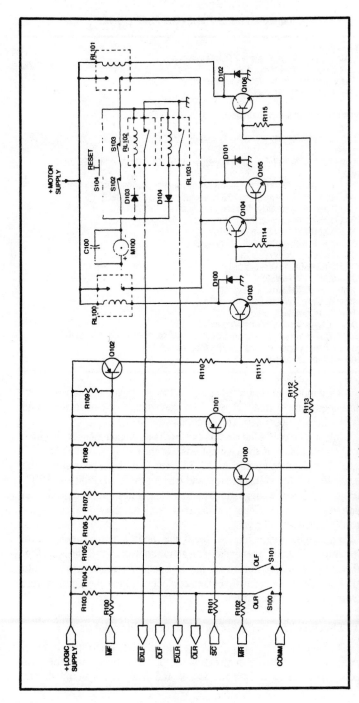

Fig. 6-11. Direction and speed-control power circuit for a manipulator system.

Table 6-2. Parts List for the Circuit of Fig. 6-12.

Specifications:
    All inputs and outputs are TTL compatible
    +LOGIC SUPPLY = 5.5V
    Motor is rated at 12VDC, 5A stall current
    +MOTOR SUPPLY=12V

Recommend Parts Specifications:

RL100,RL101—SPDT (or DPDT) relay; 12VDC, 75mA coil; 120VAC, 10A contacts
RL102,RL103—SPST (or SPDT) relay; 12VDC, 10mA coil; 120VAC, 1A contacts
D100,D101,D102,D103,D104—Any small rectifier diodes, such as 1N4001
Q100,Q101,Q102—Any small PNP transistor, such as 2N3906
Q103,Q106—Any small NPN transistor, such as 2N3904
Q104,Q105—NPN power transistor, such as 2N3055
R100,R102,R111,R115—5.6K, ¼W resistor
R103,R104,R105,R106—22K, ¼W resistor
R101—910 Ohm, ¼W resistor
R107,R109—56K, ¼W resistor
R108—9.1K, ¼W resistor
R110,R113—560 Ohm, ¼W resistor
R112—75 Ohm, ¼W resistor
R114—750 Ohm, ¼W resistor
S100,S101—normally open microswitch
S102,S103—normally closed microswitch with contact rating of 5A or more
S104—normally open momentary pushbutton switch
C100—0.1 uF capacitor

All the inputs and outputs are active-low, including the variable duty cycle speed-control input at $\overline{SC}$. The remaining I/Os have the same functions already described in connection with Fig. 5-11.

Table 6-2 is a parts list for this circuit. The component specifications have been calculated on the assumptions that the motor is rated at 12 VDC and has a stall/start-up current of 5A, and that the inputs should be compatible with the voltage levels and current-drive capabilities of TTL logic circuits. Use the suggestions for selecting transistors and the design equations presented in Chapters 5 and 6 to work out the specifications for motors having different ratings.

As an engineering exercise, see if you can fit a motor speed control circuit into the "smart" manipulator direction control circuit shown in Fig. 5-10. Hint: The driver transistors for the speed control circuitry go into the line between the normally closed contacts on the relays and system common. You take it from there.

A Two-Direction, Variable-Speed Circuit for Drive/Steer Motors

The circuit shown in Fig. 6-12 is a variable duty cycle speed control worked into a standard two-direction motor control scheme. The two-direction portion of the circuit is adapted from the system already

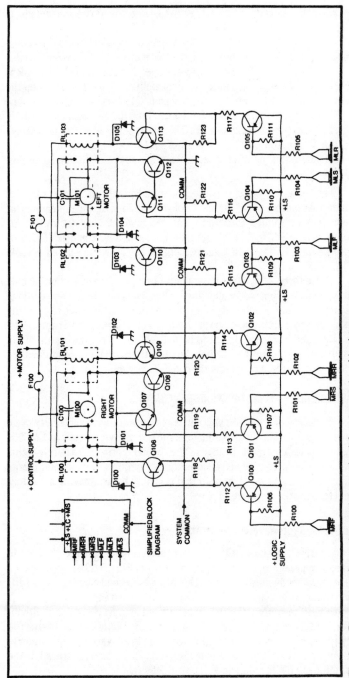

Fig. 6-12. Direction and speed-control power circuit for a two-motor, drive/steer system.

165

described in connection with Fig. 5-13. See the operating details in Chapter 5.

The speed-control portion of the circuit used a Darlington, inverting-type waveform amplifier for each of the two motors. The inverted variable duty cycle waveform for controlling the speed of the RIGHT MOTOR is picked up at the $\overline{\text{MRS}}$ signal input. Preamplifier Q101 then steps up the current level for driving the Darlington pair, Q107 and Q108. By the same token, the signal input point for controlling the speed of the LEFT MOTOR is at the $\overline{\text{MLS}}$ connection. The preamplifier in this case is Q104, and the Darlington pair is made up of Q111 and Q112. The direction-control inputs and amplifier schemes have already been discussed for Fig. 5-13.

The circuit is quite versatile, being adaptable to a wide range of drive/steer motor specifications. Table 6-3 summarizes specifications for the components under three different design conditions: for 6VDC, 1A motors (Table 6-3B), 12VDC, 5A motors (Table 6-3C), and 12VDC, 10A motors (Table 6-3D). Parts that are identical in all cases are listed in Table 6-3A.

These three sets of motor parameters and parts specifications should cover most parabot/robot designs. When considering systems having specifications outside the range illustrated here, you will have to apply the design principles and equations presented in this chapter and Chapter 5.

Incidentally, the specifications in Table 6-3 make the inputs of the circuit compatible with TTL logic levels. Alternate input conditions are considered in a few examples at the conclusion of this chapter.

VARIABLE DUTY CYCLE WAVEFORM GENERATORS

There are two general approaches to generating variable duty cycle waveforms. One way is to use analog-type timer devices, such as LM555s or their dual counterpart, LM556s. The second approach is a purely digital one, using variable-modulus counters. Both approaches have their merits and drawbacks, so study this section carefully before making a commitment to using one or the other.

The Analog-Type Waveform Generators

Figure 6-13 shows two kinds of analog-type variable duty cycle waveform generators. The first, shown in Fig. 6-13A, generates a pulse-width modulated waveform (PWM). The second circuit (Fig. 6-13B) is a frequency-modulated (FM) version of the same thing.

In both instances, Z100-A is connected as a free-running multivibrator, and its output triggers a monostable multivibrator, A100-B. The SPEED ADJUST control in Fig. 6-13A, however, varies the duration of the output pulse, and the frequency is fixed. This, you might recall, is the definition of a PWM motor speed control scheme.

Table 6-3. Additional Parts Lists for the Circuit of Fig. 6-12.

Parts Common to All Configurations:

C100,C101—0 1 uF capacitor
Q100,Q101,Q102,Q103,Q104,Q105—Any small PNP transistor, such as 2N3906
D100,D101,D102,D103,D104,D105—Any small rectifier diode, such as 1N4001

Additional Parts for Use With a 6VDC, 1A Stall-Current Motor (See Fig. 6-12 and Table 6-2A)

Q106,Q109,Q110,Q113—Any small NPN transistor, such as 2N3904
Q107,Q111—Any medium-power NPN transistor, such as 2N2907
Q108,Q112—Any small, power transistor, NPN, such as 2N4013
R100,R102,R103,R105—36K, ¼W resistor
R101,R104—4.6K, ¼W resistor
R106,R108,R109,R111—360K, ¼W resistor
R107,R110—46K, ¼W resistor
R112,R114,R115,R117—3.3K, ¼W resistor
R113,R116—360 Ohm, ¼W resistor
R118,R120,R121,R123—33K, ¼W resistor
R119,R122—3.6K, ¼W resistor
RL100,RL101—SPDT (or DPDT) relay; 6VDC, 12mA coil; 120VAC, 1A contacts
F100,F101—1A slow-blow fuse

Additional Parts for Use With 12VDC, 5A Stall-Current Motors

(See Fig. 6-12 and Table 6-2A)

Q106,Q109,Q110,Q113—Any small NPN transistor, such as 2N3904
Q107,Q111—Any medium-power NPN transistor, such as 2N2222
Q108,Q112—Any NPN power transistor with collector current meeting or exceeding 5A, such as 2N3055
R100,R102,R103,R105,R107,R110—10K, ¼W resistor
R101,R104—10K, ¼W resistor
R106,R108,R109,R111—100K, ¼W resistor
R112,R114,R115,R117—820 Ohm, ¼W resistor
R113,R116—68 Ohm, ½W resistor
R118,R120,R121,R123—8.2K, ¼W resistor
R119,R122—680 Ohm, ¼W resistor
RL100,RL101—SPDT (or DPDT) relay; 12VDC, 50mA coil; 120VAC, 5A contacts
F100,F101—5A slow-blow fuse

Additional Parts for Use with 12VDC, 10A Stall-Current Motors
(See Fig. 6-12 and Table 6-2A)

Q106,Q109,Q110,Q113—Any small NPN transistor, such as 2N3904
Q107,Q108,Q111,Q112—NPN power transistor, such as 2N3055
R100,R101,R102,R103,R104,R105,R107,R110—5.6K, ¼W resistor
R106,R108,R109,R111—56K, ¼W resistor
R112,R114,R115,R117—560 Ohm, ¼W resistor
R113,R116—36 Ohm, ½W resistor
R118,R120,R121,R123—5.6K, ¼W resistor
R119,R122—360 Ohm, ¼W resistor
RL100,RL101—SPDT (or DPDT) relay; 12VDC, 75mA coil; 120VAC, 10A contacts
F100,F101—10A slow-blow fuse

In Fig. 6-13B, the SPEED ADJUST control influences the basic operating frequency of the circuit, and the width of the output pulse is fixed. This, of course, defines an FM motor speed control.

Some of the parts values in these circuits have been specified for you, while other have not. This gives you some degree of freedom over the specifications, and yet helps you get your own designs started on the right track.

When using the PWM version in Fig. 6-13A, for example, a fixed frequency of 100 Hz serves quite well in virtually all speed control circuits. Adhering to that particular operating frequency pretty well dictates the values of C100, R102 and R100, as shown on the schematic.

It is up to you, however, to select the values of C102, R103 and R104, based on your own choice of maximum and minimum operating duty cycles, D_{max} and D_{min}. The equations accompanying that diagram show how to calculate the values of C102 and R104. It is up to you to get things started by selecting a value for R103 from the family of standard potentiometer values: 5K, 10K, 50K, etc. Take an educated guess. You will know the guess is right if the values for the other two components are reasonable ones.

Suppose, for example, you want to be able to vary the duty cycle between 0.7 and 1 (70 percent to 100 percent). If you then select 5K as a value for the SPEED ADJUST control, C102 must have a value of about 0.55 μF. You won't be able to find a 0.55 μF capacitor, so use the closest standard value of 0.47 μF. Then use that figure to calculate the value of R104—about 13.5K. A standard 5 percent-tolerance resistor value of 15K will have to do for R104.

Don't spend too much time playing around with these values, attempting to get them to work out of the money. Tolerances of the components throughout the entire circuit will throw off your calculations as much as 10 percent anyway. The only real test for the correct values is trying the circuit with the motor system. If that on-line test does not work out to your satisfaction, you can go back to the drawing board, adjusting and recalculating the values as necessary.

If you want to change the basic operating frequency of this PWM generator, just change the value of R100. Decreasing the value of that resistor will increase the base frequency, while increasing the value of R100 will lower the base frequency.

The FM version of the variable duty cycle circuit (Fig. 6-13B) assumes a fixed T_{on} time of 50 ms. The value of the SPEED ADJUST resistor, R100, should be selected from the family of standard values; once you do that, you are in a position to calculate the values for C100, R101 and R102.

Try working out an example, using the same specifications cited in the previous, PWM example: R100=5K, and the range of duty cycles is 0.7 to 1. The value of C100, according to the equations accompanying the FM circuit, is about 6 μF. Unfortunately, that's a nonstandard value, so

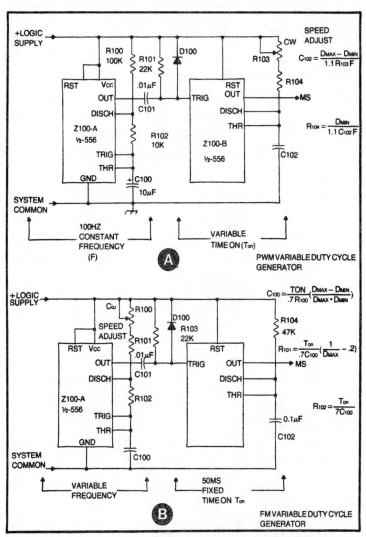

Fig. 6-13. Analog-type variable duty cycle waveform generators. (A) PWM type. (B) FM type.

pick the nearest standard size, 4.7 μF. Use that value for calculating R101 and R102. R101 thus turns out to be about 12K, and R102 works out to about 1.5K. Use those values, and you're ready to check the FM motor speed control circuit. If you choose to do so, you can alter T_{on} from its prescribed value of 50 ms by changing the values of R104, C102 or both.

The outputs of both variables duty cycle generators in Fig. 6-13 are active-high waveforms. The significance of this fact is that they can be

169

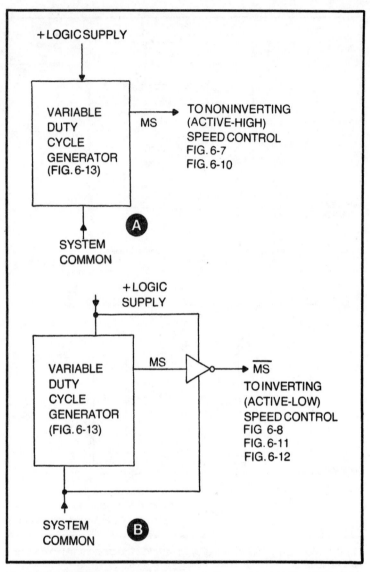

Fig. 6-14. Connecting waveform generators to the speed control inputs of motor power circuits. (A) Interfacing with noninverting power switching circuits. (B) Interfacing with inverting power switching circuits.

used directly with noninverting motor-speed driver circuits, as shown in Fig. 6-14A. If you have an inverting driver circuit, then, the waveform from these generators must be inverted first. See the simplified diagram in Fig. 6-14B.

A Complete Parabot Drive/Steer System With Analog Speed Control

Figure 6-15 is a complete two-direction, variable speed control circuit for a pair of drive/steer motors. The components used for driving the motor relays and the motors are specified in Table 6-4 for low-power applications. Specifically, the motors are to be rated at less than 1A of stall current. The whole circuit is adequate for a little parabot—a microbot, if you will. And it makes a good project for an experimenter who doesn't have much experience assembling electronic circuits.

The circuit is included in this section because it is a prime example of applying analog-type speed controls in a direct fashion. You will find two of the PWM controls here. Z100-A and Z100B make up the speed control for the RIGHT MOTOR, while Z101-A and Z101B do the same job for the LEFT MOTOR. In both of these control circuits, the "A" parts are the 100-Hz, fixed-frequency oscillators, and the "B" parts are the speed pulse width (T_{on}) monostable multivibrators. The duty cycle for both waveform generators is adjusted manually by means of the RIGHT SPEED and LEFT SPEED controls. The duty cycle control range is between 0.6 and 1 in both cases.

Because most motors will come to a complete stop whenever they are operating at a duty cycle of 0.6, the circuit gives the operator full speed control range. Using the terminology of control systems, this is an *infinite* speed control. That doesn't mean the motors can be run at an infinite speed, but rather than the user can select an *infinite number* of speeds between stop and full speed.

The waveforms from the two variable duty cycle waveform generators are active-high, so they are fed to noninverting motor-speed driver circuits. Therefore, the speed control waveform for the RIGHT MOTOR, labeled MRS, goes to a Darlington pair of transistors Q100 and Q101. Since the motors are specified as having stall currents of 1A or less, this two-transistor driver scheme is adequate for the job. In a similar fashion, waveform MLS—the active-high variable duty cycle waveform for the LEFT MOTOR speed control—sets the speed of the LEFT motor through the Darlington pair of transistors, Q102 and Q103.

The motor-direction control portion of the circuit is a straightforward application of the latching switch control described in connection with Fig. 5-17. Setting up the directions of the motors is a two-step process. First set the appropriate pattern of settings for the switches labeled RF, RR, LF and LR. Then execute the command by depressing the LOAD pushbutton, S101.

Figure 6-16 shows a suggested layout for the control panel, as well as a function table for the direction-control switches. These control components can be enclosed in a small project box and connected to the main circuitry and mainframe through an umbilical cord.

Again, this can be a good project for a novice experimenter. Its only drawback is that it doesn't interface very well with automatic or processor-controlled logic systems. The fact that the speeds of the motors

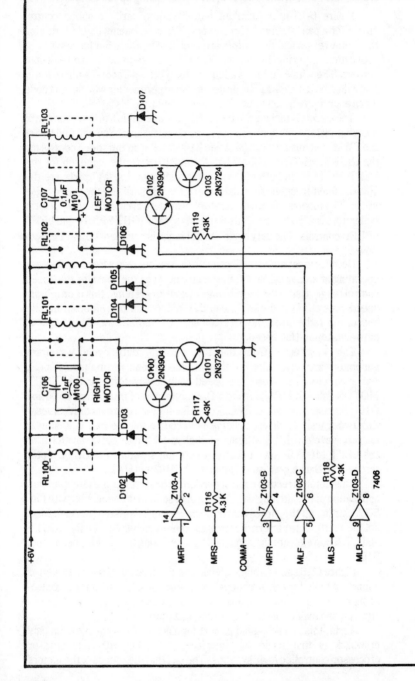

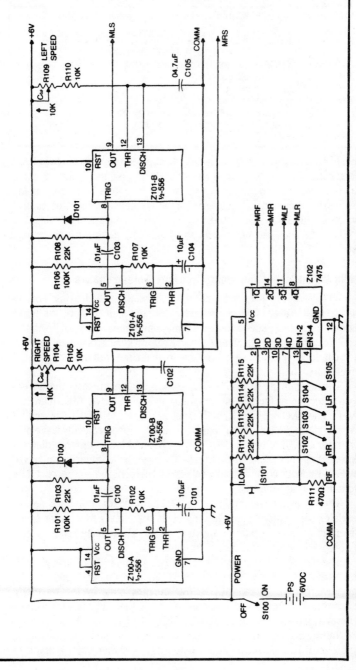

Fig. 6-15. A complete motor control system for a parabot using two-direction, variable-speed drive/steer motors.

Table 6-4. Parts List for the Circuit of Fig. 6-15.

M100, M010—Any 6VDC, permanet motor having stall current less
        than 1A
RL100, RL101, RL102, RL103—SPDT (or DPDT) relay; 6VDC, 12mA coil;
        120VAC, 1A contacts
D100 through D107—Any small rectifier diode, such as 1N4001
Z100, Z101—LM556 dual timer
Z102—7475 quad D-type latch
Z103—7406 hex, open-collector inverter
PS—6VDC power source, such as 4 C-cells in series
The remaining parts are specified on the diagram.
Use ¼W resistors.

have to be adjusted with mechanical potentiometers precludes automatic speed control from an electronic system. Of course the speed-adjusting variable resistors could be driven by little motors, but that would needlessly compound the complexity of the system. The direction-control switches could always be replaced with a source of digital logic levels, but those mechanical, analog speed-setting potentiometers are something else.

Getting Digital Control with Analog-Type Duty Cycle Generators

It is possible to get digital control over motor speed, using analog-type, variable duty cycle waveform generators, if you are willing to trade off the inifinite speed control feature inherent in Fig. 6-15. The basic idea is to use two or more analog-type waveform generators, each set for a different duty cycle, and select the one to go on line with the motor speed driver transistors.

The diagram in Fig. 6-17 illustrates this idea. In this case, the system has a selection of four different speeds: STOP, SLOW, MEDIUM and FAST. The SLOW and MEDIUM speeds are set at 80 percent and 90 percent duty cycles for a pair of analog speed-control waveform generators, VDC100 and VDC 101. Those blocks on the diagram represent one of the circuits in Fig. 6-13. As described shortly, the STOP and FAST speeds come about by selecting 0 percent and 100 percent duty cycles from other sources.

The key element in this circuit is a 4:1 multiplexer or digital selector. It delivers one of four different duty cycle waveforms to a motor-speed driver, as selected by address inputs A and B. Those inputs come from a pair of speed-selector switches, S100 and S101.

The four signal inputs to Z102 go to pins labeled 1CO through 1C3. Notice that 1C0 is connected to SYSTEM COMM, 1C1 goes to the waveform output of the SLOW (80 percent duty cycle) waveform generator, IC2 goes to the output of the MEDIUM speed waveform generator, and IC3, is fixed at the +LOGIC SUPPLY VOLTAGE.

Setting both of the speed-select switches to logic 0 thus sends the logic level at 1C0 of Z102 to the MS output. The waveform at that point is a steady logic-0 level, defining an active-high duty cycle of 0 percent. That means STOP as far as the motor speed is concerned.

Setting the switches for binary 1 (SS1=0 and SS0 =1 feeds the signal present at 1C1 to the motor. That creates the SLOW, 80 percent duty cycle speed control situation.

A switch pattern for binary 2 selects the signal at 1C2, which is the MEDIUM, 90 percent duty cycle waveform from VDC101. And setting both switches to 1 (binary 3) feeds a constant logic-1 level to the output of the data selector, defining the FAST, 100 percent duty cycle waveform.

The operator can thus select any one of the four speeds by setting the speed-select switches to the appropriate pattern of 1s and 0s. That set of patterns is summarized in the function table accompanying the diagram.

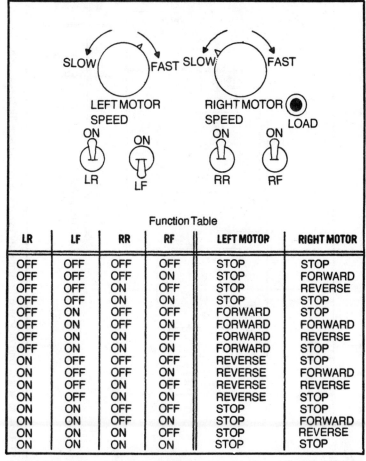

Function Table

LR	LF	RR	RF	LEFT MOTOR	RIGHT MOTOR
OFF	OFF	OFF	OFF	STOP	STOP
OFF	OFF	OFF	ON	STOP	FORWARD
OFF	OFF	ON	OFF	STOP	REVERSE
OFF	OFF	ON	ON	STOP	STOP
OFF	ON	OFF	OFF	FORWARD	STOP
OFF	ON	OFF	ON	FORWARD	FORWARD
OFF	ON	ON	OFF	FORWARD	REVERSE
OFF	ON	ON	ON	FORWARD	STOP
ON	OFF	OFF	OFF	REVERSE	STOP
ON	OFF	OFF	ON	REVERSE	FORWARD
ON	OFF	ON	OFF	REVERSE	REVERSE
ON	OFF	ON	ON	REVERSE	STOP
ON	ON	OFF	OFF	STOP	STOP
ON	ON	OFF	ON	STOP	FORWARD
ON	ON	ON	OFF	STOP	REVERSE
ON	ON	ON	ON	STOP	STOP

Fig. 6-16. Function table and suggested switch panel layout for the complete parabot control system shown in Fig. 6-15.

175

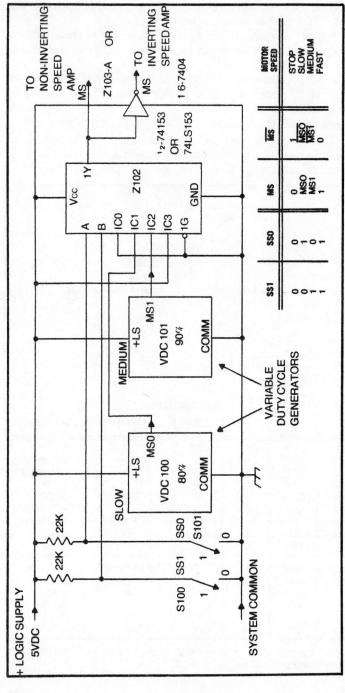

Fig. 6-17. How to select one of four different duty cycle waveforms in a digital fashion.

176

The circuit in Fig. 6-18 is a refinement of this four-speed selection scheme. Actually, it is a composite speed/direction control circuit for a single motor. Table 6-5 is the parts list for Fig. 6-18.

The circuit uses the PWM principle to generator SLOW and MEDIUM duty cycle waveforms. To do this, two monostable multivibrators, Z101-A and Z101-B, generate the T_{on} pulse intervales. These two monostables share a common, fixed-frequency, 100-Hz clocking oscillator, Z100. The SLOW speed duty cycle is trimmed by means of the LOW SPEED ADJUST control, and the MEDIUM speed in trimmed with the MEDIUM SPEED ADJUST control. In a working system, these two controls can be trimmer resistors mounted on the printed-circuit board with all the other small parts.

Z102 is the same 4:1 data multiplexer described for the general circuit in Fig. 6-17. The inverter at its output makes the selected waveform suitable for use with inverting speed control amplifiers.

Another big difference between this circuit and the one described in connection with Fig. 6-17 is this one uses logic-level speed selection. You can connect switches to inputs D0 and D1 to select the desired motor speed, but those D inputs can also come from any other sort of TTL data source, including the output port of a bus-oriented microprocessor scheme.

If you were to build the speed-select and direction-select control circuits separately, you would need four data inputs. You would need two inputs for selecting the speed and two more for setting up the direction the motor is supposed to turn. The idea would work just fine, but applying some elementary principles of data compression makes it possible to control both the speed and direction of the motor with just three inputs—D0 through D2, in this case.

Recall from Chapter 5 that most motor-direction selection schemes operating from two direction-select inputs result in two different STOP configurations. In other words, there is some redundancy in the usual approach to selecting the direction a motor will turn. This redundancy is eliminated in this circuit by letting the speed control portion of the system generate the STOP command. As a result, there are just three control

Table 6-5. Parts List for the Circuit of Fig. 6-18.

Z100—LM555 timer
Z101—LM556 dual timer
Z102—74LS153 4:1 data multiplexer
Z103—7404 hex inverter
Z104—7400 quad 2-input NAND gate
D100, D101—Any small rectifier diode, such as 1N4001
R103, R105—10K printed-circuit (trimmer) resistors

Other part values are specified on the diagram.
All resistors are ¼W

inputs that generate eight possible operating conditions. Those conditions are spelled out for you in Table 6-6. If you were to use separate control inputs for the speed and direction circuits, you would end up with 16 control states, many of them redundant. The circuit shown in Fig. 6-18 can be interfaced with a single-motor control circuit, such as the manipulator control of Fig. 6-11.

A Complete Drive/Steer Motor Control With Digital-Select Inputs

The circuit shown in Fig. 6-19 extends the basic notions just described to a two-motor drive/steer system. The controts allow completely independent direction and speed control for each of the two motors, and there are small, printed-circuit potentiometers for calibrating the SLOW and MEDIUM speeds for the motors.

Here is a summary of definitions for the output signals:

$\overline{\text{MLF}}$—An active-low logic level that causes the LEFT MOTOR to run in its forward direction. It is selected by inputs D0, D1 and D2.

$\overline{\text{MLR}}$—An active-low logic level that causes the LEFT MOTOR to run in reverse. It is selected by inputs D0, D1 and D2.

$\overline{\text{MLS}}$—An active-low speed control signal for the LEFT MOTOR drive circuitry. Four duty cycles can be selected by inputs D0 and D1:

    ☐ STOP (0 percent duty cycle)
    ☐ SLOW (user adjustable by R203)
    ☐ MEDIUM (user adjustable by R206)
    ☐ FAST (100 percent duty cycle)

$\overline{\text{MRF}}$—An active-low logic level that causes the RIGHT MOTOR to run in its forward direction. It is selected by inputs D3, D4 and D5.

$\overline{\text{MRR}}$—An active-low logic level that causes the RIGHT MOTOR to run in reverse. It is selected by inputs D3, D4 and D5.

$\overline{\text{MRS}}$—An active-low speed control signal for the RIGHT MOTOR drive circuitry. Four duty cycles can be selected by inputs D3 and D4:

    ☐ STOP (0 percent duty cycle)
    ☐ SLOW (user adjustable by R209)
    ☐ MEDIUM (user adjustable by R212)
    ☐ FAST (100 percent duty cycle)

Table 6-6. Function Table for the Circuit of Fig. 6-18.

D2	D1	D0	MOTOR RESPONSE
0	0	0	STOP
0	0	1	SLOW REVERSE
0	1	0	MEDIUM REVERSE
0	1	1	FAST REVERSE
1	0	0	STOP
1	0	1	SLOW FORWARD
1	1	0	MEDIUM FORWARD
1	1	1	FAST FORWARD

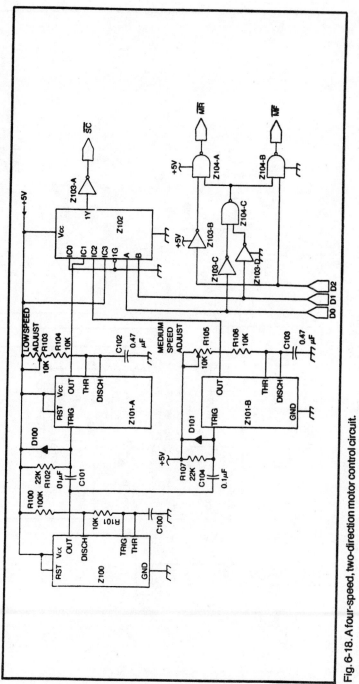

Fig. 6-18. A four-speed, two-direction motor control circuit.

179

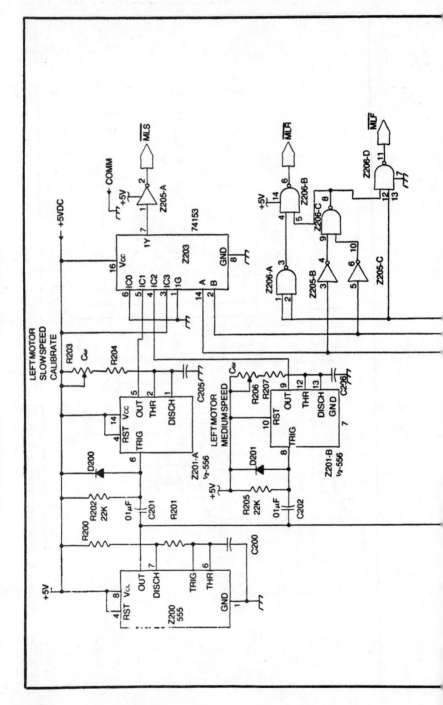

180

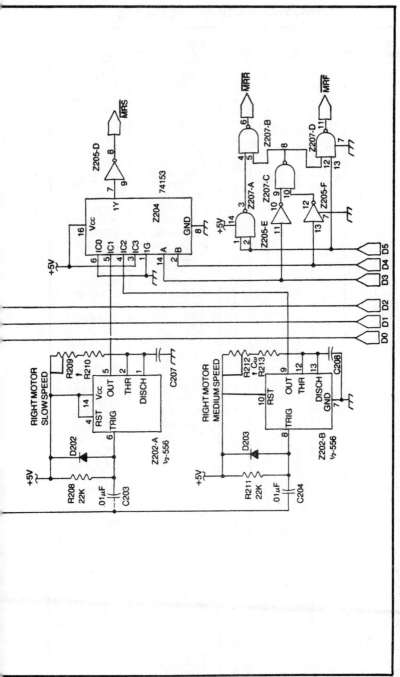

Fig. 6-19. Control circuit for a four-speed, two-direction drive/steering motor system.

181

Table 6-7. Function Table for the Circuit of Fig. 6-19.

D5	D4	D3	D2	C1	D0	RIGHT MOTOR	LEFT MOTOR
0	0	0	0	0	0	STOP	STOP
0	0	0	0	0	1	STOP	SLOW REV
0	0	0	0	1	0	STOP	MED REV
0	0	0	0	1	1	STOP	FAST REV
0	0	0	1	0	0	STOP	STOP
0	0	0	1	0	1	STOP	SLOW FOR
0	0	0	1	1	0	STOP	MED FOR
0	0	0	1	1	1	STOP	FAST FOR
0	0	1	0	0	0	SLOW REV	STOP
0	0	1	0	0	1	SLOW REV	SLOW REV
0	0	1	0	1	0	SLOW REV	MED REV
0	0	1	0	1	1	SLOW REV	FAST REV
0	0	1	1	0	0	SLOW REV	STOP
0	0	1	1	0	1	SLOW REV	SLOW FOR
0	0	1	1	1	0	SLOW REV	MED FOR
0	0	1	1	1	1	SLOW REV	FAST FOR
0	1	0	0	0	0	MED REV	STOP
0	1	0	0	0	1	MED REV	SLOW REV
0	1	0	0	1	0	MED REV	MED REV
0	1	0	0	1	1	MED REV	FAST REV
0	1	0	1	0	0	MED REV	STOP
0	1	0	1	0	1	MED REV	SLOW FOR
0	1	0	1	1	0	MED REV	MED FOR
0	1	0	1	1	1	MED REV	FAST FOR
0	1	1	0	0	0	FAST REV	STOP
0	1	1	0	0	1	FAST REV	SLOW REV
0	1	1	0	1	0	FAST REV	MED REV
0	1	1	0	1	1	FAST REV	FAST REV
0	1	1	1	0	0	FAST REV	STOP
0	1	1	1	0	1	FAST REV	SLOW FOR
0	1	1	1	1	0	FAST REV	MED FOR
0	1	1	1	1	1	FAST REV	FAST FOR
1	0	0	0	0	0	STOP	STOP
1	0	0	0	0	1	STOP	SLOW REV
1	0	0	0	1	0	STOP	MED REV
1	0	0	0	1	1	STOP	FAST REV
1	0	0	1	0	0	STOP	STOP
1	0	0	1	0	1	STOP	SLOW FOR
1	0	0	1	1	0	STOP	MED FOR
1	0	0	1	1	1	STOP	FAST FOR
1	0	1	0	0	0	SLOW FOR	STOP
1	0	1	0	0	1	SLOW FOR	SLOW REV
1	0	1	0	1	0	SLOW FOR	MED REV
1	0	1	0	1	1	SLOW FOR	FAST REV
1	0	1	1	0	0	SLOW FOR	STOP
1	0	1	1	0	1	SLOW FOR	SLOW REV
1	0	1	1	1	0	SLOW FOR	MED REV
1	0	1	1	1	1	SLOW FOR	FAST REV
1	1	0	0	0	0	MED FOR	STOP
1	1	0	0	0	1	MED FOR	SLOW REV
1	1	0	0	1	0	MED FOR	MED REV

Table 6-7. Function Table for the Circuit of Fig. 6-19.

D5	D4	D3	D2	D1	D0	RIGHT MOTOR	LEFT MOTOR
1	1	0	0	1	1	MED FOR	FAST REV
1	1	0	1	0	0	MED FOR	STOP
1	1	0	1	0	1	MED FOR	SLOW FOR
1	1	0	1	1	0	MED FOR	MED FOR
1	1	0	1	1	1	MED FOR	FAST FOR
1	1	1	0	0	0	FAST FOR	STOP
1	1	1	0	0	1	FAST FOR	SLOW REV
1	1	1	0	1	0	FAST FOR	MED REV
1	1	1	0	1	1	FAST FOR	FAST REV
1	1	1	1	0	0	FAST FOR	STOP
1	1	1	1	0	1	FAST FOR	SLOW FOR
1	1	1	1	1	0	FAST FOR	MED FOR
1	1	1	1	1	1	FAST FOR	FAST FOR

to suit your own needs. The only requirement is that the driver circuit accept TTL logic voltage and drive currents in an active-low format.

The outputs are completely determined by the logic levels present at the data inputs, D0 through D5. The logic levels at these inputs can come from a set of toggle switches for purely parabot-class operation or from any other data source—including an output port from a microprocessor control—for possible robot-class operation.

In any event, the complete function table for the circuit is shown in Table 6-7. While Table 6-7 shows 64 different combinations of input logic levels, 21 of them are redundant. There are just 41 unique sets of motor responses. If you were to use two inputs for setting the direction of each motor and two more for speed selection, you would end up with a total of eight inputs to the circuit. That would yield a total of 256 possible switch combinations but would not change the number of unique sets of motor responses. There would be 215 redundant conditions. So don't worry about data compression! It has already been handled quite nicely.

Z200 is the master frequency generator for this PWM speed control scheme. Note that it triggers all four pulse generators.

The values specified on the schematic diagram are common to just about any duty-cycle arrangement you want to use. The values specified on the accompanying parts list (Table 6-8) assume you want the master oscillator to run at 100 Hz and you want to be able to calibrate the pulse generators for duty cycles between 0.6 and 1. You can use the design procedures described earlier to select values suited to your own needs.

Figure 6-20 shows how the circuit can be connected to the two-motor driver circuit shown in Fig. 6-12. That makes the whole thing look simple. The catch is that you have to build the circuits in Fig. 6-12 and 6-19 first.

Of course, the circuit of Fig. 6-19 doesn't have to be interfaced with any particular power circuit shown in this book. Make up one of your own

Table 6-8. Parts List for the Circuit of Fig. 6-19.

Z200—LM555 timer
Z201, Z202—LM556 dual timer
Z203, Z204—74153 dual 4:1 multiplexer
Z205—7404 hex inverter
Z206, Z207—7400 quad 2-input NAND gate
D200, D201, D202, D203—Any small rectifier diode, such as 1N4001
R200—100K, ¼W resistor
R201, R204, R207, R210, R213—10K, ¼W resistor
R202, R205, R208, R211—22K, ¼W resistor
R203, R206, R209, R212—10K printed-circuit (trimmer) potentiometer
C200—10μF, 35WVDC electrolytic capacitor
C201, C202, C203, C204—0.01 μF capacitor
C205, C206, C207, C208—0.47 μF capacitor

Part values shown on the schematic are common to all TTL configuration.

Additional parts shown on the parts list provide a 100Hz fixed frequency for PWM operation. All four waveform generators can be trimmed for duty cycles between 0.6 and 1.

Where do the input logic levels for D0 through D5 come from? You can make a switch control panel such as the one shown in Fig. 6-21. That works nicely for a complete parabot drive/steering scheme. Later chapters in this book show how these control levels can be picked up from the output of a microprocessor system.

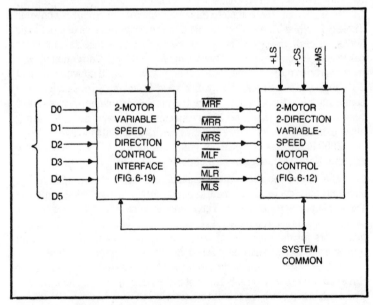

Fig. 6-20. How to interface the control circuit shown in Fig. 6-19 with the drive/steer power circuit shown in Fig. 6-12.

184

PURELY DIGITAL SPEED CONTROLS

There are almost always some distinct advantages in selecting a digital circuit in favor of an analog version that performs much the same task. The question of whether or not a digital version of some circuit is better than its analog counterpart depends on how well the advantages of one trades off against its own disadvantages.

Figure 6-22 is a basic variable duty cycle waveform generator that has been worked out in a digital format. One of its advantages is that it generates any one of 16 different PWM duty cycle waveforms, depending on the binary pattern at the select inputs, S0 through S3. Using the analog-type waveform generators described previously, you would need 16 pulse-generating sections and a four-line to 16-line decoder to sort them. But on the other hand, about one-half of the duty cycles available from the digital version are below 50 percent and are thus virtually useless

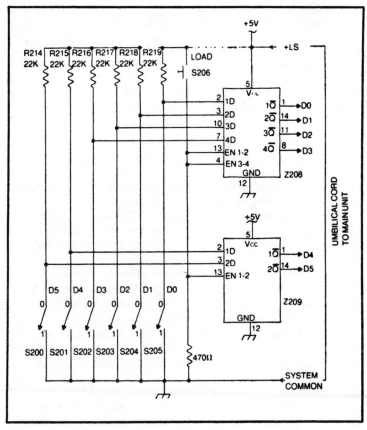

Fig. 6-21. A latching-type switch control for the drive/steer logic shown in Fig. 6-19.

185

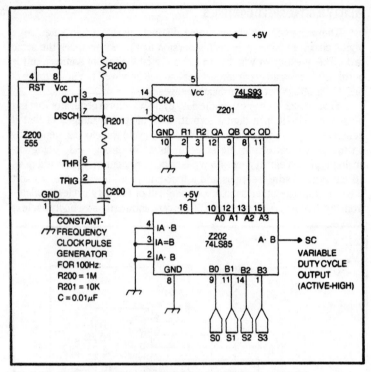

Fig. 6-22. A basic, all digital variable duty cycle waveform generator.

for motor speed control applications. Most DC motors stop running at duty cycles below something on the order of 60 percent.

Even so, that still leaves about eight selectable duty cycles from just three IC devices. Unlike with an analog generator, you cannot tweak in an infinite number of different speeds, but the ones that are available and useful do the job rather well.

Being a PWM system, it operates from a fixed-frequency oscillator, Z200. Setting this oscillator for a frequency of 100 Hz seems to work out nicely for most parabot/robot speed control systems, but of course you are free to tinker with the frequency to see the results.

The constant-frequency clocking pulses from oscillator go to a four-bit binary up counter, Z201. This counter is constantly enabled, so it continuously counts out the binary equivalent of decimal 0 through 15. The counting rate is always equal to the frequency of the oscillator that clocks it.

The third IC device in this circuit is a four-bit magnitude comparitor. This comparitor, Z202, has two sets of inputs: A0 through A3, and B0 through B3. The device constantly compares the values of the binary bits present at those inputs, and figures out whether the A inputs are larger than, equal to or less than the B inputs. For this particular application, only the A-less-than-B output has any significance.

The B inputs to the comparitor are the speed-select inputs. The A inputs are the counting patterns from the counter circuit, Z201. So the comparitor is constantly comparing the count from the counter with whatever speed-select pattern happens to be present at the B inputs.

In principle, at least, the A-less-than-B output responds with a logic-1 level whenever the A input bit pattern has a binary value less than the pattern at the select inputs. If, for example, you apply the binary equivalent of decimal 12 to the select inputs (where S0 is the least-

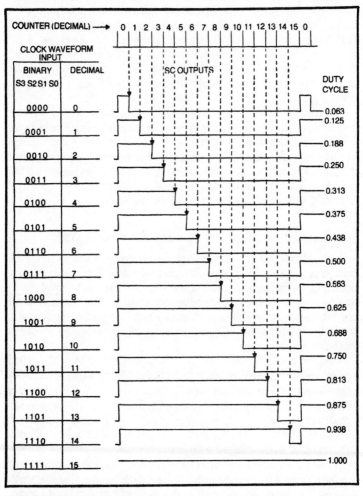

Fig. 6-23. Waveforms available from the digital variable duty cycle waveform generator shown in Fig. 6-22.

187

significant bit), the SC output will stay at logic 1 as long as the counter is counting out numbers less than 12. Once the count passes 12, however, the SC output goes to logic 0 until the counter passes 15 and returns to start over again at 0.

The larger the value of the binary number entered at the select inputs, then, the longer the waveform from SC remains in a logic-1 state. See the waveforms in Fig. 6-23.

Before some sharp, mathematically inclined reader gets too excited about a "small" technical detail, I must point out that the comparitor, Z202, is specifically and clearly shown as a 74LS85. Note also that the three cascading inputs at pins 4, 3 and 2 on that IC are all connected to ground. Using that LS version and grounding those three inputs makes the A-less-than-B output respond to both A less than B as well as A equals B. That *A-less-than-or-equal-to-B* feature is unique to the LS version of this IC device. The 7485 version cannot do the job in the same way. In short, the SC output of the circuit is at logic 1 as long as the counter is generating binary numbers less than or equal to the binary value present at the select inputs.

The waveforms in Fig. 6-23 show all possible 15 duty cycles available from this circuit. Notice that the selectable duty cycle range is between 0.063 (6.3 percent) and 1.000 (100 percent). But since duty cycles below 60 percent cannot make a motor turn, the S3 select input is normally fixed at logic 1. This procedure limits the range of duty cycles between 0.563 and 1.000, but that's the most useful range, and they can be selected from that range, using just three select inputs: S0, S1 and S2.

DIGITAL MOTOR DIRECTION/SPEED CONTROL FOR MANIPULATORS

The circuit shown in Fig. 6-24 can be any motor control scheme that calls for digitally selected motor direction and speed. Table 6-9 is the parts list for this circuit. Output $\overline{MS}$ is the variable duty cycle waveform for driving some inverting-type speed control transistors, while $\overline{MR}$ and $\overline{MF}$ are active-low motor-direction control signals. Actually, the format of these three signals is identical to most of the schemes presented throughout this chapter.

Digital inputs D0, D1 and D2 select the duty cycle of the speed-control waveform. Note that they go to the three lower-order select inputs of the four-bit magnitude comparitor, Z202. As described in connection with Fig. 6-22, this comparitor, in conjunction with a 100-Hz oscillator (Z200) and a counter (Z201) generates an active-high, variable duty cycle waveform. The larger the value of the binary number selected at inputs D0, D1 and D2 is, the longer the duty cycle becomes.

The paralleled logic inverters at the output of the comparitor invert the duty cycle waveform to make it compatible with active-low motor drivers. Two inverters are used in this parallel configuration to ensure adequate drive current.

Table 6-9. Parts List for the Circuit of Fig. 6-24.

Z200—LM555 timer
Z201—74LS93 4-bit binary counter
Z202—74LS85 4-bit comparitor
Z203—74LS00 quad 2-input NAND gate
Z204—74LS04 hex inverter
Z205—74LS10 triple 3-input NAND
R200—1 Meg, ¼W resistor
R201—10K, ¼W resistor

The remaining logic circuitry is responsible for generating the motor-direction logic levels at $\overline{MR}$ and $\overline{MF}$. Whether the motor is ultimately run in a forwward or reverse direction depends on the four-bit binary pattern at the data inputs. Table 6-10 summarizes the overall operation of the circuit.

Whenever the data inputs are fixed such that D1, D2 and D3 are all at logic 1, while D0 is at logic 0, Table 6-10 that the motor responds by running in a forward direction at speed SSE. What does SSE mean? Check the legend accompanying Table 6-10. It means a duty cycle of 0.939—almost full speed, but not quite.

The circuit does use some data compression to reduce the number of required data inputs. Without the benefit of the data-compression logic, you would need three inputs just for motor speed control, and two more for

Table 6-10. Function Table for the Circuit of Fig. 6-24.

	INPUTS			
D3	D2	D1	D0	MOTOR RESPONSE
0	0	0	0	STOP
0	0	0	1	REVERSE SS9 (minimum speed)
0	0	1	0	REVERSE SSA
0	0	1	1	REVERSE SSB
0	1	0	0	REVERSE SSC
0	1	0	1	REVERSE SSD
0	1	1	0	REVERSE SSE
0	1	1	1	REVERSE SS9 (maximum speed)
1	0	0	0	STOP
1	0	0	1	FORWARD SS9 (minimum speed)
1	0	1	0	FORWARD SSA
1	0	1	1	FORWARD SSB
1	1	0	0	FORWARD SSC
1	1	0	1	FORWARD SSD
1	1	1	0	FORWARD SSE
1	1	1	1	FORWARD SSF (maximum speed)

Motor Speed (SS) Designations

SS9 = 0.625 duty cycle
SSA = 0.688 duty cycle
SSB = 0.750 duty cycle

SSC = 0.813 duty cycle
SSD = 0.875 duty cycle
SSE = 0.938 duty cycle
SSF = 1.000 duty cycle

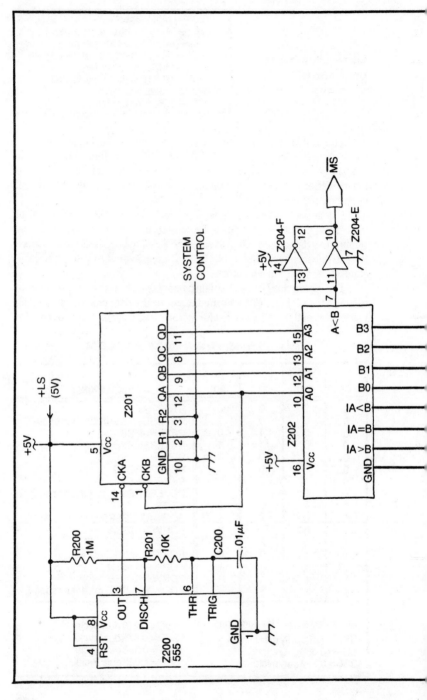

190

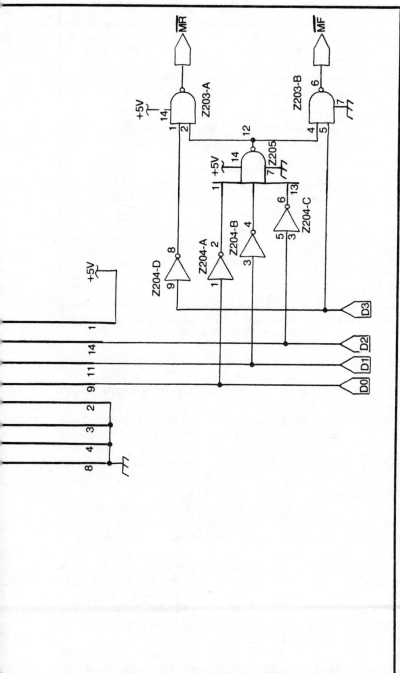

Fig. 6-24. A fully digital eight-speed, two-director motor control circuit.

191

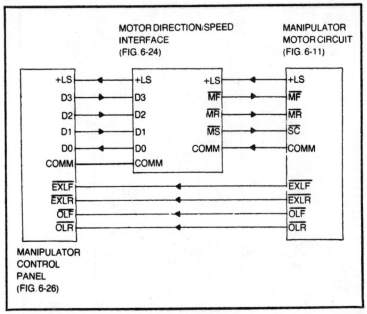

Fig. 6-25. How to interface the control circuit shown in Fig. 6-24 with a control panel (Fig. 6-26) and a manipulator motor circuit (Fig. 6-11).

motor direction control. The result would be 32 possible combinations of data inputs and 14 redundant motor responses. The four-input scheme does the same job with one less input and only one redundant (STOP) motor response.

Figure 6-25 is a block diagram that shows one way to connect this digital motor control to a manipulator motor drive circuit (such as the one in Fig. 6-11) and a parabot-class switching panel (Fig. 6-26). The overall scheme in that block diagram represents a rather complete manipulator motor control. If you aren't sure of the meaning of some of the labels, review the discussion connected with Fig. 6-11.

Of course, it is entirely possible to replace the suggested switch panel with a four-bit, bidirectional I/O port for a microprocessor system. You will find more details concerning that option later in the book.

For the time being, check the suggested switch circuit in Fig. 6-26. The parts list is Table 6-11. The data inputs come from a set of toggle switches, S300 and S303. The logic levels set at these switches can be entered into a latch (z300) by depressing the LOAD pushbutton. This portion of the circuit is practically identical to several switch-input schemes already described in this chapter.

Once you get the desired control pattern at the switches and depress the LOAD pushbutton, the bit pattern is set to the speed-control logic

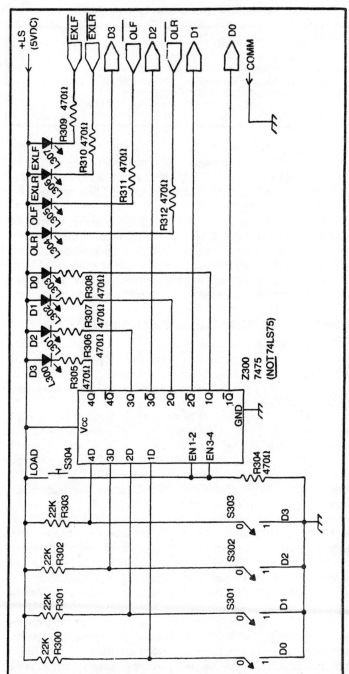

Fig. 6-26. Suggested switch input scheme for a full-control manipulator motor system.

193

Table 6-11. Parts List for the Circuit of Fig. 6-26.

Z300—7475 quad D latch (do NOT substitute 74LS75)
L300 through L307—Any standard 10mA LED
R300 through R303—22K, ¼W resistor
R304 through R312—470-Ohm, ¼W resistor
S300 through S303—SPST toggle switch
S304—normally open, momentary pushbutton switch

circuitry and, at the same time, appears at a set of four LEDs, L300 through L303. These LEDs indicate the control code being sent to the motor control system. While the LOAD pushbutton is open, you are free to set up the next control pattern without affecting the LEDs and the motor control system in any way.

The four remaining LEDs, L304 through L307, merely indicate the status of the limit switches of the manipulator. In all cases, the LED lights whenever its limit switch is energized. The idea is to give the operator a positive idea about the position of the manipulator with regard to its normal operating and extreme limits of motion. This entire switching circuit can be connected to the remaining system via a long umbilical cord.

AN EIGHT-BIT SPEED AND DIRECTION CONTROL FOR DRIVE/STEER SYSTEMS

The fully digital, variable duty cycle waveform generator can be readily expanded to control a pair of motors used for parabot/robot drive and steering operations. If the data-compression logic is also included for the direction controls, you end up with one of the most useful drive/steer systems available anywhere.

The waveform generator for this system is illustrated in Fig. 6-27. The appropriate parts list is Table 6-12. Note that there are eight data inputs, D0 through D7. The four lower-order data input bits, D0 through D3, control the speed and direction of the LEFT MOTOR, while the four higher-order bits take care of operations for the RIGHT MOTOR. The nomenclature for the six output signals, three for each motor, is consistent with that used throughout this chapter.

If you have been following all the preceeding discussions carefully, you shouldn't have any trouble figuring out how this circuit works. Generally speaking, it is simply a doubled-up version of the circuit described in Fig. 6-24. The only point of departure from that general view is that this circuit uses just one master clock oscillator, Z200, to run both duty cycle counters, Z201 for left-motor speed control and Z202 for right-motor control.

Table 6-13 is an abbreviated version of the function table for the circuit shown in Fig. 6-27. The circuit has eight separate data inputs, which means there are actually 256 possible motor-control operations. A complete function table—all 256 possible combinations—is shown in the

Table 6-12. Parts List for the Circuit of Fig. 6-27.

Z200—555 timer
Z201, Z202—74LS93 4-bit binary counter
Z203, Z204—74LS85 4-bit magnitude comparitor
Z205, Z206—74LS04 hex inverter
Z207—74LS10 triple 3-input NAND gate
Z208—74LS00 quad 2-input NAND gate
R200—1 Meg, ¼W resistor
R201—10K, ¼W resistor
C200—0.01μF capacitor

Appendix titled "Complete Function Table for Eight-Bit Drive/Steer, Variable-Speed Motor Control."

Using the abbreviated version, however, is first a matter of recognizing that D0 through D3 control the speed and direction of the LEFT MOTOR, while D4 through D7 control the RIGHT MOTOR in exactly the same fashion. Realizing this, it is possible to think in terms of controlling the two motors separately; and that's where Table 6-13 comes into the picture.

Table 6-13 applies to right-motor operations if index n is equal to 0. That way, the binary bits carry designations D3, D2, D1 and D0. Using the table for left-motor operations is a matter of letting index n equal 4. Incidentally, if you do not remember exactly how to translate the SS

Table 6-13. Abbreviated Function Table for the Circuit of Fig. 6-27.

INPUTS					MOTOR RESPONSE
		Binary			Left Motor if n=4
Hex	D_{n+3}	D_{n+2}	D_{n+1}	D_n	Right Motor if n= 0
0	0	0	0	0	STOP
1	0	0	0	1	REVERSE SS9
2	0	0	1	0	REVERSE SSA
3	0	0	1	1	REVERSE SSB
4	0	1	0	0	REVERSE SSC
5	0	1	0	1	REVERSE SSD
6	0	1	1	0	REVERSE SSE
7	0	1	1	1	REVERSE SSF
8	1	0	0	0	STOP
9	1	0	0	1	FORWARD SS9
A	1	0	1	0	FORWARD SSA
B	1	0	1	1	FORWARD SSB
C	1	1	0	0	FORWARD SSC
D	1	1	0	1	FORWARD SSD
E	1	1	1	0	FORWARD SSE
F	1	1	1	1	FORWARD SSF

For a complete Eight-bit listing, see Appendix, "Complete Function Table for Eight-Bit Drive Steer, Variable-Speed Motor Control."

195

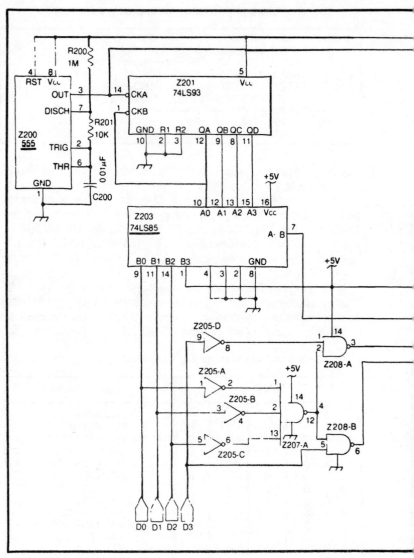

Fig. 6-27. An eight-speed, two-direction motor control unit for two-motor drive/steer systems.

motor-speed designations into actual duty cycle figures, refer back to the legend in Table 6-10.

It is no accident that the circuit has eight control inputs that are TTL compatible. Looking ahead to future developments, the circuit can be directly interfaced with an output port of a typical eight-bit microprocessor system.

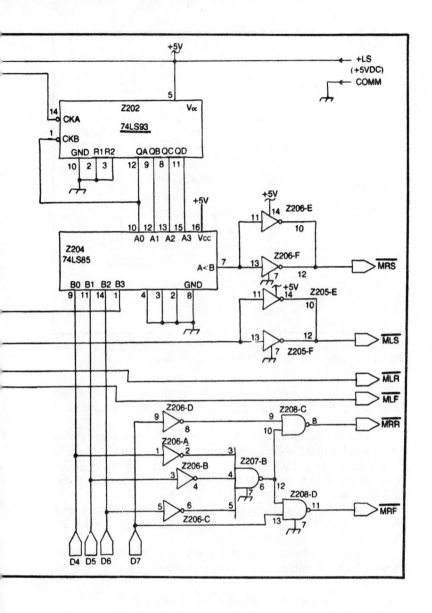

Figure 6-28 shows how the circuit can be interfaced with the power circuit described in Fig. 6-12. Once you have built the two circuits, they aren't difficult to interconnect.

The eight-bit control circuit doesn't necessarily have to be used with the driver in Fig. 6-12, however. It can be connected to any of the two-motor speed/direction power circuits that happen to use active-low

197

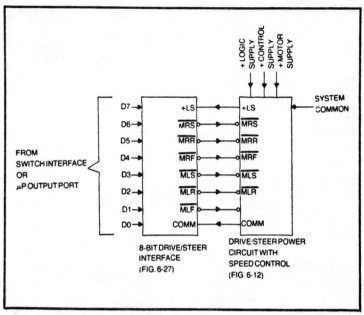

Fig. 6-28. Block diagram of interconnections between the eight-bit control circuit (Fig. 6-27) and the matching motor power circuit (Fig. 6-12).

inputs. If you happen to be using some power circuits that have active-high (noninverting) inputs, simply insert an inverter at the appropriate place in Fig. 6-27.

If, for example, the variable duty cycle inputs to your power circuit are supposed to be active-high, simply insert inverters between the pin 7s of the comparitors (Z203 and Z204) and the inputs of their respective paralleled output inverters (Z206-E and 206-F for the output of Z204, and Z205-E and 205-F for the output of Z203).

If you need active-high motor-direction control signals, place logic inverters just before the $\overline{MLR}$, $\overline{MLF}$, $\overline{MRR}$ and $\overline{MRF}$ outputs in Fig. 6-27. That will upright each one of them.

It takes a lot of talent to control the motor system from the eight-bit circuit in a manual fashion. It can be done, however, and if you use the

Table 6-14. Parts List for the Circuit of Fig. 6-29.

Z300, Z301—7475 quad D latch
L300 through L307—Any standard, 10mA LED
R300 through R307—22K, ¼W resistor
R308 through R316—470-Ohm, ¼W resistor
S300 through S307—SPST toggle switch
S308—normally open, momentary pushbutton switch

198

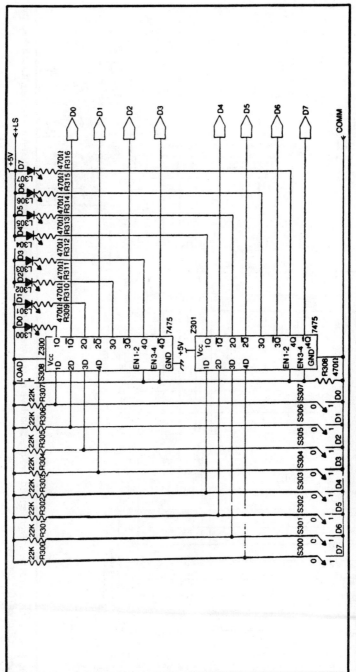

Fig. 6-29. Suggested source of control data for the eight-bit motor control system.

199

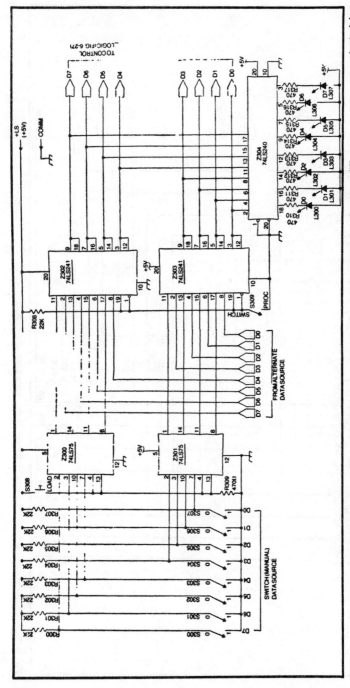

Fig. 6-30. A data source for an eight-speed, two-direction drive/steer system. The data can originate from two difference sources, as selected by the PROC/SWITCH control.

200

Table 6-15. Parts List for the Circuit of Fig. 6-30.

Z300, Z301—74LS75 quad D latch
Z302, Z303—74LS241 3-state octal buffer/driver (non-inverting)
Z304—74LS240 3-state octal buffer/driver (inverting)
L300 through L307—Any standard, 10mA LED
R300 through R308—22K, ¼W resistor
R309 through R317—470-Ohm, ¼W resistor
S300 through S307, S309—SPST toggle switch
S308—normally open, momentary pushbutton switch

circuit shown in Fig. 6-29 as a switch-control input, you will have the makings of a complete parabot drive/steer system with variable speed control. Table 6-14 is the parts list for the circuit of Fig. 6-29.

Using that switching circuit, you can first set up the desired set of eight-bit motion-control logic levels at the toggle switches, and then put the information into action by momentarily depressing the LOAD pushbutton. The system will then respond to the code and display it on the LEDs as well. Changing the operation is a matter of setting up the new bit patterns at the toggle switches and putting it into action by depressing the LOAD button again. You have to count on a lot of practice to get close control over this parabot drive/steer scheme, but it can be done with highly satisfying results.

If you have any plans for eventually interfacing the system with a microprocessor output port, you can still play around with the motors from the eight toggle switches by using the circuit shown in Fig. 6-30. The outputs are connected to the data inputs to a control circuit such as the one shown in Fig. 6-27; in this case, however, you have the option of manual switch control or automatic control, with the data inputs coming from an alternate source. Table 6-15 is the parts list for the circuit shown in Fig. 6-30.

If you want to run the system from the toggle switches, S300 through S307, simply set the PROC/SWITCH control, S309, to its SWITCH position. And there it is: complete manual control. Set the PROC/SWITCH control to its PROC position, however, and the motor-control data will have to come from the alternate set of data inputs—presumably a microprocessor output port or similar data source.

This selectable data-source input circuit can give you a couple of nice advantages. First, you can play with the motor control system long before you have a chance to build any sort of microprocessor control for it. Second, it provides a manual override scheme that gives you full manual control over the motor system in the event something goes wrong with the microprocessor hardware or software. With regard to that latter feature the PROC/SWITCH control can be viewed as a panic button

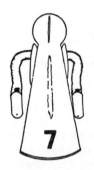

Some Internal Input Sensing Mechanisms

7

Except for the simplest kinds of parabot machines, there is always some need for monitoring the status of mechanisms within the machine. On a very primitive level, for instance, you need to know whether or not a motor is turning when it is supposed to be turning. And if that system incorporates some motor speed controls, you need to know whether or not the motor is turning at the speed the command signals prescribe for it.

If you have any plans for using some automatic controls for manipulators, there is a need for sensing the position of all the moving parts. Monitoring response rate might be important, too.

Then there is the matter of building self-reliant machines—machines that can monitor critical parameters within themselves to take action aimed at self preservation. If nothing else, that boils down to using battery voltage sensors and battery recharger current monitors.

This chapter offers a survey of techniques for sensing motor speed, mechanical positions and the status of battery systems. Virtually all these sensing schemes must be backed up with voltage amplifiers. By the time you complete your study of sensing mechanisms and amplifiers presented here, you will be in a position to deal with a great variety of different sensing schemes.

MOTOR SPEED SENSING

There are two different mechanisms for monitoring the speed of motors in parabot/robot systems. One is basically optical in nature, while the other is wholly electromechanical. Neither is an obvious choice in a general sense, and your final choice depends on the nature of your own system, how you want the motor-speed parameters to be represented and, finally, your own subjective feelings on the matter.

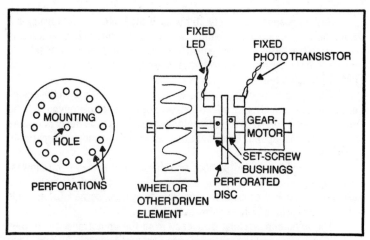

Fig. 7-1. General mechanical arrangement for optical motor speed sensing.

Optical Motor Speed Sensing

Figure 7-1 illustrates the general scheme for monitoring motor speed in an optical fashion. In a manner of speaking, this system actually "looks" at the speed of the motor in much the same way humans do.

The basic idea is to construct a disc that has a series of evenly spaced holes drilled around its outer edge. This perforated disc is then mounted to the shaft of the motor so that it spins at the same rate.

An LED is mounted on one side of the disc, and a phototransistor is mounted on the other side. They are arranged so that light from the LED will fall directly onto the lens of the phototransistor whenever one of the perforations in the disc comes between them. As the disc spins, it effectively chops the light from the LED, and the phototransistor switches off and on at a rate that is proportional to the speed of the motor. It is the frequency of the rectangular waveform from the phototransistor that provides some indication of how fast the motor is running.

$$f = Ms(n)/60 \qquad \text{Equation 7-1}$$

where f = the frequency of the waveform from the phototransistor, Ms = the motor speed in rpm, and n = the number of holes around the perimeter of the disc.

The larger the number of holes, n, the more precise the system will be. At one extreme of the matter, you might drill 360 holes—one for every degree of arc. Even if the motor is turning as slowly as 1 rpm, then, you can expect a frequency from the phototransistor of 6 Hz. The significance of that result is that the system requires only about one-sixteenth of a second to find out whether or not the motor is actually turning at that very slow rate.

But suppose you use only 16 holes. If the motor is still turning a 1 rpm, the frequency from the phototransistor (according to Equation 7-1) is only about 0.267Hz. It will take nearly 4 seconds for the system to tell you whether or not the motor is turning. A motor problem lasting for 4 seconds before it is detected can spell a lot of trouble.

Two specifications must be determined before settling on the design of this sort of optical motor speed sensor. You must decide for yourself the slowest speed you want to sense and the maximum time delay allowable for sensing that speed. Keep both numbers realistic and be prepared to make some compromises.

For a medium-sized parabot or robot, you might set the slowest working speed of a drive motor at 30 rpm, or about 1 revolution every 2 seconds. As far as the maximum allowable delay is concerned, 0.1 seconds is a good figure for a high-performance machine.

The minimum working frequency from the phototransistor is thus 1/0.1, or 10 Hz, and the motor speed at that frequency is 30 rpm. Plug those figures into Equation 7-1 and rearrange it to solve for n (the number of holes):

$$n = 60(10Hz)/30\,rpm = 20\,holes$$

Because the holes should be evenly spaced, they should be drilled at 360/20, or 18-degree intervals. And at this point, the mechanical design of the perforated disc is well underway. Equations 7-2 through 7-5, along with the diagram in Fig. 7-2, make up a complete design format.

$$n = 60f_{min}/Ms \qquad \textbf{Equation 7-2A}$$

or

$$n = 60/(Mst_{max}) \qquad \textbf{Equation 7-2B}$$

where n = the minimum number of holes, f_{min} = the lowest desired sensing frequency, t_{max} = the longest desired delay time to sensing, and Ms = motor shaft speed in rpm.

$$a = 360/n \qquad \textbf{Equation 7-3}$$

where a = the angle between holes in degrees, and n = the number of holes.

$$r_{1min} = nd/\pi \qquad \textbf{Equation 7-4}$$

where r_{1min} = minimum radius of hole placement, n = number of holes, and d = diameter of the holes.

$$d_{2min} = 2r_1 + d \qquad \textbf{Equation 7-5}$$

where d_2 = the overall diameter of the disc, r_1 = radius of the hole placement, and d = diameter of the holes.

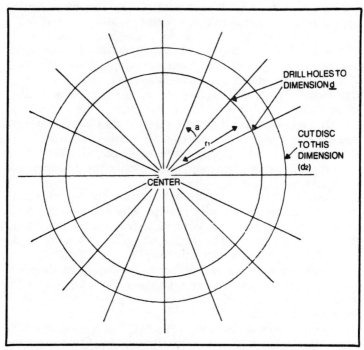

Fig. 7-2. How to lay out a perforated disc for optical motor speed sensing.

Equation 7-2A lets you calculate the number of holes necessary for generating a minimum motor speed output frequency or sensing time delay. If you can live with a 0.1-second time delay at a minimum motor speed of 30 rpm, Equation 7-2B leads to using a minimum of 20 holes. But if you can settle for sensing a minimum motor speed of 45 rpm at a 0.1-second delay, you need 13-1/3 holes. How can you drill one-third of a hole? Don't bother. Compromise for the next-higher integer value—14 holes in this particular instance.

Equation 7-3 works out the number of degrees between the evenly spaced holes. If you need 16 holes, that equation shows that they have to be arranged 22.5 degrees apart.

We like to keep things as small as possible in this parabot/robot business, and Equation 7-4 helps you do that. It provides the minimum possible distance between the center of the disc and the centers of the perforations. Most LED/phototransistor devices are about 3/16-inch in diameter. That is the d term in the equation. Still assuming you need 16 holes, it turns out that they should be centered at least (16) (3/16)/3.14 = 0.96 inches from the middle of the disc. You're in good shape if you round up to 1 inch.

Equation 7-5 works out the minimum diameter of the entire disc assembly. Using the figure of 1 inch for the radius of the hole centers and still assuming the holes are three-sixteenth of an inch in diameter, the disc ought to be about $2(1) + 3/16$, or 2-3/16 inches in diameter.

To make your own template for constructing a disc for this photosensing motor speed monitor, first determine the maximum system delay you want (t_{max}) and the minimum working rpm of the motor (Ms). Then apply Equation 7-2A to calculate the number of holes. If you end up with a fractional part of a hole, round the number up to the next-higher whole number.

Then mark the center of a sheet of paper. That's where the mounting hole of the disc will go. Use Equation 7-3 to calculate the number of degrees between each hole. Score those angles on your paper with an ordinary protractor. Draw long angle lines, radiating from the center mark to the angle markings you just made.

Next, use Equation 7-4 to calculate the distance from the center of the disc to the centers of the holes you will be drilling along the outer perimeter. Set a compass to that length and score a circle that is, of course, centered on your original center mark. The places where this circle intersects the angle lines are where the holes should be drilled.

Finally, work out Equation 7-5 to get the overall diameter of the disc. If you want to include that circle on your template, divide the result by 2 in order to come up with the radius. Set the drawing compass to that radius and score the larger circle.

There it is: a finished template for the disc. Tape it to a sheet of thin, but durable, aluminum and cut out the circle pattern. Center-punch the positions of all the holes, including the shaft-mounting hole in the center. Drill the holes, and the job is completely done.

If you're lucky, you'll be able to mount the disc directly to the motor shaft, leaving plenty of shaft for mounting the device the motor is supposed to drive. But what if you aren't so lucky? What if there isn't some convenient place on the motor shaft for mounting the disc? That is where you are going to have to apply some mechanical ingenuity. Look around for any rotating part of the gear motor assembly that might be a pickoff point for turning the disc. Perhaps you will find it convenient to pick up the motor speed at some higher-speed point in the system—a point ahead of the speed-reducing geartrain. Equations 7-2 through 7-5 still apply, but you have to work with the higher rpm figure, rather than the lower one at the final stage of the geartrain.

As a last resort, give some thought to mounting the disc on its own shaft and bearing assembly. That shaft can then be mechanically ganged to the main motor shaft with a gear or belt assembly. Things get a bit complicated, but sometimes no alternatives exist.

If you find you must use a separate gearing arrangement for the disc assembly, remember to take into account any possible change in gearing speed. Actually, the Ms term in Equations 7-1 through 7-5 should reflect the rpm of the *disc*, not necessarily that of the gear motor. But as you have seen in some specific examples, life is a lot simpler if you can mount the

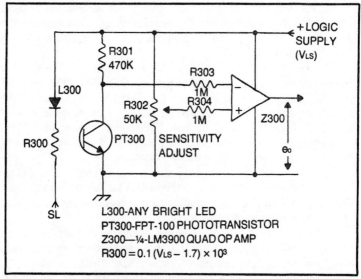

Fig. 7-3. A circuit for operating an optical motor speed sensing system. See also Fig. 7-23.

disc directly to the gear motor shaft and make the safe assumption that the two turn at precisely the same rate.

Figure 7-3 shows a simple circuit for operating the electrical part of this type of motor speed sensor. The LED is turned on whenever input $\overline{SL}$ is a logic 0, or grounded. Pulling $\overline{SL}$ up close to the +LOGIC SUPPLY level turns off the LED. While it is true that the LED must be turned on to pick up the chopped-light signals in this scheme, there is some merit to the notion of being able to turn off the LED during intervals when it isn't necessary to monitor the motor speed.

The value of the limiting resistor of the LED, R300, depends on the +LOGIC SUPPLY voltage level. The equation for R300, accompanying the diagram in Fig. 7-3, assumes you want to use a standard LED current level of 10 mA.

The phototransistor, PT300, changes conduction with the amount of light reaching it through the disc assembly. If the system is operating in complete darkness and the peak wavelengths of the LED and phototransistor are fairly well matched, the phototransistor switches to its ON state whenever it "sees" light through one of the perforations in the disc. The phototransistor will then switch OFF whenever a space between the perforations interrupts the light from the LED. Ideally, then, the voltage waveform at the collector of the phototransistor switches between V_{LS} and zero as the disc turns.

Conditions are rarely ideal, however, and you should expect something less than a perfect rectangular waveform at the collector of the phototransistor. Ambient lighting and stray reflections from the LED can introduce a lot of "noise" into the optical system. Consequently, the waveform at the collector of the phototransistor is ragged and rides on some DC offset level.

These less-than-ideal situations call for conditioning the signal in some fashion. The first step is to provide a SENSITIVITY ADJUST control, R302, and a voltage amplifier, Z300. The idea is to desensitize the circuit, putting most of the unwanted lighting effects below the level set at the SENSITIVITY ADJUST control. The larger, chopped-light signal should rise above the noise level, and the task of the amplifier is to bring the wanted signal back up to a workable voltage level.

The circuit cannot be adjusted properly until it is installed in its final form. Run the motor, making sure the LED is turned on, and adjust the SENSITIVITY ADJUST until you see a reasonably clean, rectangular waveform from the output of the amplifier.

As shown in Fig. 7-3, the amplifier inverts the chopped-light signal from the phototransistor. If you follow the phasing of the overall system, then, you will find the output of the amplifier rising up toward V_{LS} whenever light is striking the phototransistor. Whenever the light is blocked, the output of the amplifier should go close to zero.

There is bound to be some offset in the output signal of the amplifier. That's the nature of such amplifiers. You will find a complete discussion of this matter later in this chapter, notably in those sections on sensing amplifiers of all kinds.

If you have trouble getting an adequate signal-to-noise ratio from this circuit, there is a chance that the optical properties of the phototransistor and LED do not match very well. The FPT-100 phototransistor offers the greatest optical sensitivity in the infrared part of the spectrum. If the LED you choose is heavily filtered in favor of shorter, therefore, visible wavelengths, you will have to pick an LED having no passive filter. Of course you can avoid this particular problem by buying a matched set—LED and phototransistor—at the outset.

Serious problems concerning ambient lighting can be solved by using an AC version of the amplifier circuit. The circuit in Fig. 7-4 suggests one approach. In this case, the chopped-light signals at the collector of the phototransistor are coupled through capacitor C300 to the operational amplifier. Any DC component of conduction, mostly caused by steady background lighting, is blocked out of the system at that point.

It takes some tinkering with some of the component values to get this circuit working under the conditions you want to build into your motor speed-sensing scheme. R303, for example, sets the DC offset level at the output of the LM3900 operational amplifier. That is altogether important when attempting to interface the output of the circuit with some digital logic circuits. R304 and R305 establish the maximum voltage gain of the operational amplifier, while R302 sets the actual gain within that range.

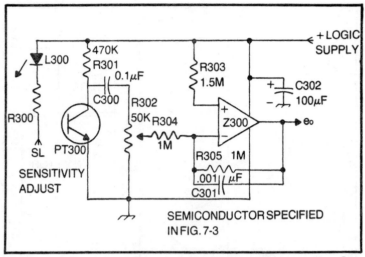

Fig. 7-4. An AC version of the optical motor speed sensing amplifier. See also Fig. 7-24.

In the context of the discussion to this point, you should resort to using the AC version of the circuit only when attempts to block out ambient light in a mechanical fashion have failed and the DC circuit in Fig. 7-3 proves useless. However, there are some very good reasons, other than ambient lighting problems, for using the AC-coupled version. Those reasons are described later in this chapter.

Electromechanical Motor Speed Sensing

The optical sensing scheme just described seems to be very popular among parabot/robot experimenters these days. But Fig. 7-5 is a very attractive alternative to the whole idea. This is a purely electromechanical technique for monitoring the speed of a motor.

The scheme is based on the principle that a permanent-magnet motor can be used as either a motor or a generator. It works as a motor whenever DC power is applied to its commutator rings and armature coil. If you spin the shaft and armature by some external, mechanical means, however, you end up with an AC voltage at the commutator rings. The nice feature inherent in using a DC motor as a generator is that it produces an AC waveform that has both a voltage and frequency component proportional to the shaft speed.

All you need do to get the system running is to work out a belt or gearing arrangement that lets the main motor—the one whose speed you want to monitor—drive the shaft of the generator. You then have the option of working with the output frequency or voltage of the generator.

On the practical side of this matter, these "generators" are flooding the surplus marketplace these days. Many firms sell a plastic bag with

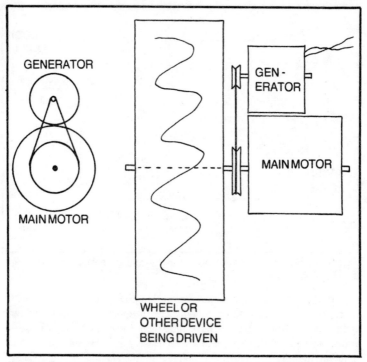

Fig. 7-5. General mechanical arrangement for electromechanical motor speed sensing.

three or four of them in it for just a couple of dollars. Actually, they were originally intended to be drive motors in inexpensive cassette and cartridge tape players. They're inexpensive, widely available and quite durable and easy to use. It is difficult to make a convincing case *against* using them as suggested here.

Figure 7-6 shows one kind of circuit that can be used for interfacing the output of the generator to some other circuitry. In this instance, the e_o output rises up to a more positive level whenever the speed of the generator reaches a certain threshold level. Whenever this motor/ generator set is running more slowly than prescribed by the SENSITIVITY ADJUST, the signal at e_o drops toward a minimum level. In a sense, this is a speed threshold sensing circuit that works according to the amount of voltage from the generator.

Rectifier diode D300 does the job of a half-wave rectifier circuit. A portion of the half-wave signal is picked off at R300 and filtered by C300. The result is a fairly smooth DC voltage across C300 and at the inverting input of the operational amplifier. That DC voltage level is supposed to be proportional to the speed of the generator shaft, but there is actually some tapering of the voltage level due to the time-constant character of the circuit.

If you want to deal with the frequency component of the signal, rather than the voltage component, just change the position of capacitor C300. Put it in series with the cathode of D300, rather than in parallel with the lower section of R300. Set up the SENSITIVITY ADJUST properly, and you will see an MSS output made up of a constant-voltage, rectangular waveform. The frequency of that waveform is, indeed, proportional to the frequency of the motor that is driving the little generator; none of the nonlinearity associated with the voltage-sensitive version exists.

There is little point in describing design equations for the motor/generator speed sensing scheme. The problem is that experimenters generally do not have access to the necessary specifications for the small motors—the ones used backwards as generators.

So just pick out one of the little motors and gang it to your main motor assembly as suggested in Fig. 7-5. If you have an oscilloscope available, connect it to the terminals on the generator and monitor both the voltage and frequency at the same time. Run the system and see what voltages and frequencies you get at maximum speed and any other speed you think is relevant to your system.

An oscilloscope isn't absolutely necessary. It is possible, for instance, to get some idea about the voltage response by monitoring it with an ordinary AC voltmeter. The frequency can be measured with a digital frequency meter, but if you don't have access to that particular sort of instrument either, you will find that you can get along quite well. Just read through the rest of this chapter to see how well you can work out a good system based on a very minimal amount of test data.

The only potentially troublesome feature of the whole idea is getting the shaft of the generator ganged to that of the gear motor system. Get that one figured out for your own particular case, and the job is more than half done.

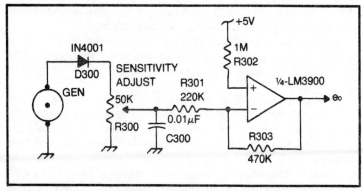

Fig. 7-6. A circuit for monitoring the amplitude of a signal from an electromechanical motor speed sensing arrangements. See also Figs. 7-25 and 7-26.

SENSING MECHANICAL POSITIONS

A parabot or robot using separate drive and steering motors should include some scheme for sensing the actual position of the steering mechanism—if you want some automatic or semi-automatic steering features, that is. So even if your project does not include any sort of manipulator, but you are using separate drive and steering motors, you will benefit from some of the discussions in this section.

The need for sensing mechanical positions, however, is paramount to automatic *and* semi-automatic manipulator designs. A machine really cannot make a good judgment concerning *what* to do with its manipulator if it has no provisions for sensing *where* the various components are located relative to others. If we exclude the more esoteric position-sensing schemes, such as ultrasonic or laser-beam proximity sensors, the practical ideas that remain fall into two general categories—discrete-position and continuous-position sensors of a purely electromechanical variety.

Discrete-Position Sensing

Figure 7-7 shows two different approaches for accomplishing the same general task: sensing the position of a moveable frame relative to a fixed one. Both schemes use switches that are fastened to the fixed frame and energized by the relative position of some activating mechanism on the moveable frame.

In Fig. 7-7A, the switches are ordinary roller-lever micro-switches. While these switches are usually available with a SPDT format, assume for the sake of the discussion that the switches go to their ON position whenever the lever is pressed down.

The moveable frame has an extension that rolls over the levers and depresses them whenever it makes mechanical contact. As pointed out shortly, it is important to arrange the spacing of the microswitches and the size of the extension from the moveable frame such that any movement causes a make-before-break situation between two adjacent switches. Perhaps putting it more simply: At least one switch must be energized at any given moment during the motion of the moveable frame.

For example, Fig. 7-7A shows that S1 and S2 are ON, while S3 and S4 are OFF. If the direction of motion of the moveable frame happens to be toward S3 and S4, switch S3 will be energized next, and certainly before S2 is de-energized.

The setup in Fig. 7-7B uses magnetic reed switches in place of microswitches. Here, a magnet is attached to the moveable frame in such a way that it energizes the reed switches laying directly below it. As the moveable frame moves, the magnet energizes different pairs of reed switches.

Again, it is important to space the reed switches and select the length of the magnet so that one switch remains energized until the next in line is also energized. This creates the necessary make-before-break situation.

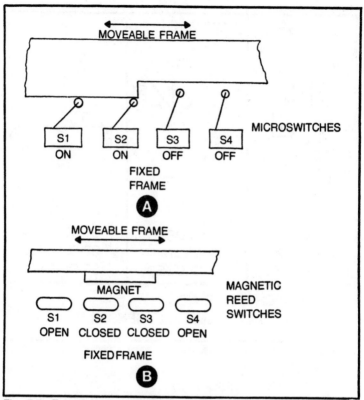

Fig. 7-7. Two discrete position-sensing schemes. (A) Using roller-lever microswitches. (B) Using magnetic reed switches.

Comparing the two schemes shown in Fig. 7-7, the reed-switch version has the advantage of having no physical connection between the elements of the moveable frame and the switches. You have to make a trade-off between durability and cost. Plain magnetic reed switches—those having their glass bodies fully exposed—are less expensive than roller-lever microswitches. But having glass components exposed to the potentially rugged environment of a manipulator is a cause for some concern.

The ultimate in durability is a reed-switch scheme that uses switches enclosed in a rugged plastic case. This kind of magnetic switch is commonly used for security systems that sense the opening or closing of a door. The enclosed reed switches can cost two or three times as much as a microswitch, but they will survive all but the most catastrophic events in the life of the manipulator.

Both examples of Fig. 7-7 illustrate a setup for sensing the position of a frame that is moving in a linear, back-and-forth fashion. The same

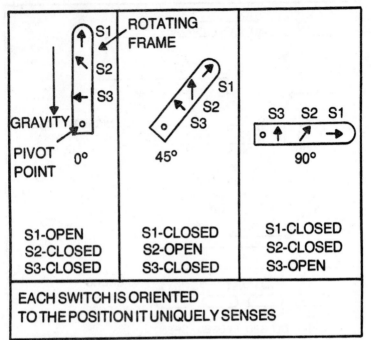

Fig. 7-8. Using mercury tip switches to sense angles of rotation.

general idea can be applied just as well to a frame that is rotating. Simply arrange the switches in a circular pattern.

The discrete position-sensing arrangement illustrated in Fig. 7-8 uses mercury-filled tip switches to sense the rotational position of a frame. These switches are open only when they are oriented straight up and down, relative to the local gravitational field. There are a couple of degrees of tolerance in any direction away from straight up, but the general idea is that the mercury contacts are closed whenever the switch is tipped away from true vertical.

To see how these tip switches can sense the angular position of a rotating frame, note the little arrows in the diagram. They point toward the upright, contact-opening position for each of three switches. So when the frame is set upright (at an angle of 0 degrees), only switch S1 is open. The other two switches are tipped off their center axis and are thus closed.

As the frame rotates to 45 degrees from its vertical position, switch S2 opens and S1 closes. Switch S2 is now the only one that is pointing straight upward.

And finally, switch S3 opens as the frame rotates to 90 degrees off its vertical position. Switches S2 and S3 are closed because neither is in its particular contact-opening position.

Using a series of mercury tip switches, each arranged at a particular sensing angle along the length of a rotating frame, is a rather simple idea.

Just three such switches as illustrated here are usually inadequate, however.

The problem is that a tip switch closes whenever it is tipped some small amount from its vertical position. That angle of tolerance varies somewhat, and you'll have to connect an ohmmeter to one of them and tip it around to find out what that angle of tolerance is.

Once you know the tolerance of the open-connection angle, you can begin getting a realistic idea about how many of these switches you will need to do the job. The basic idea is to set the angles of enough switches so that at least one will open before the preceding one closes. That's getting back to the make-before-break requirement.

So if you find your mercury tip switches close at about 10 degrees in any direction from true vertical, you should count on using enough switches to create a change in the switching pattern at no less than 10 degrees. The switches would have to show a tolerance of 45 degrees to make the example in the diagram work properly. That isn't a very realistic tolerance: It's way too much. Here is an equation that will help you get some perspective on the number of tip switches for a particular job:

$$n = A/T \hspace{2cm} \textbf{Equation 7-6}$$

where n = minimum number of tip switches required, A = maximum angle of excursion of the frame (degrees), and T = angular tolerance of the tip switches (degrees).

To see how this all works, suppose you want to be able to sense angles of rotation anywhere between 0 degrees and 90 degrees. The tip

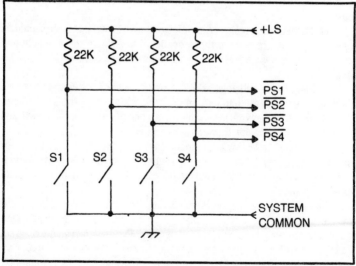

Fig. 7-9. A circuit for interfacing discrete position-sensing switches with some control circuitry.

215

SWITCH CONDITIONS				OUTPUT BITS			
S1	S2	S3	S4	PS1	PS2	PS3	PS4
OFF	OFF	OFF	OFF	1	1	1	1
OFF	OFF	OFF	ON	1	1	1	0
OFF	OFF	ON	OFF	1	1	0	1
OFF	OFF	ON	ON	1	1	0	0
OFF	ON	OFF	OFF	1	0	1	1
OFF	ON	OFF	ON	1	0	1	0
OFF	ON	ON	OFF	1	0	0	1
OFF	ON	ON	ON	1	0	0	0
ON	OFF	OFF	OFF	0	1	1	1
ON	OFF	OFF	ON	0	1	1	0
ON	OFF	ON	OFF	0	1	0	1
ON	OFF	ON	ON	0	1	0	0
ON	ON	OFF	OFF	0	0	1	1
ON	ON	OFF	ON	0	0	1	0
ON	ON	ON	OFF	0	0	0	1
ON	ON	ON	ON	0	0	0	0

you are using show an angular tolerance of 5 degrees. Using Equation 7-6:
$n = 90/5$, or 80 switches. And this doesn't allow for some switches that
might not close until the tip angle is somewhat greater than 10
degrees—that sort of thing can happen, you know. So it's a good idea to
increase the number of switches by a few, just to make up for any of them
that might be a bit out of tolerance. Use maybe 20 of them in this particular
case.

What about the angular orientation of the tip switches? Each should be
oriented A/n degrees from the preceding one. Using the example just
cited, the 20 tip switches should be arranged at increments of 90/20, or 4.5
degrees.

Holding that small angular difference for some 20 switches is a task
calling for some close and careful craftsmanship. Once the job is done,
though, the entire switch assembly can be enclosed in a durable case and
mounted to the moveable frame as a separate item.

Obviously the tip-switch, angular-sensing scheme is impractical for
sensing relatively large angles of motion. It is a viable alternative,
however, when the angular displacement is going to be on the order of 20
or fewer degrees.

Figure 7-9 shows an electrical arrangement for preconditioning the
switch signals. The circuit applies equally well to all three discrete
position-sensing schemes described thus far. The switches on the diagram
can thus represent microswitches, magnetic reed switches or mercury tip
switches. The four signal outputs are TTL and CMOS compatible,
assuming, of course, the +LS voltage is also set to the necessary level.

Table 7-1 summarizes the logic function of the circuit. The circuit and function table can both be extended to include any necessary number of switches and a corresponding number of $\overline{PS}$ signals.

In any discrete position-sensing scheme, it is important to build in a make-before-break feature. This has been emphasized several times already in this section, and now it's time to explain why.

The diagram in Fig. 7-10 shows how *not* to arrange a discrete position-sensing mechanism. And while it relates specifically to one using magnetic reed switches, the same general idea applies to the other scheme as well.

The table accompanying the diagram shows the responses of the switches as the moveable frame and its magnet moves from position P0 to P5. The logic-1 and logic-0 levels are those that would come from a preconditioning circuit such as that in Fig. 7-9.

When the magnet is at position P0 and the movement toward switch S1 is just beginning, not one of the switches is energized and the logic outputs show all logic-1 levels. There is nothing wrong so far—an output of three ones can signify the moveable frame is at its starting position.

The magnet then moves to point P1 over switch S1. Switch S1 is thus closed, nicely represented by the logic-0 level at PS1 and ones at PS2 and PS3. Still, so far so good, but here comes the problem.

As the magnet moves away from P1 and to position P2, it lets S1 open before it closes S2. All three switches are open again, and the signal

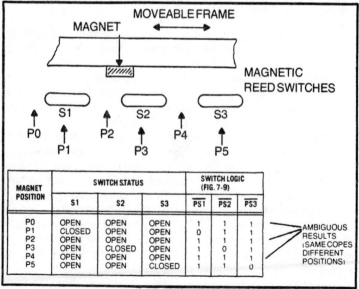

Fig. 7-10. An example of how not to set up a discrete position-sensing scheme. The problem is that the zero-position code appears at two places other than the zero-position point of the manipulator.

217

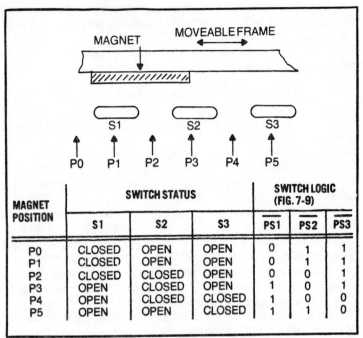

MAGNET POSITION	SWITCH STATUS			SWITCH LOGIC (FIG. 7-9)		
	S1	S2	S3	$\overline{PS1}$	$\overline{PS2}$	$\overline{PS3}$
P0	CLOSED	OPEN	OPEN	0	1	1
P1	CLOSED	OPEN	OPEN	0	1	1
P2	CLOSED	CLOSED	OPEN	0	0	1
P3	OPEN	CLOSED	OPEN	1	0	1
P4	OPEN	CLOSED	CLOSED	1	0	0
P5	OPEN	OPEN	CLOSED	1	1	0

Fig. 7-11. How to set up a discrete position-sensing scheme to exhibit the necessary make-before-break feature.

outputs are all at logic 1. That is the pattern of ones and zeros the system is supposed to provide when the magnet is back at position P0. But here it is at position P3!

Just looking at the data outputs, the system cannot tell the difference between having the moveable frame at position P0 and P2. Being able to detect the difference between those two positions can be critical to the proper function of any mechanical positioning system.

The same sort of ambiguity arises as the frame moves the magnet to position P4. In fact, the only time this arrangement is doing a decent job is when the magnet is directly over one of the three reed switches. Otherwise, a whole lot of confusion can occur. What good is a position-sensing scheme that is reliable over just 50 percent of its working range?

The way around the problem is to make the magnet longer, set the reed switches closer together, or both. Do whatever is necessary to make sure one switch remains energized until the next one along the line is energized. Figure 7-11 shows how the scheme works when it is set up properly to incorporate the make-before-break format.

The table accompanying the improved setup clearly shows that there is no ambiguity with regard to the pattern of ones and zeros from the switch circuit and the position of the magnet over the reed switches. To be sure, the output bit patterns are identical for positions P0 and P1, but that is of no

218

real concern. Those two positions are at the extreme end of the path of the magnet. The main point is that the beginning codes appear only at the beginning, not at other places as indicated in the table of Fig. 7-10.

A Further Note About Discrete Position-Sensing Schemes

Discrete position-sensing schemes do have their place in modern parabot/robot projects. In any particular case, though, making a commitment to using them or not should be based on some intelligent considerations. Those considerations include some trade-offs among the desired level of resolution, cost, complexity and reliability.

In position sensing, *resolution* is a measure of the number of different positions the scheme can sense within a given length of travel. A system using four sensing switches over a space of 12 inches can reliably sense no more than five positions (one point for each switch and an additional one where no switches are energized). About 3 inches of "slop" are between switches in this example, and the implication is that the system has to move at least 3 inches before it can sense any change in position.

This might be a bit too coarse for many position-sensing applications. The obvious way to improve the resolution is to use more switches over the same amount of space. Suppose you use 12 switches, evenly spaced,

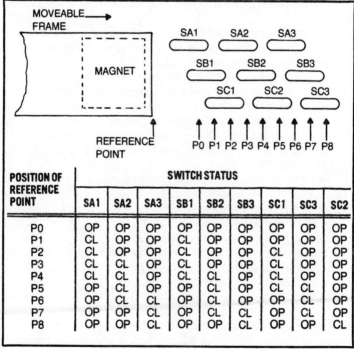

POSITION OF REFERENCE POINT	SWITCH STATUS								
	SA1	SA2	SA3	SB1	SB2	SB3	SC1	SC3	SC2
P0	OP	OP	OP	OP	OP	OP	OP	OP	OP
P1	CL	OP	OP	CL	OP	OP	OP	OP	OP
P2	CL	OP	OP	CL	OP	OP	CL	OP	OP
P3	CL	CL	OP	CL	OP	OP	CL	OP	OP
P4	CL	CL	OP	CL	CL	OP	CL	OP	OP
P5	OP	CL	OP	OP	CL	OP	CL	CL	OP
P6	OP	CL	CL	OP	CL	OP	OP	CL	OP
P7	OP	OP	CL	OP	CL	CL	OP	CL	OP
P8	OP	OP	CL	OP	OP	CL	OP	OP	CL

Fig. 7-12. Improving the resolution of a discrete position-sensing scheme.

for picking up the position of a moveable frame over a 12-inch distance of travel. That amounts to a marked improvement in system resolution over the simpler, four-switch version. The frame then only has to move 1 inch in order to detect a change in position. The four-switch scheme can resolve five positions, while the 12-switch system can resolve 13 positions.

The point of this discussion is to lead you to the general proposition that the resolution of a position-sensing scheme is proportional to the number of discrete sensing elements you use. The larger the number of discrete sensing elements included in a given setup, the higher the resolution is.

When striving for higher resolution with discrete-position sensing devices, you are bound to encounter a *packing-density* problem. No two objects can occupy the same space at the same time, and you will have trouble trying to fit two ¾-inch switches into a 1-inch space.

The way around this particular difficulty is to stagger the positions of the sensing elements. Figure 7-12 suggests this approach when using magnetic reed switches. The same idea holds equally well for micro-switches.

The actual pattern of switch openings and closings depends on the physical dimensions of all the components and, in this case, the strength of the magnet as well. In any event, the scheme works as long as the pattern of switch settings is different for every specified position of the moveable frame.

The same idea can be modified for use with rotating frames. If you find you cannot pack enough switches along the radius of turn, stagger the placement of the switches along two or more radii.

No matter how clever you are when it comes to packing a lot of switches into a small space to get a higher resolution, however, another potential problem still exists. There is a practical limit to the number of discrete-sensing devices you should use. When your design calls for something more than eight or 10 switches for sensing the position of a single element in the system, you have to begin wondering if there is a better way to go about the job.

Switches can cost a lot of money, and the cost of your system can begin getting out of hand if you have to use a lot of switches for position-sensing tasks. While the wiring for a single switch is just about as simple as any wiring job can be, multiply that by some large number of switches and you come up with a big wiring job.

Then don't forget that all that switch information has to be processed somewhere farther down the line. You can tie up a lot of electronics, just sorting out the signals from all those switches.

So discrete-position sensing should be used sparingly. There can be no question that it is quite suitable for small and simple systems, but more complex position-sensing systems and systems that use a lot of position sensing of high resolution call for an alternate approach to the whole matter—continuous position sensing.

Continuous Position Sensing

Achieving a fairly good level of positional resolution with discrete-device position sensors create some problems. In fact, most of the disadvantages inherent in descrete position sensing stem from a need for reaching higher levels of resolution.

Continuous position sensing gets around the resolution problem, as least with the sensing mechanism. The continuous-position sensing device used most often through this book is a simple potentiometer. The moveable frame rotates the shaft of the potentiometer, generating a continuous change in its resistance. In a sense, a potentiometer offers an infinite number of different sensing positions, each providing a level of resistance that is different from all the others. In theory, at least, a potentiometer has *infinite resolution*.

One device (a potentiometer) will be doing the job of an infinite number of discrete devices (position-sensing switches)? That's terrific. It's almost too good to be true. There has to be a catch somewhere!

Yes, there is a catch. While using a potentiometer to sense the position of a moveable frame generates a change in resistance that has virtually infinite resolution, the varying resistance has to be divided into discrete steps somewhere along the line. No data handling system—analog or digital—can work properly with an infinite amount of information. The continuous signal has to be divided into a manageable number of discrete elements.

Fortunately, the process of breaking up a continuous position-sensing signal can be held off until everything has been translated into an electrical format. It is far easier to divide an analog signal into 16 different voltages in a purely electrical fashion than it is to work out a mechanical scheme for picking up 16 different sets of on/off signals from some switches. Because most of the control systems are digital in nature, the process of breaking up a continuous signal into discrete elements calls for an analog-to-digital (A/D) conversion process. That process and its circuitry are described in detail in the "Data Formatting" section of this chapter.

For our immediate purposes, the objective is to translate rotary and linear motion into a varying resistance. The problem of using that varying resistance to generate binary signals representing some positions will be handled later.

Most potentiometers are made of *carbon* that shows a varying resistance between either end of the material and the wiper arm that slides along it. Most of the discussions about these potentiometers will assume they have a linear taper: Their resistances change linearly with changes in shaft rotation. While potentiometers showing a nonlinear, audio-taper might prove useful in some position-sensing operations, they are not considered here.

Also, it is convenient to assume that the shaft of a potentiometer can be rotated through an arc of about 300 degrees. Most potentiometers can handle that much rotation. Bearing in mind that a

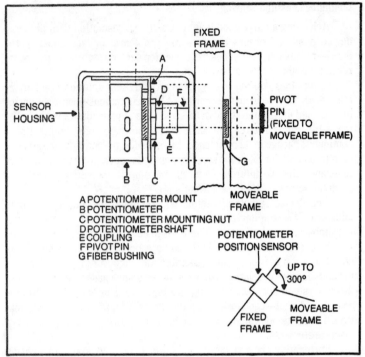

Fig. 7-13. Mechanical arrangement of a potentiometer for sensing the position of a rotating manipulator.

typical potentiometer has a carbon-composition resistance element (as opposed to wire-wound), has a linear taper (as opposed to an audio taper) and can be rotated through an arc of 300 degrees (as opposed to a 10-turn pot), it will be possible to get through the following discussions without getting tangled in needless technicalities.

Sensing the Position of Rotating Elements

As position-sensing devices, potentiometers lend themselves more naturally to situations involving a rotational motion. This is especially true if the angle of rotation falls within the usual 300-degree limit imposed by the rotational arc of the shaft of the potentiometer.

The basic idea is to fix the main body of the potentiometer to a fixed frame of reference, and fasten the shaft of the pot to the moveable frame at the point of rotation. This rather simple idea is shown in Fig. 7-13.

The main body of the potentiometer is attached to the fixed frame through a potentiometer mounting bracket and sensor housing. The pot is fastened to the mounting just as though the mounting were a front panel assembly. The shaft of the potentiometer is ganged to the main pivot pin for the moving mechanism by means of a screw-type coupling.

As the moveable frame rotates with respect to the fixed frame, the shaft of the potentiometer turns. The angle of turn of the pot shaft is a clear indication of the angle between the fixed and moveable frames.

Incidentally, expressions "fixed frame" and "moveable frame" are merely relàtive terms. What is called the fixed frame in this case might be moving with respect to an entirely different frame of reference. That doesn't change anything as far as this particular setup is concerned, however. The pot still generates a varying resistance that reflects the angular difference between the two frames shown here. It is not relevant which frame is moving, or if both are moving with respect to another frame of reference.

Sensing Linear Positioning With a Potentiometer

Potentiometers respond quite naturally to rotary motion, but unfortunately, many manipulator schemes exhibit motions that have linear, back-and-forth components. Getting a potentiometer to respond to linear positioning is a matter of mechanically translating that linear motion into rotary motion. The position of the shaft of the potentiometer is then a reflection of the position of the moving frame.

Figure 7-14 shows one such scheme. The main body of the potentiometer is secured to the fixed frame of reference. A shaft mounting

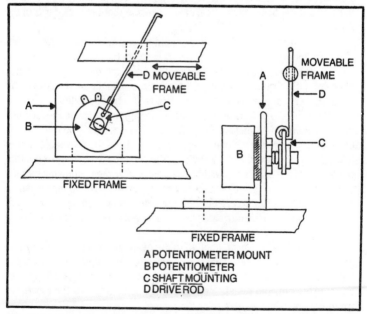

Fig. 7-14. A simple mechanical arrangement for translating linear motion into rotary motion. The objective is to sense the linear position of a manipulator with a potentiometer.

device and one end of a drive rod are fastened to the shaft of the potentiometer.

The drive rod, perhaps a length of sturdy steel wire, is crimped tightly to the shaft mounting element. The free end of the drive rod, however, slides freely through a slot in the moveable frame. So as the moveable frame moves back and forth with respect to the fixed frame, the drive rod is forced to move and rotate the shaft of the potentiometer.

Figure 7-15 is a dimensional schematic of this system. The full distance of the linear excursion is represented by distance S. L is the length of the drive rod, D is the right-angle distance between the shaft of the potentiometer and the moveable frame, and *theta* (Θ) is the angular excursion of the potentiometer shaft.

The potentiometer should be arranged at the midpoint of the linear excursion. Doing so, you have the benefit of using the maximum amount of variable resistance in the most effective manner. And, incidentally, mounting the pot in this position simplifies the design equations.

The equations accompanying the diagram are all based on some straightforward applications of geometry and trigonometry. They work if you stay with the notion that the pot is centered at the midpoint of linear excursion.

Knowing how far the moveable frame will move is important to getting started. That's usually no big problem. You probably have already worked out the basic mechanical design, so use that term, S, for L in the equation.

It would seem that distance D could be any arbitrary or convenient distance. Indeed, you can set that pot as far from the moving frame as you choose. But you will find that the smaller D is, the more the resistance changes with a given amount of motion. In a manner of speaking, that means the sensor becomes more sensitive as D is made smaller.

For the sake of working out a specific example, suppose the full span of linear motion is 3 inches. For one reason or another, you find the pot cannot be mounted any closer than 1.5 inches from the moveable frame. Term D, in other words, is 1.5. Solving the equation for L, it turns out that the length of the drive rod should be at least 2.12 inches. Allowing for play and connections to the potentiometer shaft, setting L to 2.5 inches isn't a bad idea. It is better to err on the long side, leaving some extra length protruding through the moveable frame. You can cut off the extra bit when the project is done.

The equation for *theta*, applied to the example just cited, provides an angle of about 90 degrees. That means you will be using about 90 out of the 300 degrees normally available from the potentiometer. If you happen to use a 1Meg pot, the resistance will change some 300K as the moveable frame covers its entire span.

While you consider this linear position sensing device, it is important to realize that the change in resistance is not exactly proportional to the change in position of the moveable frame. The shaft of the potentiometer

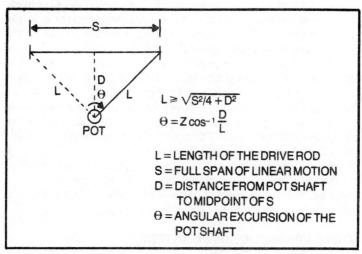

$$L \geq \sqrt{S^2/4 + D^2}$$

$$\Theta = Z \cos^{-1} \frac{D}{L}$$

L = LENGTH OF THE DRIVE ROD
S = FULL SPAN OF LINEAR MOTION
D = DISTANCE FROM POT SHAFT
    TO MIDPOINT OF S
Θ = ANGULAR EXCURSION OF THE
    POT SHAFT

Fig. 7-15. A dimensional schematic for the simple manipulator position sensor in Fig. 7-14.

turns more for each unit of linear motion near the extremes of that motion than it does while the drive rod is nearly straight up from the potentiometer shaft. The sensing system, to put it another way, is more sensitive at the ends of travel than it is toward the middle. This doesn't pose a serious problem, but you should be aware of the situation.

One problem does call for serious consideration: The drive rod follows a circular path that has a radius equal to the length of the rod. The rod thus sweeps out an arc on the side of the moveable frame that is opposite the position of the potentiometer. That area must be kept free of anything that might snag the drive rod; in fact, it should be protected in an enclosure. For experimenters wanting to cram as many devices as possible into a small space this position-sensing scheme poses something of a problem.

The mechanical schematics in Fig. 7-16 illustrate the operation of a slightly different sort of linear position sensor. The main sensing device is still an ordinary potentiometer, but in this case, the amount of space required for operating the sensor is about one-half that of the former suggestion.

This setup uses two drive rods, generally of equal length. One end of a rod, r_1 is securely fixed to the body of a potentiometer. The second rod, r_2, is fixed to the shaft of that potentiometer. The shaft of the pot thus reflects the *angular difference* between the two rods. The end of rod r_1 is fixed to some point on the moveable frame, while the end of the other rod is fixed to an element of the fixed frame.

The figures show the response of the system as the moveable frame travels from its minimum to maximum position. Notice that the shaft of the

225

potentiometer undergoes an angular change that approaches 180 degrees. The scheme uses a good share of the available resistance from the potentiometer, and it doesn't require as much operating space as the sensor described in connection with Figs. 7-14 and 7-15.

There is just one problem with this particular sensing scheme: The lengths of the two rods must be equal. If they are not exactly the same length, the system binds up as it approaches the 90-degree point. Rather than using a stiff steel wire for the drive rods, you can use narrow strips of aluminum. And to get around the equal-length problem, just make one of the mounting holes at one end of the rods an oblong hole. That way, you can adjust the two for equal lengths as you set up the system.

The change in resistance is not a perfectly linear one with respect to the motion of the moveable frame, but this rarely poses any serious problems. The system tends to be more accurate near the two extreme limits of travel than through its middle region of travel.

SENSING INTERNAL CURRENTS AND VOLTAGES

All but the simplest kinds of parabots have some need for sensing at least one current or voltage level within themselves. Current sensors can be used for monitoring battery charging current or for detecting stall conditions in DC motors. Voltage sensors can do a variety of tasks, including watching over the voltage level of critical batteries in the system.

Voltage Sensors

Voltage sensors do not have to be anything more elaborate than some sort of *voltage amplifier circuit*. Experimenters often insert such amplifiers into a system, not being fully aware that they are installing a voltage-sensing device. There is little point in dwelling on the use of voltage amplifiers as voltage-sensing devices. Just bear in mind a couple of basic requirements.

First, the voltage sensor must have an input impedance that is far greater than the device it monitors. Unless this is the case, the sensor becomes a significant part of the load, thus altering the characteristics of the voltage source. Second, the voltage sensor should not inject too much noise into the signal, nor should it distort the signal in some undesirable fashion.

Neither of these conditions is difficult to meet in this day of high-quality, IC operational amplifiers. We will leave further discussion of the matter to specific examples cited later in this book.

Current Sensors

It is far more difficult to sense current levels than it is to sense voltage levels. Most current-sensing schemes convert the current into a voltage parameter as soon as possible, thus translating an essentially difficult current-sensing task into a relatively simple voltage-handling task.

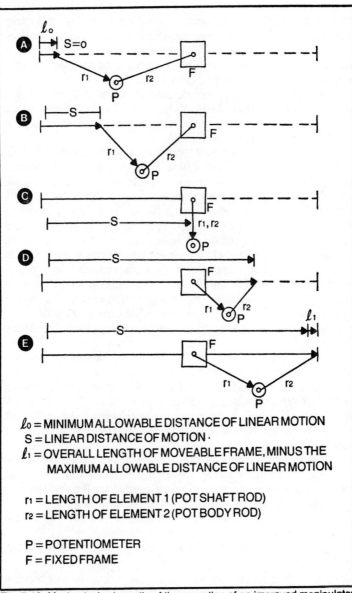

ℓ_0 = MINIMUM ALLOWABLE DISTANCE OF LINEAR MOTION
S = LINEAR DISTANCE OF MOTION
ℓ_1 = OVERALL LENGTH OF MOVEABLE FRAME, MINUS THE
       MAXIMUM ALLOWABLE DISTANCE OF LINEAR MOTION

r_1 = LENGTH OF ELEMENT 1 (POT SHAFT ROD)
r_2 = LENGTH OF ELEMENT 2 (POT BODY ROD)

P = POTENTIOMETER
F = FIXED FRAME

Fig. 7-16. Mechanical schematic of the operation of an improved manipulator position sensor.

There are two fundamental approaches to sensing current levels. One is to insert some sort of resistance in series with the current path, and convert the current level to an IR-type voltage drop. The other approach is to pick up the magnetic field that inevitably surrounds a current-carrying conductor, and translate that field into a voltage level. This is a tricky procedure if the current to be monitored is a slowly changing DC level, as opposed to a pulsating DC or AC current level. Inserting a resistance in series with the current-carrying conductor is by far the most common means of sensing current levels.

Figure 7-17 illustrates a procedure for developing an IR drop across a series-connected resistor. By Ohm's law, the voltage developed across that resistor is proportional to the value of the resistor and the amount of current flowing throught it.

As simple as this procedure appears in principle, you must be fully aware of two potential problems with it. First, the resistor steals voltage from the load device that is being operated from the current source. If the scheme is being used for monitoring current to a DC motor, it steals voltage from the motor when that motor needs it the most—during high-current start and stall intervals. The larger the amount of current being sensed, the more voltage the sensor takes from the load. A second problem closely associated with the series-connected resistor is that it can dissipate more power than you might want expended for current-monitoring purposes. The power dissipated by the resistor is proportional to the square of the current flowing through it. If that resistor happens to have a value of 5 ohms and it is passing 10A of current, then, its power dissipation is 500W. This would be a terrible waste of valuable battery power.

These two disadvantages must be tolerated to some extent, but their effects can be minimized by using current-sensing resistors that have very low values. If, for example, the 10A circuit just described is being monitored with a 0.1-ohm resistor, its voltage drop is 1V and its power dissipation is only 10W. And if the value of that resistor is reduced to 0.01 ohms, the voltage at 10A is 100 mV and the power dissipation is a relatively small 1W.

Of course, you cannot expect to use resistors that are too small for the current levels being monitored. Certainly the efficiency of the sensing scheme improves as the resistor values become smaller, but then the problem of amplifying the small IR drops becomes a problem.

Equation 7-7 will help you get some perspective on the value of the resistor you are to use.

$$R = \Delta e_r / \Delta i \qquad \textbf{Equation 7-7}$$

where R = the value of the sensing resistor, Δe_r = the difference between the maximum and minimum voltage to be dropped across the sensing resistor, and Δi = the difference between the maximum and minimum currents you want to measure.

Suppose, for instance, you have a DC voltage amplifier that can handle input voltages between 100mV and 1V. You want the sensing resistor to respond to currents between 0.5A and 10A. According to Equation 7-7, the value of the current-sensing resistor should be

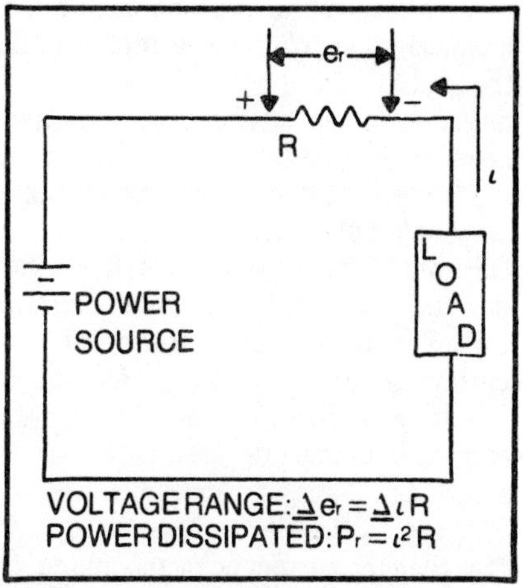

Fig. 7-17. A circuit for translating a current level into a proportional voltage level.

VOLTAGE RANGE: $\underline{\Delta} e_r = \underline{\Delta} \iota R$
POWER DISSIPATED: $P_r = \iota^2 R$

$(1V - 0.1V) / (10A - 0.5A) = 0.095$ ohms. The maximum power dissipation of that resistance will be 10^2 x 0.095, or 9.5W. A 0.1-ohm, 10W wire-wound power resistor will do the job. Don't worry about the error introduced by rounding off the calculated value; you can compensate for that by scaling the gain of the amplifier circuit.

Just be sure that the Δe_r term in Equation 7-7 is selected so that it is less than about 10 percent of the voltage specified for the load device—the motor, battery or whatever is connected to the current source. Most devices, you see, can operate satisfactorily with a supply voltage that is no more than 10 percent down from the nominal rating.

Figure 7-18 shows an entirely different approach to sensing the amount of current flowing through a conductor. The idea is to sense the magnetic field that is generated by that current, translating it into a voltage by means of a *Hall-effect device*.

The Hall-effect scheme overcomes virtually all the difficulties associated with the series-resistor sensor. There is one very significant practical drawback, however: Hall-effect sensors are fairly expensive. It is difficult for most experimenters to justify the cost of a Hall current sensing scheme when the low-cost resistive sensor is available. Considering that some experimenters are willing to give a new idea a try, regardless of the cost, and looking toward the possibility that the cost of Hall devices will drop to a more attractive point in the future, a brief discussion of the Hall current-sensing scheme is in order.

A basic Hall device is little more than a slice of semiconductor material. Unlike most semiconductors, this one has no pn junctions. As shown in Fig. 7-18, a control current I_c, is applied to the device so that it flows from one end to the other. An external magnetic field, B, falling onto the device at right angles to the flow of control current creates a voltage, V_H, that is proportional to both the strength of that magnetic field and the amount of control current.

The control current is normally fixed at some constant level; 50 mA is a typical value. With the current level thus fixed, the Hall-effect voltage then varies with the amount of magnetic field. If the angle between the direction of the magnetic field and control current is not exactly 90 degrees, the Hall voltage is lowered by the sine value of that angle.

Fitting this into the context of sensing current levels within parabot/robot machines, the idea is to run the conductor carrying the current level to the sensed parallel to the control current through the Hall device. The magnetic field generated around the current-carrying conductor influences the value of the Hall voltage. The larger the amount of current being monitored, the larger the Hall voltage becomes.

It's a simple idea, really. The only problem, aside from the high cost of good Hall-effect devices, is that the Hall voltage is rather small. The simplest Hall-sensor arrangements show a sensitivity of about 0.5 mV/A. This means the Hall voltage changes just 500 microvolts for each 1A change in the current being monitored. So if you want to monitor current levels between 0.5A and 10A, that particular Hall-effect arrangement will produce output voltages in the range of 25 μV to 500 μV. That's a pretty small DC signal, and you're going to have a lot of trouble finding a DC amplifier that can boost the voltage to a workable level without drowning out the signal in noise.

Fortunately, there are two tricks for handling the small-signal situation. The first, and more advisable of the two, is to wrap the current-carrying conductor around a ferrite toroidal core and fix the Hall sensor in the center of that core. The ferrite material will not change the current level being monitored, but it certainly will increase the strength of the magnetic field it generates.

Commercial versions of this field-concentrating setup show sensitivities on the order of 1 mV/A and as much as 1 V/A. Such systems cost more than $150.

If the Hall-voltage output is still too small to be handled properly with any available DC amplifier, the Hall-effect people have a second trick up their sleeve: Pulse the control current to get a pulsating Hall voltage and use an AC amplifier to boost the voltage level. Low-level, high-gain AC amplifiers are far easier to build and use than DC amplifiers of the same calibre. You can obtain further information about commercial Hall-effect current sensors by writing to *F.W. Bell Inc.*, 4949 Freeway Drive East, Columbus, OH 43229.

AMPLIFIERS FOR INTERNAL SENSING SYSTEMS

Just about all the sensing devices described in this chapter and elsewhere in this book should be used in conjunction with an amplifier of some sort. The role of that amplifier can vary according to the nature of the sensor, the nature of the interfacing circuit it feeds and, indeed, the desires of the experimenter. Because of this wide spectrum of possible roles, the amplifiers have not yet been covered in any real detail.

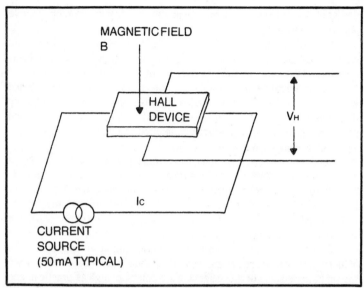

Fig. 7-18. General layout of a Hall-effect current-sensing system.

These sensing amplifiers are going to perform several different kinds of tasks. Sometimes they will perform just one task, while at other times they will be expected to do two or more different kinds of jobs simultaneously. One such task is to raise small-signal voltage levels to a point where those signals can be properly treated by standard control and logic circuitry. A second task is matching vastly different electrical specifications that happen to exist between the sensing device and the circuitry it is supposed to feed. Then, too, the circuits will sometimes be called upon to clip, clamp or otherwise modify the character of the input signal.

For the most part, these discussions of sensing amplifiers deal with DC amplifiers. It simply turns out that DC amplifiers are more useful in parabot/robot machines than AC amplifiers are.

You will also notice that I have a personal preference for the LM3900 quad Norton operational amplifier package. Of course other kinds of op amps can work equally well—perhaps better in some instances. But any experimenter versed in the use of operational amplifiers should have no trouble applying devices of his or her own preference. I strongly suggest avoiding operational amplifier packages that normally call for dual, plus-and-minus power sources. Such packages add needless complexity to the power supply design of the system.

General-Purpose Amplifiers

Figure 7-19 is a schematic diagram for a general-purpose, noninverting DC amplifier. The input and output are both referenced to ground

231

potential, and it is assumed that the inputs and outputs are both positive with respect to that point.

The circuit features a NULL ADJUST potentiometer that lets you null out any DC offset at the output whenever the input signal is at zero. This is one-time adjustment that can be a small, printed-circuit pot.

You will notice that two resistors are labeled R_i and two are labeled R_f. This isn't a printing error. Rather, it means that the resistors having the same labels have the same values. Violate that principle here, and you're going to have trouble getting the circuit to work according to the equations accompanying the diagram.

According to those equations, the voltage gain of the amplifier, A_v, is equal to the ratio of the value of the R_f resistors to that of the R_i resistors. Therefore, if you want an amplifier to have a voltage gain of 10, work out the values of those resistors so that the R_f values are 10 times the R_i values. And if you want a noninverting, nonamplifying buffer circuit, just set the R_i values equal to the R_f values.

The diagram and its equations imply nothing at all concerning the actual values of those resistors. As a rule of thumb, it's a good idea to keep the input impedance of these sensing amplifiers as high as practical, and that means using rather large values for R_i. Selecting the values of R. to be 100K or more will make things work out rather well. So if you want that amplifier with a voltage gain of 10, you can do that by setting R_i to 100K and R_f to 10 times that value, or 1 Meg. And there it is : a gain-of-10 voltage amplifier.

The second equation in Fig. 7-19 can be used for calculating the value of the output voltage, e_o, as a function of the input voltage level, e_i. If you have picked the ratio of R_f to R_i at 10, the output voltage will be 10 times the input voltage.

Of course, the value of the output voltage cannot exceed the supply voltage to the amplifier circuit, +LOGIC SUPPLY. If that happens to be a 5V source and you have set up the circuit for a gain of 10, don't expect the output voltage to be 50V whenever you apply 5V at the input. The highest positive output voltage swing you can get in this case is 5V.

When thinking in terms of a sensing amplifier, you are not usually in a position where you can simply pick a value for the voltage gain out of the air. Equation 7-8 can help you:

$$A_v = e_{omax} / e_{imax} \hspace{2cm} \textbf{Equation 7-8}$$

where A_v = voltage gain of the amplifier, e_{omax} = maximum desired output voltage level, and e_{imax} = maximum input voltage level.

For example, supposed you are working with a sensor voltage that swings between 60mV and 1.2V, and you want to be amplified as much as possible without clipping or distortion of any other kind. The e_{imax} is thus 1.2V, and the maximum signal output is equal to the LOGIC SUPPLY voltage of—say, 5V. The equation thus shows the necessary gain to be $5/1.2 = 4.17$. According to the gain equation in Fig. 7-19, then the value of R_f should be about 4.17 times greater than the value of R.

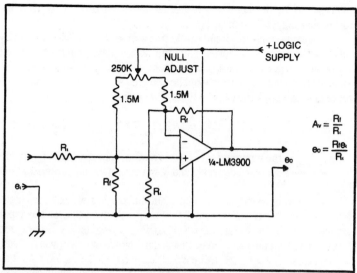

Fig. 7-19. Schematic for a noninverting, general-purpose voltage amplifier circuit.

You might have some trouble finding one resistor that is 4.17 times greater than another. The basic, workable approach is to round off one of the resistor values such that it decreases the gain of the amplifier from your original calculation. If you round off the increase the gain, you risk clipping the signal whenever the input reaches its maximum level.

Suppose you have selected a value of 100K for R_i and, as calculated earlier, you want the circuit to have a gain of 4.17. R_f should be equal to 100K times 4.17, but you will have trouble finding a 417K resistor. Fiddling around with crazy series/parallel combinations of resistors is a waste of effort. So round down the value of R_f to the standard value of 390K. The voltage gain of the circuit is now 390K/100K, or 3.9—but that isn't bad. It's close enough for most applications. Also bear in mind that you can trim out any offset with the NULL ADJUST control.

It is always a good idea to set up your circuit on a breadboard, checking the values to make certain it all works the way you expect. Simply apply some test voltages and monitor the voltages at the output.

In the case of the circuit shown in Fig. 7-19, you should avoid the temptation of using variable resistors to give it a variable gain. The resistors labeled R_f must have the same values if the NULL ADJUST is to do any good and you want to avoid having a troublesome offset at the output. If you make one of the R_f resistors a variable resistor, you should do the same for the other R_f resistor. And on top of that, they both have to be adjusted to the same resistance value. That's a tricky operation at best.

So maybe you won't get the gain and signal level you want within 10 percent. You won't need high precision at this point, despite general impressions to the contrary. It can all be taken care of by the data formatting and processing further down the line.

The same amplifier circuit can be transformed into an AC amplifier by inserting capacitors in series with the nongrounded input and output terminals. To get a reasonably faithful reproduction of the input signal, the value of the input capacitor should be selected so that its reactance is no more than one-tenth the value of R_i. In other words:

$$C_{imin} = 1/(2\pi f_{min} R_i) \qquad \textbf{Equation 7-9}$$

where C_{imin} = the minimum value of the series input capacitor, f_{min} = the lowest expected input frequency, and R_i = value of the input resistor (see Fig. 7-19).

So if you have chosen R_i to be 100K and the lowest expected input frequency is 100 Hz, Equation 7-9 shows that the value of the input capacitor should be at least 159 μF. Rounding upward (since the equation turns out a minimum value) to get a standard capacitor value, you find 220 μF will do the job for you. If the circuit calls for a AC-coupled output as well, you'll be in good shape most of the time by using the same capacitor value there.

As yet another example of how this basic, general-purpose amplifier can be used, consider a situation where the signal from a sensing device is larger than the signal level that can be applied safely to some logic or processing circuitry. A case in point occurs whenever the sensing device is operating at 12 VDC and the processing circuitry is powered from a 5V source. If there is no DC offset in the sensor signal, getting the sensor signal level down to a 5V format is a simple matter of applying it to a passive voltage divider, as illustrated in Fig. 7-20.

For most applications, R_2 ought to have a value on the order of 470 ohms. That value makes the impedance of the e_i point compatible with the input requirements for TTL circuitry. The equation accompanying the diagram then shows how to calculate the value of R_1 based on the largest input signal (e_s) you expect and the largest signal the proceeding circuitry can tolerate (e_i).

So if the sensing circuit operates from a 12 VDC source and the signal is to be made compatible with a circuit that operates from a 5 VDC source, fix R_2 at 470 ohms and solve for R_1: 470 (12V−5V)/5V = 658 ohms. Round up to the next higher standard value, just to make sure e_i isn't too large, and the value of R_1 turns out to be 680 ohms.

This process assumes the output of the divider is to drive some low-impedance TTL circuitry, the input signal has 0V as its minimum point and the signal source can tolerate a fairly low output impedance. If your own situation fails to meet one or more of these conditions, you can fix the situation by connecting the e_i signal point from the voltage divider shown in Fig. 7-20 to the input point of the general-purpose operational amplifier circuit shown in Fig. 7-19.

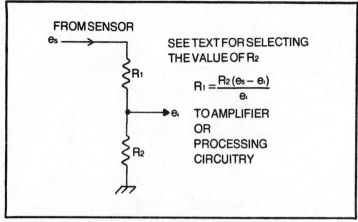

FROM SENSOR

e_s ⟶

SEE TEXT FOR SELECTING
THE VALUE OF R_2

R_1

$$R_1 = \frac{R_2 (e_s - e_i)}{e_i}$$

➤ e_i TO AMPLIFIER
OR
PROCESSING
CIRCUITRY

R_2

Fig. 7-20. A voltage divider for reducing the level of input signals.

By matching this divider to the operational amplifier circuit, the value of R_2 can be as much as 10 times greater than the value of R_i you select for the amplifier circuit. So if you are working with 100K values as suggested for R_i, resistor R_2 in the voltage divider can go as high as 1Meg. This generally eliminates any problems with overloading the sensing device.

If you use 1Meg for R_2 in the divider circuit, matching a 12V sensor signal with a 5V amplifier and logic circuit is a matter of selecting R_1 by 1 x $10^6 (12V - 5V)/5V = 1.4$ Meg. The next-higher standard resistor value in this case is 1.5Meg.

The sensor signal level is stepped down to a 5V format. If you want no signal gain from the amplifier, simply use 100K for all the resistors labeled R_i and R_f. The circuit thus operates as a noninverting buffer that can translate a high-voltage, high-impedance sensor signal into a lower-voltage signal of any output impedance you choose. What's more, using the divider in conjunction with an op amp lets you null any DC offset that might be present in the sensor signal.

All things considered, the strict assumptions imposed by using the voltage divider by itself are eliminated by coupling it to a buffer amplifier as just described here. This is an especially attractive circuit in cases where you aren't sure what the loading specifications for the sensing device might be. The amplifier is not very complicated and it does not cost very much, so why not make a policy of coupling all sensor signals to your interface circuitry in this fashion?

Buffering Sensor Signals Not Referenced to Ground

There are rare, but important, situations where a signal from a sensing device is not directly referenced to ground potential. Notice that one side of the signal input in Fig. 7-19 is tied to ground, which cannot always be done. A case in point is one where the system is supposed to monitor the current through a conductor, using the series-connected

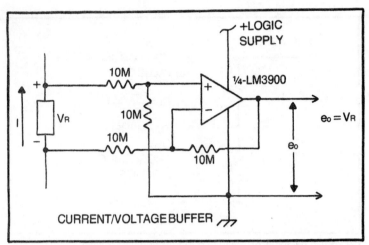

Fig. 7-21. A buffer circuit that establishes a ground reference for a voltage signal.

resistor described earlier in this chapter. In such cases, neither side of that resistor is connected to system ground.

The circuit of Fig. 7-21 shows how to deal with the situation by means of a special noninverting, unity-gain buffer circuit. V_R is some varying DC voltage level that has no direct reference to the common ground connection of the system—it's floating somewhere above ground potential. By the time the signal gets through the buffer circuit, however, the output (e_o) is, indeed, referenced to ground potential.

The output voltage follows V_R rather closely, within the tolerance of the values of the resistors. Those voltages, however, are usually rather small and are not compatible with any processing circuitry further down the line. The output of this circuit is generally applied to a general-purpose amplifier such as the one already described in connection with the circuit of Fig. 7-19.

The primary purpose of the buffer circuit is to give a sensor voltage a ground-reference character that it otherwise does not have. After that, the signal is treated as an ordinary source of sensor information.

Some Notes About Voltage Comparators

Generally speaking, a *voltage comparator* is an amplifier circuit that amplifies only the difference between two input voltages. One of those input voltages is usually considered a reference voltage, while the second input is the signal, itself. The idea is to compare the signal voltage with the reference voltage and come up with a difference signal.

Such an operation can perform two kinds of operations that can be essential in parabot/robot sensing mechanisms. In one instance, the comparator can monitor some DC voltage level, and generate an output

236

signal only when that DC input signal reaches a particular threshold point. That threshold point is fixed by the reference input of the circuit. Another application for voltage comparators is to clip off the top of a signal in order to amplify just the lower portion. This kind of operation is normally used for pulse-shaping purposes.

It is also possible to adjust a voltage comparator so that it amplifies only the top portion of an input signal. Such a circuit can remove low-level noise or a DC component from a sensor signal.

Actually, the general-purpose amplifier shown in Fig. 7-19 can perform most of these voltage-comparitor operations. It is the NULL ADJUST feature in that circuit that makes this possible. While the comparison range is somewhat limited by the NULL ADJUST potentiometer when its value is 250K as shown in Fig. 7-19, it is possible to extend the adjustment range by a substantial amount by simply using a 1Meg pot in its place.

In an electronic system that can become as complex as that typifying experimental parabots and robots, it's a good idea to use the same kind of circuit wherever it can possibly do the job. The more sameness in the system, the easier it is to set up things and keep them running. If you know how one kind of circuit works and you use it wherever possible, life is a lot easier in the long run.

But in the event you find you have a need for a wide-range voltage comparator circuit, here are two that can do the job. The comparator shown in Fig. 7-22A removes the portion of the input signal that is below or less positive than, the reference level, and outputs an amplified and noninverted version of the input signal existing above, or more positive than, the reference voltage level.

The amount of amplification is rather close to the ratio of the values of R_f/R_i. As in the case of the other amplifiers described in this chapter, R_i should be at least 100K. So to amplify the upper portion of a signal 10 times, set R_i to 100K, R_f to 1Meg, and adjust the REFERENCE ADJUST potentiometer for the clipping point.

The equation accompanying the circuit of Fig. 7-22A should be used with some reservation. According to that equation, if e_{ref} is larger than e_i, the output swings negative, or below ground potential. This cannot happen, however, when the amplifier is operated from a single, positive supply. So whenever that equation calls for a value of e_o that is less than zero, you can count on the circuit outputting a level of 0V (or rather close to it).

As long as the input signal level is less than the reference level, the amplifier shown in Fig. 7-22A delivers a near-zero output voltage. Otherwise it outputs a positive voltage, up to the +LOGIC VOLTAGE level.

The circuit shown in Fig. 7-22B turns around the comparison operation, making it possible to scrape the top off an input waveform and amplify just the bottom (less positive) portion of it. This is a signal-

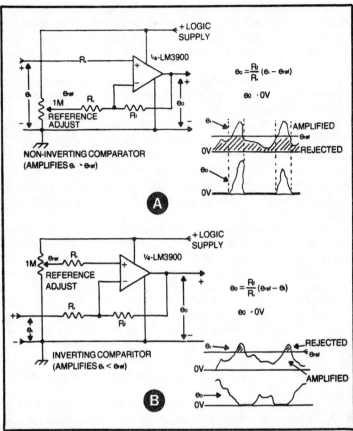

Fig. 7-22. Voltage comparitor circuits. (A) A circuit for amplifying signals that are more positive than a reference point set by REFERENCE ADJUST. (B) A circuit for amplifying signals that are less positive than the reference point.

inverting amplifier that generates a near-zero output voltage as long as the signal input level is more positive than the reference voltage level. As the input signal level falls below the reference level, however, the output signal begins rising toward the positive supply voltage.

Either of the circuits in Fig. 7-22 can be used as a threshold-detecting amplifier by making the value of R_f at least 10 times greater than R_i. The threshold, or trigger point, is adjusted by means of the REFERENCE ADJUST control.

SOME SPECIFIC SENSOR AMPLIFIER CIRCUITS

It is one thing to understand how an amplifier circuit works; applying that knowledge calls for an even greater degree of insight. This section pulls together much of the information already presented in this chapter,

238

showing specific examples of circuits for sensing motor speeds, the positions of manipulators, currents and voltages in a linear or analog format. The first part of Chapter 8 takes up the subject of translating those analog signals into a digital format.

Amplifiers and Signal Conditioners for Optical Motor Speed Sensors

Figure 7-23 shows a circuit that transforms the light-generated pulses from a phototransistor into TTL-compatible pulses that have the same phase as the light source. Whenever light is falling into the phototransistor, PT, the output of the circuit, labeled LS, snaps up to a logic-1 level. And whenever anything interrupts the light source, such as the metal between the perforations in a spinning disc, the light to the phototransistor is interrupted, and the output of the circuit drops to a logic-0 level.

Diode D1 and resistor R9 are largely responsible for making the LS output compatible with most TTL devices. The diode ensures that the output of the operational amplifier circuit pulls down very close to ground potential whenever the output of the amplifier is at its minimum level. Resistor R9 further guarantees a well-defined logic-0 level for the Schmitt-trigger logic inverter. It is that *logic inverter*, incidentally, that makes the phase of the LS output follow the on/off phasing of the light pulses reaching the phototransistor.

The circuit is a fairly simple one, and the values of the resistors have been selected to give the circuit a voltage gain of about 5. If the tops are clipped from the incoming signal because of this rather high voltage gain,

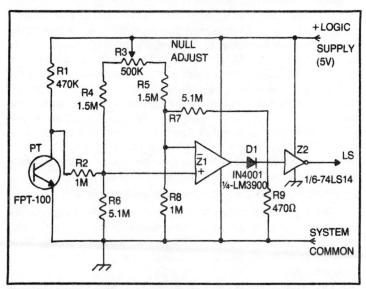

Fig. 7-23. A DC amplifier/signal conditioner for optical motor speed sensing schemes.

239

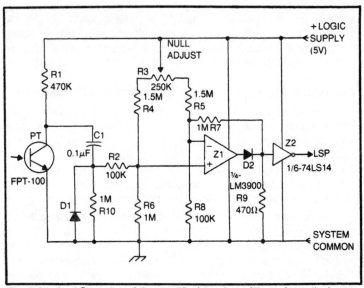

Fig. 7-24. An AC version of the amplifier/signal conditioner for optical motor speed sensing schemes.

there's no harm done. It is the *frequency* of the pulses, and not their amplitude, that is relevant in this case. Once the speed-sensing system is set up and running, set the NULL ADJUST control to produce a clean and reliable rectangular waveform from LS.

The proper operation of the circuit of Fig. 7-23 assumes that ambient lighting is not a real problem. If you are having some trouble separating the light pulses at the phototransistor from ambient lighting, however, try the circuit shown in Fig. 7-24. This circuit is basically an AC version of the circuit just described. In this case, however, capacitor C1 blocks any gradual changes in ambient lighting and passes only the relatively fast changes in light level created by the spinning disc. The gain of the amplifier is greater, too. Note that it runs close to 10. This circuit also has a TTL-compatible output and is adjusted in a manner identical to the simpler, DC version already described. If you are having any trouble understanding the central purpose of these circuits or the function of the amplifiers, see the discussions that concern Figs. 7-3, 7-4 and 7-19.

Signal Conditioners for Electromechanical Motor Speed Sensors

Discussions built around Figs. 7-5 and 7-6 are about a scheme for sensing motor speed by having the motor mechanically ganged to the shaft of a little AC generator—a small DC motor used backward, actually. Here are a couple of circuits that are suitable for monitoring the signals from these generators.

The circuit shown in Fig. 7-25 assumes the experimenter is mainly interested in monitoring the amount of voltage from the generator. In this

240

case, the output of the generator is first rectified by a standard full-wave bridge rectifier assembly, BR1. This produces a full-wave, pulsating DC waveform.

Capacitor C2 plays the role of a filter capacitor that smooths out most of the changes in the waveform, leaving behind just a bit of ripple on top of a DC voltage level. It is this DC voltage level that is proportional to the speed of the motor being monitored.

A voltage divider made up of resistors R1 and R2 is used here on the assumption that the peak voltage from the little generator will be greater than the +LOGIC SUPPLY voltage to the amplifier. It is difficult to say exactly what that peak voltage level will be, and the only way to get that figure is by hooking up the generator and running it from its motor system. If it happens that the peak output voltage is less than the +LOGIC SUPPLY level, you can omit the voltage divider, running the positive side of the full-wave bridge assembly directly to resistor R3.

If the divider circuit must be used, however, the equation for R1 that accompanies Fig. 7-25 will help you get things going. The equation assumes that R2 is fixed at 10K. So if you find the generator turning out peak voltages on the order of 12V and you are using a +LOGIC SUPPLY voltage of 5V, that equation shows that R1 should be close to 10(12V−5V)x10³/12V, or 5.8K. See the discussion for Fig. 7-20 if you aren't sure about the purpose and application of this voltage divider arrangement.

The amplifier is set for a gain of 1. That means the signal from output GS is equal to the voltage from the divider circuit, give or take a bit of offset created by the NULL ADJUST setting. All things considered, the voltage from GS is proportional to the speed of the motor that is driving the generator, which is the purpose of the whole idea.

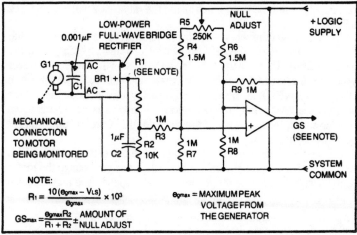

NOTE:
$$R_1 = \frac{10(e_{gmax} - V_{LS})}{e_{gmax}} \times 10^3$$

$$GS_{max} = \frac{e_{gmax}R_2}{R_1 + R_2} \pm \text{NULL ADJUST}$$

AMOUNT OF

e_{gmax} = MAXIMUM PEAK VOLTAGE FROM THE GENERATOR

Fig. 7-25. A circuit that monitors the amplitude component of an electromechanical motor speed sensor.

241

Figure 7-26 is designed for experimenters who want to monitor the motor speed in terms of *frequency*, rather than an analog voltage level. In this case, the waveform from the generator is rectified in a half-wave fashion by diode D1. This process creates a series of positive pulses that have a frequency and amplitude proportional to the speed of the motor. The output of the rectifier is not filtered in this circuit, simply because the pulse frequency must remain intact. Capacitor C1, being on the AC side of the rectifier, merely filters out voltage spikes that are created by the commutating effect of the generator. That capacitor is far too small to have any significant effect on the waveform.

The voltage divider circuit between the rectifier and operational amplifier serves exactly the same purpose as the divider of Fig. 7-25. It simply ensures that the signal applied to the noninverting input of the amplifier does not exceed the power supply rating.

The output of the amplifier circuit is a series of pulses that have a fairly constant amplitude. Diode D2 and resistor R10 work together to make that rectangular waveform compatible with TTL circuitry. Schmitt-trigger inverter Z2-A inverts the waveform, preparing it for a pulse generator circuit made up of Z3, Z2-B, C2 and R11.

Before reaching this pulse generator, the waveform shows pulses having a width that varies with the pulse frequency. The pulse generator is a simple sort of one-shot multivibrator that guarantees the same pulse width, regardless of the frequency of the waveform. That pulse width is determined here by the value of C2. See the equation for C2 accompanying the diagram.

The pulse width, t_{go}, should be more than 10 μs, but less than one-twentieth the period of the signal from the generator. In most instances, you can save yourself a lot of trouble by simply assuming an output pulse width of 10 μs, which means the value of C2 will be about 0.02 μF.

Once you get the scheme set up and running, run the motor at its full speed and adjust the NULL ADJUST control for a clean, 5V signal from the output of Z2-A. Then run the motor down to its slowest speed—whatever speed you feel is too slow to be considered a slow speed. If the waveform at the output of Z-2A is still a clean 5V signal, you're in good shape. If the signal at that test point drops out before the motor reaches its minimum running speed, though, readjust the NULL ADJUST control until you get that clean signal again.

Sometimes you will have to compromise the setting of the NULL ADJUST, trimming it back and forth a little bit as you change the motor speed from full speed to the minimum useful speed. The signal should appear at the output of Z-2A through the entire speed range.

Amplifiers for Manipulator Position-Sensing Potentiometers

An earlier section in this chapter described the use of potentiometers for sensing both the rotary and linear positioning of mechanical elements in

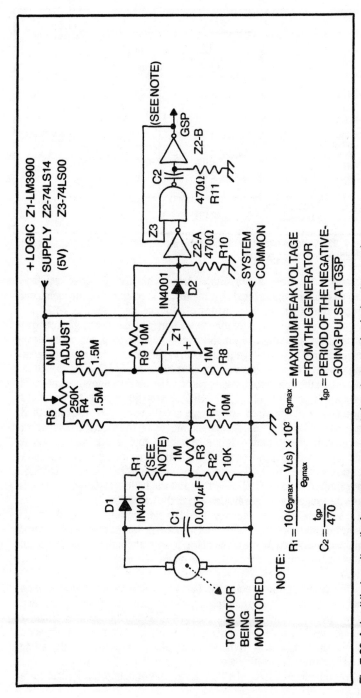

Fig. 7-26. A circuit that monitors the frequency component of an electromechanical motor speed sensor.

243

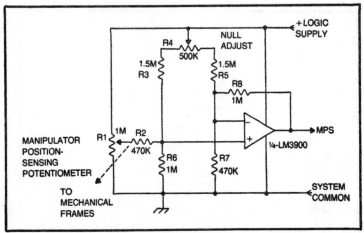

Fig. 7-27. A simple circuit for conditioning the signal from a potentiometer-type manipulator position sensor.

a manipulator. Here is a circuit that transforms that change in resistance into an analog voltage level. Ultimately, it is responsible for monitoring the position of a manipulator in term of a voltage level. See Fig. 7-27.

Resistor R1 is the potentionmeter that is mechanically ganged to the manipulator system. The amplifier is a general-purpose, noninverting amplifier that has a voltage gain slightly greater than 2. An astute reader might be wondering why the amplifier has a gain of 2 when the circuit shows that the voltage at the wiper arm of the sensing potentiomenter, R1, can range from 0 to full LOGIC SUPPLY voltage. Wouldn't that saturate the amplifier at any wiper-arm voltage greater than $V_{LS}/2$, causing half the response to be lost? Yes, that would indeed be the case if the wiper arm were adjustable through its entire range. But recall from an earlier discussion of these position-sensing schemes that the sensing potentiometer rarely travels through an angle greater than 180 degrees. Its setting is thus limited to some fraction of the total available resistance range—and so is the range of voltages available from the wiper arm.

The sensing potentiometer should be mechanically arranged so that it shows ground potential at the wiper arm whenever the manipulator is in its minimum position. As the manipulator moves, then, the wiper arm should move up toward, but never reach, the +LOGIC SUPPLY point as shown in Fig. 7-27.

So to set up this scheme, first arrange the sensing potentiometer so that its wiper arm is at ground potential whenever the manipulator is in its "zero" position. Then adjust the NULL ADJUST in the amplifier circuit for the smallest possible voltage at the MPS output. As you move the position of the manipulator away from its starting point, you should see the voltage at MPS rising in a positive direction. That voltage should just reach about its maximum +LOGIC SUPPLY level as the manipulator reaches the

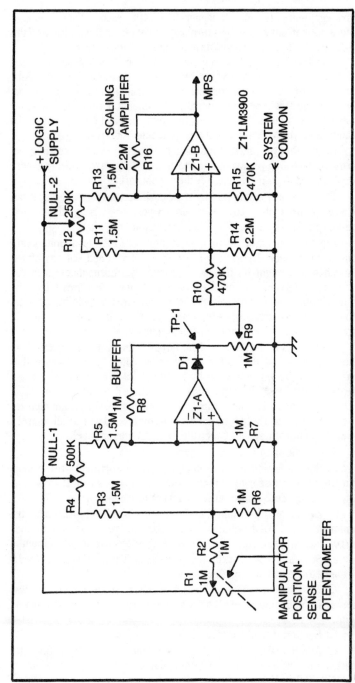

Fig. 7-28. An improved circuit for monitoring the signal from a potentiometer that reflects the position of a manipulator.

maximum limit of travel. If it happens that MPS reaches its peak voltage before the manipulator reaches its maximum extent of travel, trade off the NULL ADJUST setting against its minimum-position setting.

The idea is to get a nice representation of the position of the manipulator as a voltage level at MPS. Bear this in mind while making these adjustments: If the potentiometer is sensing purely rotary motion, the MPS signal will be proportional to the amount of rotation; but with a scheme that is monitoring the position of a back-and-forth element of a manipulator, the MPS signal will change at different rates.

In the event you find that you cannot get the scheme properly aligned, no matter how long you fuss around with it, you'll have to resort to a more sophisticated amplifier circuit. See the one shown in Fig. 7-28.

This is a more flexible version of the position-sensing amplifier. It can be adapted to just about any potentiometer position-sensing scheme, and the only price to pay is that of an additional amplifier section.

The first amplifier section, built around Z1-A, is a unity-gain, noninverting buffer. Its main purpose is to let you null out any offset resistance that might be present whenever the manipulator is in its minimum-travel position. Unlike the earlier version of this circuit, there is no need to make a careful mechanical alignment. It should be close to showing zero resistance when the manipulator is at its point of minimum travel, but it doesn't have to be right on the money.

The real key to the success of this circuit is the "volume" control, R9. Once the output of the buffer amplifier is nulled, you can vary the sensitivity of the whole system by means of R9. That resistor can be adjusted so that the MPS output shows a full range of variation from 0V to +LOGIC SUPPLY as the manipulator is moved from its minimum to maximum position. There is no need for fussing around with a null control, making compromises between the two extremes of travel.

Here's the one-time setup procedure:

☐ Set the manipulator to its minimum position of travel—to the point that will cause the sensing potentiometer to show its smallest amount of output voltage. Zero volts is nice, but not really necessary.

☐ Set the NULL-1 control until the voltage at TP-1 *just reaches* its minimum level. Do not overadjust to the point where you are still turning the control and the voltage at TP-1 has already reached its minimum point.

☐ Set resistor R9 to its minimum position. The idea here is to feed a 0V signal to the scaling amplifier.

☐ Set the NULL-2 control until the voltage at the MPS output *just reaches* its minimum level. Both amplifiers are properly nulled at this point.

☐ Set the manipulator to its maximum position.

☐ Adjust resistor R9 so that the voltage at the MPS output just reaches its *maximum* level. That voltage might not reach all the way to +LOGIC SUPPLY, but it will be close.

The manipulator position-sensing circuit is now completely aligned. Run a final check by moving the manipulator between its two extremes, observing the MPS output with a voltmeter. The output voltage should be a clear reflection of the position of the manipulator.

A Battery Voltage Monitor

A battery is generally considered discharged ("dead") whenever a normal load pulls its voltage down 20 percent from its full-charge voltage rating. There is thus little need to monitor voltages all the way from 0V up to full-charge voltage; in fact, it is a waste of data-generating capability to consider battery voltages below the "dead" level. The voltage monitor suggested in Fig. 7-29 is a comparator-type amplifier circuit for this very reason.

The part of the circuit connected directly to the battery being monitored is a voltage divider. The assumption here is that the battery voltage is normally larger than the +LOGIC SUPPLY level. Typically, the

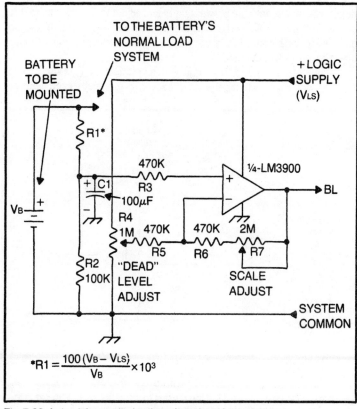

Fig. 7-29. A circuit for monitoring the voltage level from a battery.

247

+LOGIC SUPPLY voltage is on the order of 5V, while most batteries to be monitored are 6V or 12V batteries. The voltage divider, made up of R1 and R2, simply scales down the battery voltage to make it compatible with the amplifier/comparitor circuit.

The equation in Fig. 7-29 shows how to calculate the value of resistor R1, assuming the value of R2 is fixed at 100K. The purpose of capacitor C1 is simply to filter out short-term drops in battery voltage that might occur whenever a motor load is first turned on.

This comparator circuit works according to the explanation offered in connection with Fig. 7-22A. It amplifies only that portion of the input signal that is above the voltage level selected by the LEVEL ADJUST control.

To look at the scheme in a different way, output BL is at 0V whenever the battery voltage is below the "dead" level. There is no need to know the *exact* battery voltage at any point below this level, but as long as the battery voltage is above the "dead" level, output BL shows a voltage that is proportional to the battery voltage.

One way to calibrate this circuit is to replace the battery to be monitored with a variable power supply and adjust that supply so that its output is equal to 80 percent of the nominal voltage level. If you are planning to work with a 12V battery, for instance, the power supply should be set to 0.8 x12, or about 9.6V.

While observing the BL output with a voltmeter, adjust the LEVEL ADJUST control until that output *just* reaches its minimum level. Then set

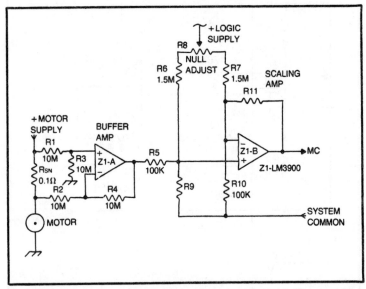

Fig. 7-30. A circuit for monitoring the current to a motor circuit. The same idea applies equally well to monotoring recharge current to a battery.

Table 7-2. Suggested Design Chart for Motor Control Sensors.

Motor Voltage Rating	Peak Motor Current	Value of R_{SN}	Suggested Power Rating of R_{SN}	Value of R9, R11	System Sensitivity
12V	12A	0.1 Ohm	15W	390K	0.4V/A
12V	1.2A	1 Ohm	2W	390K	4 V/A
6V	12A	0.05 Ohm	10W	820K	0.4V/A
6V	1.2A	0.5 Ohm	1W	820K	4 V/A

the temporary, variable power supply to a point representing the nominal, full-charge voltage of the battery and adjust the SCALE ADJUST control until the BL output *just* reaches its maximum point. That's all there is to calibrating the circuit.

Check its operation by running the power-supply voltage (the voltage from the temporary power supply) between 0V and the nominal full-charge level of the battery. BL should remain at 0V, or very close to it, until the test voltage reaches the "dead" level. At voltages above that point, output BL should rise toward its full V_{LS} level as the test voltage rises. Even though a discharged battery is still showing 80 percent of its nominal voltage specification, this circuit rescales things so that it works only with the upper 20 percent.

A Motor Current Monitor Circuit

Figure 7-30 shows a circuit for monitoring the current running through a motor circuit. The same notion applies equally well to another important current-sensing situation: monitoring recharge current for a battery.

The situation calls for using two amplifier circuits. The first is a noninverting, unity-gain buffer, and its main job is to establish a system-ground reference for the IR drop across the current-sensing resistor, R_{SN}. See the discussions for Figs. 7-17 and 7-21 if you aren't sure about what is going on here.

The voltage at the output of Z1-A is proportional to the current flowing through R_{SN}, but it is usually a voltage less then 1V. It has to be amplified by Z1-B, the scaling amplifier, to bring it up to a level that is compatible with most kinds of control circuits.

The procedures for selecting the values of R_{SN}, R9 and R11 are earlier in this chapter. Table 7-2 can make life a bit simpler, however; it suggests some values for four typical current-monitoring situations. The NULL ADJUST control should be set so that the MC output just reaches its minimum point whenever the current through the motor and R_{SN} is at some minimum point you regard as the smallest current worthy of consideration.

Formatting
Sensory Input Information

8

An experimenter experiences an unusual turn of good fortune whenever the information from a sensory mechanism and its amplifier perfectly matches the specifications and format of the internal processing system. Most of the time, an input interface circuit is needed between the sensory mechanism and the processing system.

It is difficult these days to think of an internal processing system being anything other than a digital, TTL-compatible one. Even the NMOS microprocessor ICs are designed to be used with TTL circuitry. And that certainly calls for a digital processing format.

The only exception to this general view is one calling for CMOS digital processing in very small parabot systems. CMOS logic circuits have the advantage of low power consumption and ready compatibility with a wide range of supply voltages. But CMOS logic quickly loses out to TTL/microprocessor systems whenever the machine is to have something more than a modest amount of data-processing capability. So all of the examples in this chapter, as well as most of the examples elsewhere in the book, are based on the assumption that the information from sensor mechanisms must be translated into a digital, TTL-compatible format.

As far as parabot/robot sensory information formatting is concerned, the task can be approached on two levels: generating one-bit threshold signals or multibit digital representations of sensory parameters. Here are a couple of examples that will help clarify the matter.

Suppose you want the internal processing system to know when the main battery-supply voltage drops below some critical threshold level. The actual battery voltage isn't relevant. Instead, a one-bit logic level must show a logic 1 when the battery voltage is above some threshold level and a logic 0 when the battery is running low on power.

The formatting circuit for this job thus generates either a 1 or 0 logic level, depending on the charge status of the battery being monitored. And if the signal is to be compatible with TTL processing circuitry, the logic-1 level should be a voltage of at least 2V, but not more than about 5V. The logic-0 level should be no more than 0.4V. These figures merely reflect the V_{IH} and V_{IL} specifications for standard TTL devices.

A more sophisticated multibit format is necessary whenever your plans call for monitoring some sensory parameter over a wide range of values—in instances where the amount of sensor signal is relevant to the internal processing system. Suppose, for example, you want to monitor the position of a manipulator system. If you are using a potentiometer-type sensor to pick up the position of the manipulator in a continuous fashion, you probably want to format that position with something more than a single bit. Using a four-bit format, the scheme could sense 16 different positions for that manipulator. And if that resolution isn't good enough for the job at hand, you could sense 128 different positions by using an eight-bit format.

Or perhaps you have a four-speed motor. As described in Chapter 6, such a motor requires a two-bit control input, but now you want to compare the actual speed of the motor with the speed indicated by the two-bit commands you give it. The speed sensing scheme must be able to sense at least four different motor speeds and generate a two-bit signal that reflects those actual running speeds. The processor can then compare the command signal with the speed-sensing signal to see whether or not the motor is doing what it is being told to do. That's an example of two-bit sensory data formatting.

One-bit and multibit data formatting both play valuable roles in parabot/robot technology. As you might suspect, a one-bit, threshold-sensing scheme is much simpler to design, build and use. If for no other reason than that, you should use one-bit data formatting wherever possible. Whether you ever use a more complicated multibit sensory formatting scheme depends on your expectations for the system. It is difficult to imagine a useful manipulator, for instance, that works at only two extreme limits of motion.

ONE-BIT BATTERY STATUS FORMATTING

Figures 8-1 and 8-2 are circuits for monitoring battery voltage and recharge current respectively. In both instances, they generate a one-bit signal (or word) that indicates their status.

The battery voltage monitor shown in Fig. 8-1 uses a comparator circuit to pick up and amplify battery voltage levels that are *above* a point selected by the LOWBAT LEVEL ADJUST control. As long as the battery voltage is running about that present preset point, the output of the operational amplifier will be close to +5V. The Schmitt-trigger inverter, Z301, interprets that voltage level as a logic-1 input, and delivers a clean logic-0 output at LOWBAT.

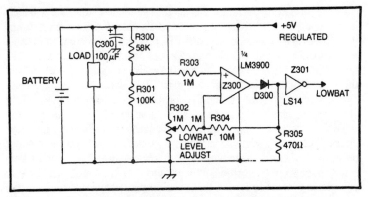

Fig. 8-1. Circuit for one-bit, threshold formatting of a battery voltage level.

Whenever the battery voltage drops *below* the point determined by the setting of the LOWBAT LEVEL ADJUST control, the output of the operational amplifier pulls down close to 0V. D300 and R305 ensure a good logic-0 level at the input of the Schmitt-trigger inverter. And when the output of the operational amplifier is at this near-zero point, you can bet the output of the inverter is at logic 1, or close to +5V.

So once you get this circuit set up and adjusted, LOWBAT will be at logic 0 as long as the battery is providing an adequate amount of voltage to the LOAD circuit. LOWBAT then snaps up to a logic-1 level whenever the battery shows a low-voltage condition. That is a good example of a 1-bit, threshold-monitoring data format.

The simplest way to calibrate this circuit is to remove the battery and LOAD, replacing them temporarily with a variable voltage source. Set that variable voltage source to the lowest battery voltage your LOAD can tolerate (generally 80 percent of the full-charge voltage). Monitor LOWBAT with a voltmeter or logic probe and adjust the level control until the output shows a logic-0 state. Slowly turn the level adjustment in the opposite direction until the LOWBAT output just rises to logic 1.

The purpose of the circuit in Fig. 8-2 is to tell a processing system whether or not the battery is still drawing charge current from an external battery charger. The ONCH (ON-CHarge) output is at logic 1 as long as there is a substantial amount of recharge current flowing into the battery. As soon as that recharge current reaches a minimum level, however, as set by the CHARGE TRIP-POINT ADJUST control, the ONCH output drops to logic 0.

The circuit uses two operational amplifiers. The first is a noninverting, unity-gain amplifier that does little more than establish a ground reference potential for the voltage—the IR drop—across sensing resistor R300.

The output of that first buffer amplifier is fed to a comparator circuit that amplifies voltage levels below the point set at the trip-point recharge status of the battery. A simplified block diagram and function table for such a system are shown in Fig. 8-3.

The first line in the function table shows a 00 combination of outputs from LOWBAT and ONCH, respectively, This particular combination tells the system that the battery voltages is above its minimum operating level and that recharge current is not flowing. This is the expected status whenever the machine is running around with a good charge remaining on the battery. It is also the status that occurs when a battery recharging cycle is finished.

The 01 combination in the second line of the function table indicates a good battery voltage level, but that recharge current is still flowing. This condition usually arises during the recharging cycle. The battery charger is supporting the voltage to the power supply, but recharge current is still flowing. The machine ought to remain connected to the battery charger as long as this particular condition persists.

The third line, the one showing a 10-bit combination, indicates a low battery voltage and no recharge current. In this case, the machine is in need of a recharge cycle, but it hasn't been connected to the battery charger yet. The condition should flag an alarm or sequence of operations that inform the machine or experimenter it is time to stop fooling around and get the charger connected to the system very soon.

The 1-1 combination of bits in the last line of the function table of Fig. 8-3 indicates a low battery voltage and a flow of recharge current. This combination often occurs about when the battery charger is first connected to the system.

This is not the place to discuss exactly how the machine should respond to these conditions. Here, the purpose is to show how such logic conditions occur. Most of the material in subsequent chapters describes what can be done by way of responses to them.

ONE-BIT MOTOR SPEED STATUS FORMATTING

If a parabot/robot machine is to exhibit any automatic qualities of behavior, it needs a method of monitoring the speed of motors—motors used for drive, steering and manipulator responses. Formatting motor status as a one-bit signal generally means looking for a minimum allowable threshold speed.

The circuits described in this section generate a one-bit signal that tells the machine or human operator that a motor is running at a speed less than some preset level. Such a scheme can be especially useful for sensing motor stall conditions.

The circuit in Fig. 8-4 monitors motor speed with an electromechanical speed sensing setup. See the discussions related to Figs. 7-5 and 7-6. In this instance, the circuit looks at the amplitude component of the signal from a motor speed generator.

The voltage appearing at the noninverting input (+) of the first amplifier is one that is proportional to the running speed of the motor being adjustment. The output of that comparator/amplifier is thus close to 0V as long as the recharge current is above the trip-point level. As the battery reaches its full charge and passes the trip-point level, though, the output of the comparator/amplifier rises toward the logic-1 level. The Schmitt-trigger logic inverter turns things around in such a way that its output is at logic 1 while the battery recharge current is flowing. It then drops to logic 0 when the battery is fully charged.

Calibrating this circuit can be a bit tricky, especially if the LOAD draws some current from the battery charger. The LOAD should be set up so that it draws a minimal current from the power supply during the recharge cycle. That can be handled by some internal logic processing, and you will find some examples along those lines later in this book. That LOAD current, you see, will add to the battery current, whether the battery is charging or discharging.

You can compensate for the LOAD current by setting the trip-point adjustment in the right manner. Here's how the adjustment goes. Set up the whole system—battery charger, battery, load and this sensing circuit—and let the battery recharge to a point you deem satisfactory for your machine. With the battery charger still connected and running, adjust the trip-point control (R306) until you see a change in the ONCH output from logic 1 to logic 0.

ONCH should remain at logic 0, even after you disconnect the battery charger from the system. Let the battery discharge for a while. Place a 1-ohm, 15W resistor across the battery terminals if you want to discharge it in a hurry. Don't hang onto the resistor with your bare fingers, though; if you are working with a high-capacity battery, the resistor is going to get pretty hot.

The battery is being discharged, but this should have no effect on the ONCH output until the battery charger is hooked up to the system again. Then the ONCH output should go to logic 1 and remain there until the recharging cycle is over.

These two circuits can be assembled on the same circuit board, providing a two-bit binary data format that indicates the running and

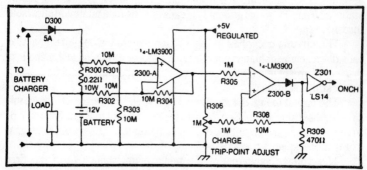

Fig. 8-2. Circuit for formatting a one-bit battery recharge current monitor.

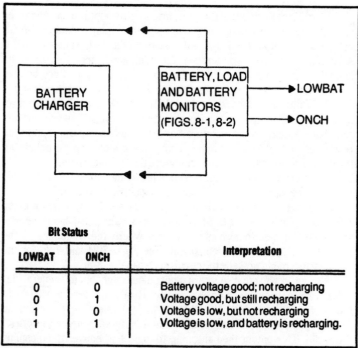

Bit Status		Interpretation
LOWBAT	ONCH	
0	0	Battery voltage good; not recharging
0	1	Voltage good, but still recharging
1	0	Voltage is low, but not recharging
1	1	Voltage is low, and battery is recharging.

Fig. 8-3. Function table and block diagram for a system that monitors battery voltage and recharge current, both generating a one-bit status signal.

monitored. The value of R300 assumes that the maximum peak voltage from the generator is 12V. If that voltage is different from 12V in your particular case, you will have to change the value of that resistor according to the discussions associated with Fig. 7-25.

The circuit of Fig. 8-4 is a noninverting, unity-gain amplifier. The output of Z300-A is likewise a positive voltage that is proportional to the running speed of the motor being monitored. The second amplifier is really a voltage comparator/amplifier. It is also a noninverting amplifier, but it has a voltage gain slightly better than 2 for input signals that exceed the threshold level adjusted by means of the MINIMUM SPEED ADJUST control.

The output of Z300-B is thus near zero volts whenever the motor is running at a speed below that set by the speed adjust control. But when the motor is running at a speed above that threshold level, the output of that same amplifier is near +5V.

The Schmitt-trigger inverter following the comparator/amplifier cleans up the signal and inverts it. $\overline{\text{MRUN}}$ is thus at logic 1 whenever the motor is running at a speed below the preset threshold level, and it shows a logic 0 whenever the motor is running at any speed above the threshold level.

To calibrate a circuit of this sort, first stop the motor to be monitored so that there is no voltage at the input of the circuit. Monitor the GS voltage at the output of Z300-A with a voltmeter and adjust the GS NULL ADJUST control until you just see the minimum output voltage of the amplifier. Then run the motor to be monitored at a speed you believe should represent its lowest operating speed—the threshold speed for this circuit. While doing that, adjust the MINIMUM SPEED ADJUST control until the MRUN output just snaps up to its logic-1 level. As you vary the speed of the motor to be monitored, you should observe the appropriate responses at the $\overline{\text{MRUN}}$ output: logic 1 for a motor speed below the threshold level and logic 0 for speeds above the threshold level.

The circuit shown in Fig. 8-5 does the same sort of job, but in this instance, the input to the threshold-sensing circuit is a frequency (rather than a voltage level) that is proportional to the speed of the motor being monitored. The source of this frequency can be an optical speed-sensing system or an electromechanical scheme where the frequency of the generator is the relevant component. See Figs. 7-24 and 7-26 for the circuitry that might precede the circuit shown in Fig. 8-5.

No matter what that source of input pulses might be, the general idea is to set a frequency threshold level and get a 0 or 1 logic level at $\overline{\text{MRUN}}$. In this particular case, MRUN will be at logic 0 whenever the incoming frequency and speed of the motor being monitored is *above* the preset threshold. This output then snaps up to logic 1 whenever the incoming frequency or motor speed drops below the desired threshold level.

Z350 is a 555-type timer wired to work as a monostable multivibrator. A negative-going pulse at the TRIG input of that device initiates a timing cycle that has an interval proportional to the values of C352, R351 and R350. If this were a conventional monostable multivibrator circuit, the OUT terminal of the timer would go to logic 1 and remain there through the timing interval. It would drop to logic 0 again until the next TRIG pulse occurred. This is not a conventional monostable circuit, however; the use of the transistor, Q350, changes the operation a bit. The OUT terminal of the timer still goes to logic 1 whenever a TRIG pulse occurs, but if another TRIG pulse occurs before the timing interval is over, the whole timing cycle is restarted and the OUT does not show a drop to logic 0.

The timing interval is set by the MINIMUM SPEED ADJUST. As long as the frequency of the pulse waveform applied to the TRIG input of the timer has a period less than the timing interval, the OUT terminal remains fixed at logic 1. The only time that output will drop to logic 0 is when the period of the incoming waveform drops below the timing interval set by the MINIMUM SPEED ADJUST.

Actually, the output of the timer does not remain at logic 0 as long as the motor is turning even slightly. Rather, it shows some alternations between logic 1 and 0. The output of the timer will go to zero and remain there, however, when the motor stops turning altogether.

The circuit includes an inverter at the output, Z351. This is not

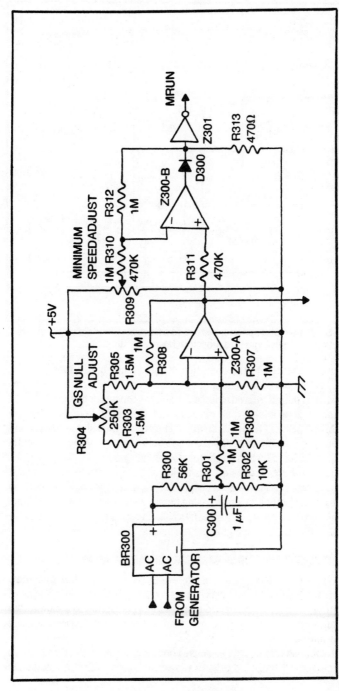

Fig. 8-4. Where motor speed is sensed as a proportional voltage level, this circuit generates a one-bit speed status signal at a prescribed low-speed threshold.

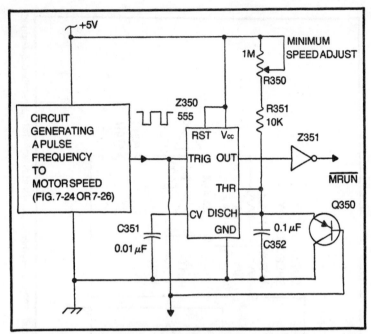

Fig. 8-5. Where motor speed is sensed as a frequency, this circuit generates a one-bit speed status signal at a prescribed low-speed threshold.

absolutely essential to the operation of the circuit, but it does give the $\overline{\text{MRUN}}$ output the same logic phase as described for the circuit of Fig. 8-4. That $\overline{\text{MRUN}}$ output is fixed at logic 0 as long as the motor being monitored is running at a speed that exceeds the threshold level set by the MINIMUM SPEED ADJUST control. The output then shows alternating ones and zeros as the motor speed drops below the threshold level, and it shows a solid logic 1 when the motor stops running.

Both of these circuits are examples of one-bit data formatting as it applies to sensing a low-speed condition for a motor. Whether or not you should do any further data processing before applying the low-speed logic levels to some internal processing mechanism depends on the nature of the system and your own feelings about the matter.

ONE-BIT FORMATTING FOR MECHANICAL CONTACT SENSING

Whenever a mechanical load exceeds the power capability of a motor, that motor stalls, or it stops turning. The preceding discussion of formatting data for motor speed monitors can be applied to a motor-stall sensing scheme which, in turn, can be quite useful for sensing contact with immovable objects.

A robot should be able to sense conditions where it is attempting to apply more mechanical force than its motors can handle. Such a scheme can

258

be used for sensing contact with objects in the external environment. The drive and/or steering motors, for instance, will stall whenever the machine runs into a heavy object (furniture, walls, people, other robots, etc.). Even system using sophisticated proximity detectors should be backed up with this sort of scheme.

It's also nice to know when a manipulator is attempting to deal with an object that is too heavy for it or squeezing something too hard. This, too, can be handled by sensing a motor-stall condition.

Figures 8-4 and 8-5 illustrate a couple of approaches to determining whether or not a motor is running at a speed above or below some preset threshold level. This, in itself, is not adequate for the present situation.

A motor should be considered in a stall condition only if it is being told to turn and it is not really turning. The circuits, as described thus far, cannot sense whether the motor is *supposed* to be turning or not; they only indicate whether the motor is turning or not.

The task here is to design some logic circuitry that compares what the motor is supposed to be doing with what it is actually doing. In Fig. 8-6, what the motor is supposed to be doing is represented by the $\overline{\text{RUN}}$ input. What it is actually doing (running or stopped) is represented by the $\overline{\text{MRUN}}$ input. The $\overline{\text{STALL}}$ output is a one-bit binary signal that shows a logic-1 level if the motor is indeed doing what it is supposed to do. If the motor is not responding as commanded, however, the $\overline{\text{STALL}}$ output shows a logic-0 level.

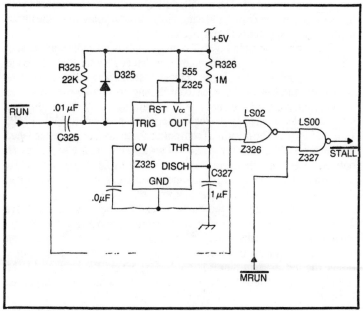

Fig. 8-6. Circuit for sensing a motor stall condition.

As far as the $\overline{\text{RUN}}$ input is concerned, it comes from some logic circuitry or control switches that generate a logic-0 level whenever the motor is supposed to run. Presumably, that same signal goes to a motor control subsystem as described in Chapter 5. A logic 1 at $\overline{\text{RUN}}$ implies the motor is being told to stop running.

The $\overline{\text{MRUN}}$ input comes from a one-bit, threshold-sensing motor speed circuit such as those shown in Figs. 8-4 and 8-5. $\overline{\text{MRUN}}$ is at logic 0 whenever the motor is running at a speed above a predetermined threshold level, and it is at logic 1 whenever the motor is running too slowly or not at all.

Z325 is a 555-type monostable timer of the conventional variety that generates a 1.1-second positive pulse whenever the $\overline{\text{RUN}}$ input shows a transition from 1 to 0. In other words, the OUT terminal of that timer shows a 1-second pulse just as the motor is told to begin running.

The purpose of the timer is to override the stall-sensing scheme whenever the motor is first told to start running. No motor can be expected to go from a full stop to full speed in zero time. The timer and NOR gate that follows it prevent the scheme from generating a stall signal during normal motor startup.

So the output of Z326 rests at logic 0 as long as the motor is *not* being told to run—while the RUN input is at logic 1. Then when the RUN input drops to logic 0, the output of Z326 remains at logic 0 for approximately 1 second, thus masking the normal stall situation that exists whenever the motor is first turned on. The modified RUN signal is then compared with the $\overline{\text{MRUN}}$ signal at Z327, producing a logic-0 output only when the motor is reasonably expected to be running, but it is not.

The circuit shown in Fig. 8-7 is an example of how this stall sensing scheme can be applied to a two-motor drive/steer system. Table 8-1 is the respective parts list. This is one of those systems where the two motors are driven independently, each contributing to both the driving and steering functions of the machine. The turning directions of the two motors are determined by the logic levels present at inputs $\overline{\text{RMF}}$, $\overline{\text{RMR}}$, $\overline{\text{LMF}}$ and $\overline{\text{LMR}}$. (The notational convention for these labels is established in Chapter 5.) The actual responses of the two motors are sensed by an optical speed-sensing system composed of LEDs L300 and L301, phototransistors PT300 and PT301 and, of course, a spinning, perforated disc at each of the two drive/steer motors.

The $\overline{\text{STALL}}$ output of this circuit remains at logic 1 as long as there are no discrepancies between what the motors are being told to do and what they are actually doing. $\overline{\text{STALL}}$ goes to a logic 0, indicating something is wrong with the drive/steer system, under these conditions: either the left or right motor is not turning when it is supposed to be turning, or either the left or right motor is turning when it is not supposed to be turning.

The latter situation rarely arises. It implies a force external to the motor itself, causing it to turn. Such a situation need not be considered, a stall condition, but it is treated that way in this circuit.

Gate Z300-A effectively ORs together the input signals that tell the right motor to turn in a forward or reverse direction, producing a logic-1 level whenever that motor is supposed to turn in either direction. The signal is inverted by Z301-A and applied to a start-up delay circuit of the kind described in connection with Fig. 8-6. The output of Z306-B, labeled RRC, is thus an active-high signal that indicates whether or not the right motor is supposed to be turning (and that the usual 1-second start-up delay interval has passed.)

PT-300 is the device that senses light pulses from L300 and a perforated disc attached to the right motor assembly. The amplifier circuitry following PT300 produces a series of pulses whenever the right motor is, indeed, turning. The low-speed threshold detector built around timer Z303-A is described in some detail in connection with Fig. 8-5.

The RMRUN signal from Z303-A is at logic 1 as long as the right motor is running at a speed above the threshold level set by R313. It begins showing drops to logic 0 as the motor runs below the threshold level, and it remains at logic 0 whenever the motor is not running at all.

This RMRUN running signal is compared with the told-to-be-running signal (RRC) at an exclusive OR gate, Z305-A. This gate is an active-low, one-bit logic comparator that generates a logic-0 output as long as its two input levels are identical, which is as long as the motor is doing what it is supposed to be doing: stopping when it is supposed to stop and running when it is supposed to run. Any discrepancy between the two signals forces the output of Z305-A to a logic-1 level.

Table 8-1. Parts List for the Circuit of Fig. 8-7.

```
L300, L301—10mA LED
PT300, PT301—FPT-100 phototransistor
Q300, Q301—Any small PNP transistor, such as 2N3906
Z300—74LS00 quad 2-input NAND gate
Z301—74LS14 hex Schmitt-trigger inverter
Z302, Z303—556 dual timer
Z304—LM3900 quad Norton operational amplifier
Z305—74LS86 quad exclusive-OR gate
Z306—74LS02 quad 2-input NOR gate
D300, D301, D302, D303, D304, D305—Small rectifier diode, such as 1N4001
R300, R302—22K, ¼W resistor
R301, R303, R308, R310, R311, R319, R320, R321—1 Meg, ¼W resistor
R304, R315—470K, ¼W resistor
R305, R316—250K printed-circuit potentiometer
R306, R307, R317, R318—1.5 Meg, ¼W resistor
R309, R312, R322, R327—100K, ¼W resistor
R313, R323—1 Meg printed-circuit potentiometer
R314, R324—10K, ¼W resistor
R325, R326—470-Ohm, ¼W resistor
C300, C301, C304, C305, C308, C311—0.01μF capacitor
C307, C309, C310, C312—0.1μF capacitor
C302, C306—1μF, tantalum capacitor
```

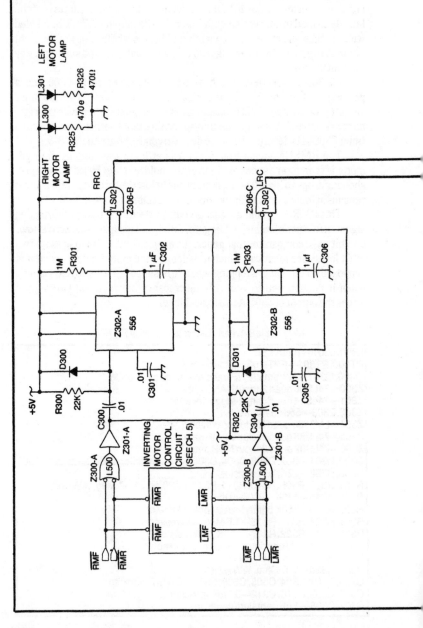

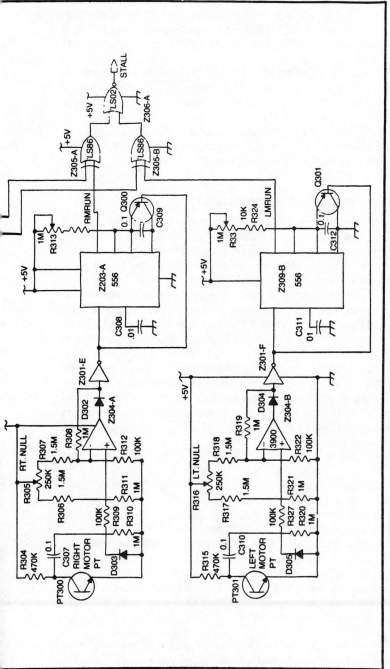

Fig. 8-7. Complete circuit for sensing stall conditions for a two-motor drive/steer system. This circuit is especially useful for sensing contact with an immovable object.

263

The same process, as applied to the left motor, is carried out by a set of circuitry that is identical to that just described for the right motor. Ultimately, the output of Z305-B shows a logic-0 output as long as the left drive/steer motor is actually doing what it is told to be doing by the LMF and LMR inputs.

The stall condition for this circuit should exist whenever either motor is not doing what it is supposed to be doing; the $\overline{STALL}$ output should be at logic 1. This particular output format is generated by the NOR gate, Z306-A.

Aside from being a practical example of a drive/steer motor stall sensing scheme, this circuit demonstrates how a number of different input parameters can be reduced to a one-bit format.

MULTIBIT FORMATTING WITH A/D CONVERTERS

Any continuous, analog voltage level can be translated into a digital number by means of an A/D (analog-to-digital) converter circuit. This means that any parabot/robot sensory parameter that can be expressed as a voltage level can be transformed into a binary number that is meaningful to some processor circuitry.

There are a number of ways to approach the process of A/D conversion and, indeed, IC devices are available for the job. The technique described here uses the simplest approach. Although it is built up from several conventional linear and digital IC devices, it proves to be the most useful in the long run.

Figure 8-8 shows the front-end section of this A/D converter scheme. It has two analog inputs: an A_{in} input representing the voltage level that is to be digitized and an E_{in} representing an error voltage from the conversion process. In both cases, you must assume that the inputs are scaled such that they range between 0V and the V_{cc} supply voltage for the system—generally +5V.

The purpose of amplifiers Z300-A and Z300-B is to determine whether the A_{in} voltage level is equal to, greater than or less than the E_{in} error voltage. These amplifiers are connected as high-gain comparators, and you should note that A_{in} is connected to the inverting input of Z300-A, but to the noninverting input of Z300-B. Likewise, the E_{in} voltage is applied to inverting and noninverting inputs.

So whenever $A_{in} = E_{in}$, both comparators are balanced, and their outputs are both very close to 0V. But what is the output configuration from the two amplifiers when A_{in} is at some voltage level greater than E_{in}? The output from Z300-A will be at 0V, and the output of Z300-B will be close to the system power supply voltage of 5V. Whenever E_{in} is greater than A_{in}, the situation is reversed: The output of Z300-A is at 5V, and the output of Z300-B is at 0V.

This setup can be described in terms of logic levels. If $A_{in} = E_{in}$, both outputs from the amplifiers are at logic 0. If A_{in} is greater than E_{in}, only the

264

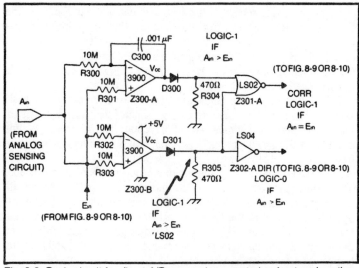

Fig. 8-8. Basic circuit for direct A/D conversion—control or front-end section.
See the options for the remaining circuitry in Figs. 8-9 and 8-10.

output of Z300-B is at logic 1. And if A_{in} is less than E_{in}, only the output of Z300-A is at logic 1.

Notice that the outputs of the comparator/amplifier circuits are connected to a two-input NOR gate, Z301-A. The output of this gate, the CORR signal, will be logic 1 is only as long as both amplifiers are generating a 0V signal—as long as $A_{in}=E_{in}$. Whenever there is any mismatch between the two input signals, one of the two amplifiers is generating a 5V output, and the NOR gate generates a logic-0 level.

Look at the purpose of the NOR gate from a slightly different view: Its job is to determine whether or not the two analog input signals are equal. It cannot tell which is greater whenever there is a mismatch between them; it only detects a match or a mismatch.

The DIR signal from a logic inverter, Z302-A, provides some information regarding the direction of the mismatch between A_{in} and E_{in}. Its output is at logic 0 only when A_{in} happens to be greater than E_{in}. It is at logic 1 at all other conditions, and cannot tell the difference between $A_{in}=E_{in}$ and A_{in} less than E_{in}.

Some valuable information is missing from both the CORR and DIR signals. Use the two together, however, and the circuit generates a complete picture of the relative magnitudes of the two analog input voltages.

You might be able to dream up some applications for the circuit shown in Fig. 8-8; it is a 2-bit analog voltage magnitude detector. Our objective here, however, is to use it as a front end, or control circuit, for a full-blown A/D converter.

265

The remaining section of the system is illustrated in Fig. 8-9. The CORR and DIR logic levels from the front-end section are applied to this digitizing part of the scheme as shown. The E_{in} signal from the resistor network in this circuit is the one applied to one of the analog inputs in the front-end section. In other words, the circuits in Figs. 8-8 and 8-9 should be used together as a unit.

The digitizing circuit shown in Fig. 8-9 is made up of four basic sections. First, there is a high-frequency oscillator composed of a series of logic inverters and a timing capacitor, C300. The frequency of this oscillator really isn't important, just as long as it runs continuously at some frequency above 10 kHz or so.

The second basic part of the circuit is a four-bit binary counter, Z303. This counter runs at a frequency determined by the oscillator circuit. Its counting takes place only as long as its G terminal is at logic 0, and that point is connected to the CORR signal from the front-end section. So whenever CORR goes to a logic-1 level (indicating a good match between the A_{in} and E_{in} analog voltages), the counter stops running and holds its last-counted, four-bit binary number.

The counter is also bidirectional: It counts upward or downward, depending on the logic level at its U/D pin. As you might well suspect, that pin is connected to the DIR signal from the front-end sections. Things are worked out in such a way that the counter counts upward whenever A_{in} is greater than E_{in} and counts downward whenever E_{in} is greater than A_{in}. The DIR signal also sets the counter for down counting whenever the analog inputs at the front end are equal, but the counter doesn't run at that time because it is disabled by the logic 1 at the CORR connection.

In short, the counter (Z303) counts upward—at a rate determined by the frequency of the oscillator part of the circuit—whenever A_{in} is greater than E_{in}. It counts downward at the same rate whenever A_{in} is less than E_{in}. And it doesn't count at all—it just holds the last-counted binary number—while $A_{in} = E_{in}$.

The third basic part of the circuit is a four-bit latch, Z304. This is simply a four-bit binary register that accepts new binary codes from the counter whenever the CORR signal is at logic 1. Whenever CORR drops to logic 0, the latch holds the last number it saw from the counter, and it holds it until the CORR signal goes to logic 1 again.

The overall effect, as far as the latch is concerned, is one where the counter can run up and down while attempting to find a match between E_{in} and A_{in}. During that running time, however, the latch is holding the last binary count that represented a good match between the two analog inputs. The only time the binary output of the latch changes is after the counter has made some changes that make A_{in} and E_{in} equal.

But perhaps the meaning of that effect is lost until understanding the function of the fourth part of the circuit: a resistor network that plays the roles of a digital-to-analog (D/A) converter. The input to this resistor network is the four binary bits from the counter circuit. Its output is an

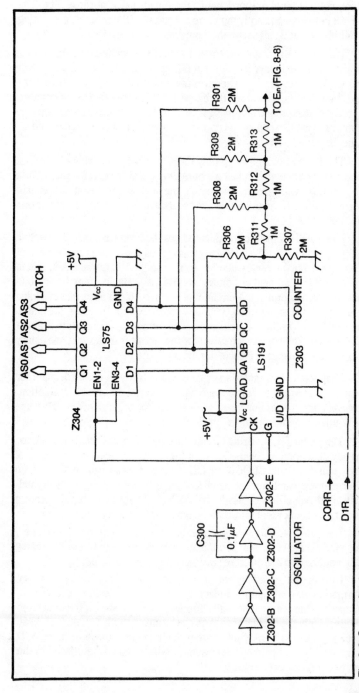

Fig. 8-9. Counter/latch circuitry for a four-bit, direct A/D converter. This circuit must be used in conjunction with the front-end control in Fig. 8-8.

267

analog voltage level proportional to the binary value of those bits. The resistor network constantly monitors the output of the counter, generating the all-important E_{in} signal for the front end.

Putting this all together, the purpose of the whole circuit is to come up with a four-bit binary number that is proportional to the voltage coming into the A/D circuit at A_{in}. Whenever there is any discrepancy between the input analog voltage level and the binary number generated by the counter, E_{in} reflects that error in an analog form. The logic circuitry then forces the counter to run up or down to try to come up with a binary number and E_{in} value that matches A_{in}.

The latch circuit picks up only those binary values where there is a good match between the analog input voltage and the error signal. While the circuit is hunting for a good match, the latch keeps the output reading steady. It filters the junk that always occurs while the system is attempting to balance itself.

If this still seems too complicated, just bear this in mind: Connecting the circuits in Figs. 8-8 and 8-9 together gives you an A/D converter that generates a four-bit binary number from outputs AS0 through AS3 that is proportional to an analog voltage level at A_{in}. Having four binary bits available at its output, the circuit can resolve 16 different input voltage levels. So if your input analog voltage ranges between 0 and 5V, the binary output will represent that voltage in 16 increments—increments of about 0.3V apiece. The conversion factor for the circuit is thus about one binary count for each change of 0.3V at the analog input.

That particular resolution or conversion factor is quite adequate for most parabot/robot sensing systems. If you need eight-bit resolution, however, or a conversion factor of about one count per 20mV, you can use the digitizing circuit shown in Fig. 8-10.

This eight-bit digitizing circuit operates from the front-end section already described in connection with Fig. 8-8. You also need a source of clock pulses at the CLOCK PULSES input. These can come from an inverter-type oscillator as used in Fig. 8-9, from a separate crystal-controlled pulse source as described later in this chapter, or from a clocking frequency that is available from most microprocessor boards.

The principle of operation is exactly the same as for the smaller, four-bit A/D conversion system. This one, however, resolves the analog input voltage into 256 discrete binary numbers, intead of just 16.

This is not at all a bad scheme for formatting analog sensory information into a multibit binary word. It is especially attractive in instances where a system calls for just one or two A/D conversion systems. Only the oscillator section can be shared with a number of these circuits, so if you think your system might require three or more A/D converters, you should consider some of the alternatives offered in the next two sections of this chapter.

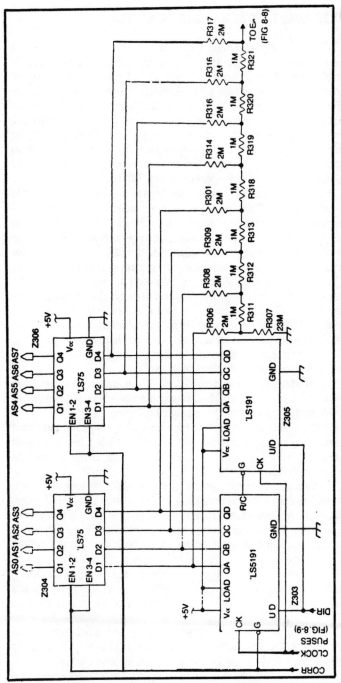

Fig. 8-10. Counter/latch circuitry for an eight-bit, direct A/D converter. This circuit must be used in conjunction with the front-end control circuit in Fig. 8-8.

269

MULTIBIT FORMATTING WITH A DIGITAL TACHOMETER

The circuits shown in Figs. 8-11 and 8-12 work together to make up a digital tachometer—a circuit that translates an incoming pulse frequency into a binary number that is proportional to that frequency. The circuit in this case generates a four-bit binary format, but if you choose to use an eight-bit format, the circuit shown in Fig. 8-11 can be coupled to the one shown in Fig. 8-13. The scheme is especially suitable for formatting sensory parameters that are inherently frequency related, such as the pulses from an optical motor speed sensor (Fig. 7-25 or 7-24) or an electromechanical motor speed sensor that responds to the frequency component of the output of the generator (Fig. 7-26).

The control portion of the tachometer circuit, shown in Fig. 8-11, runs continuously. Z300 is a free-running multivibrator that always generates zero-going pulses of about 47 μs in duration. The duration of the logic-1 intervals (and the overall operating frequency) is selected by the value of R301 and the calibration adjustment at R300. The procedure for selecting the value of R301 to suit your own system will be described later.

For the time being, it is more important to see that the circuit includes two different *one-shot multivibrators,* each generating positive pulses of about 17 μs in duration. The first one-shot multivibrator, Z301, generates its LOAD pulse the very instant the oscillator shows a drop from a logic-1 to a logic-0 level. If you are following this discussion to this point, you will realize that the frequency of the LOAD-pulse waveform from Z301-A will equal that of the free-running oscillator.

The third device in the circuit, Z301-B, generates a positive pulse identical to that from Z301-A. This second pulse, however, is generated as the LOAD pulse shows a switch from logic 1 to logic 0.

Putting the action of these three 555-type timers together, the circuit generates two 17 μs positive pulses in rapid succession, beginning with the negative-going edge of the waveform from the oscillator. See the waveforms accompanying Fig. 8-11.

The first of these two pulses is labeled LOAD, implying that it will ultimately load the tachometer reading into a register. The second pulse, labeled CLR, will then clear a counter that is keeping track of the number of input pulses which occur during a specified interval of time. Of course, the overall function of this tachometer cannot be fully explained without reference to the register and counter circuit shown in Fig. 8-12.

The waveform whose frequency is to be digitized is applied to the F_{in} connection of a four-bit binary counter, Z302. The counter thus runs at a frequency strictly determined by the frequency of the waveform being monitored.

The output of the counter is fed to the input terminals of a four-bit binary latch, Z303. That latch will pick up any number from the counter whenever the LOAD connection to that latch stands at a logic-1 level. When the LOAD waveform is at logic 0, the latch "remembers" the

270

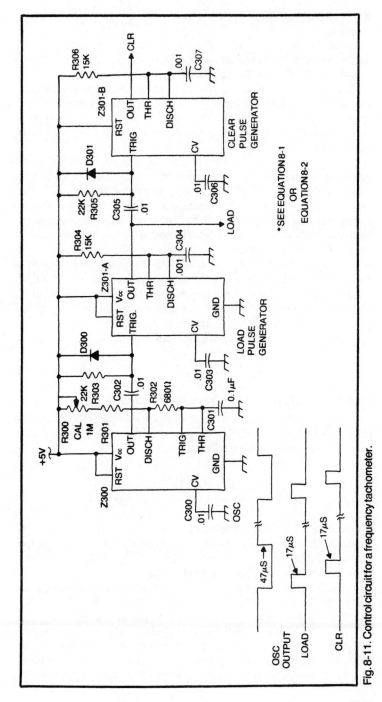

Fig. 8-11. Control circuit for a frequency tachometer.

number from the counter it saw just before the LOAD waveform dropped to logic 0.

Recall that the LOAD and CLR pulses are occurring in rapid succession and in that order—first LOAD, then CLR. That timing operation is handled by the control portion of the circuit shown in Fig. 8-11. So as far as this part of the circuit is concerned, some number from the counter is always loaded into the latch just before the counter is cleared to zero. Whatever binary number is loaded during that LOAD interval is then "remembered" at the AS outputs of the latch until the next LOAD pulse occurs.

For example, suppose the LOAD/CLR combination of pulses is taking place at a 1-Hz rate (as determined by the value of R301 in Fig. 8-11). Further suppose that the frequency to be monitored, f_{in}, is 10 Hz. Doesn't it follow, then, that the counter will count to a binary version of 10 during each of the 1-second LOAD/CLR interval? That binary number is then loaded into the latch, the counter is cleared to zero, and the circuit is ready for the next pulse-counting operation.

As long as the counter is never allowed to overflow—count beyond 15—the number appearing at the AS0 through AS3 outputs of Z303 will be a digital representation of the input frequency, f_{in}. The value of R301 in Fig. 8-11 is supposed to be carefully selected so that the counter will not overflow whenever f_{in} reaches its maximum frequency.

Here is how the value of that resistor can be selected:

$$R_{301} = \frac{16 - 35f_{imax} \times 10^{-3}}{.07f_{imax} \times 10^{-6}}$$ **Equation 8-1**

where f_{imax} = highest frequency expected to be digitized.

For example, suppose you find that a motor speed sensing circuit will generate a frequency of 100 Hz whenever the motor is running at its peak speed. Substituting that value into Equation 8-1:

$$R_{301} = \frac{16 - (35)(100) \times 10^{-3}}{.07(100) \times 10^{-6}.} = 1.79 \times 10^6$$

The value of R301, then, should be a standard resistor value of 1.8Meg. It makes no difference which way you round off the result, because the equation assumes the CAL potentiometer, R300, is set to its midpoint, thereby allowing you to adjust out any slight error.

And incidentally, Equation 8-1 further assumes you will be using the values specified for C301, R302 and R300. Change around any of the values, and you'll have to work out a different equation on your own.

After you have calculated the value of R301 and built up the circuit, you can make a final calibration adjustment and test its operation in the following manner. Attach a variable source of TTL-compatible frequencies to the f_{in} terminal. Set that frequency to the maximum point you used for calculating the value of R301, and monitor the four-bit binary number from the latch, Z303. Just how you go about monitoring that binary output depends on the kind of equipment you have available. If you have just a single-point logic probe or voltmeter, you will have to test each point individually.

In any event, feed in the highest frequency you ever expect to monitor and carefully adjust the CAL resistor until all four latch outputs are simultaneously at logic 1. If you have made this adjustment properly, the binary numbers from the latch outputs will decrease in value as you lower the input frequency to the circuit.

Of course, the circuit can be calibrated and tested in an on-line fashion. Run the motor to be monitored at its maximum rate and adjust the CAL control for a 1111 output from the latch. As you slow down the motor,

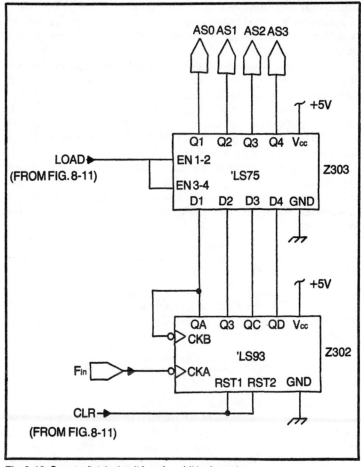

Fig. 8-12. Counter/latch circuit for a four-bit tachometer.

the value of the binary number at the output of the latch should decrease by a proportional amount.

If you do not see a consistent decrease in the value of the binary number from the latch as you lower the input frequency of the circuit, the counter is most likely overflowing; the calibration adjustment isn't correct. The binary number from the latch should always be 1111 when the input frequency is at its peak level. As you move the CAL adjustment through its full range, though, you might find that 1111 output occurring at several different points. The trick is to calibrate the system at the one point that lets the binary numbers decrease in a rational and orderly fashion as the input frequency decreases.

By way of a special hint for anyone having special trouble with this circuit, bear in mind that the period on the waveform from the free-running oscillator (Z300) should be about 16 times greater than the period of the f_{in} waveform when it is set at its maximum frequency. The counter overflows and causes a lot of confusion if the period of the calibration waveform is *more* than 16 times the input period.

The circuit, as described thus far, is sensitive to input frequencies at 16 different levels. In other words, the scheme can resolve 16 out of an infinite number of input frequencies.

If your own situation calls for a much higher level of resolution and accuracy, you need a counter/latch circuit such as the eight-bit version shown in Fig. 8-13. Coupled to the LOAD/CLR circuit in Fig. 8-11, the scheme resolves the input frequency into 256 levels.

The theory of operation is identical to that of the eight-bit version just described. There are two important differences, however. First, *the value of C301 must be increased to 1 μF* to accommodate the usual range of input frequencies. Second, the equation for determining the value of R301 is a bit different.

$$R_{301} = \frac{256 - 350f_{imax} \times 10^{-3}}{0.7f_{imax} \times 10^{-6}}$$ **Equation 8-2**

So if you have opted for the eight-bit digital tachometer and are still expecting a maximum frequency input (f_{imax}) of 100 Hz, C301 should be 1 μF and R301 is:

$$R_{301} = \frac{256 - 350(100) \times 10^{-3}}{0.7(100) \times 10^{-6}} = 3.15\text{Meg}$$

The closest standard value in this case is 3Meg.

The calibration procedures are identical to those already described for the four-bit tachometer. Just remember to calibrate for a 11111111 output from the latches when the input frequency is at its maximum point.

This tachometer scheme, both the four-bit and eight-bit versions, can be quite practical when it comes to monitoring several difference

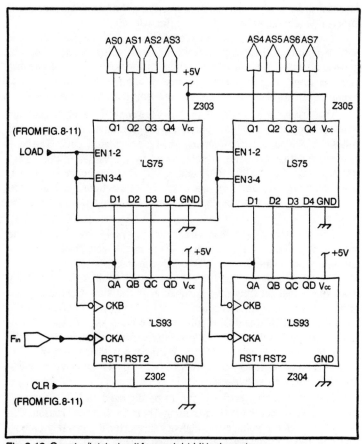

Fig. 8-13. Counter/latch circuit for an eight-bit tachometer.

frequency sources—provided those sources all have about the same frequency range. The ideas is to use the same LOAD/CLR generator circuit (Fig. 8-11) for all the sources and work out a counter/latch circuit for each source. The LOAD/CLR generator can drive up to 10 individual counter/latch circuits. If you want to drive more than 10 counter/latch circuits, simply insert a noninverting buffer/driver at the source of LOAD and CLR pulses. If your frequency sources have vastly different operating ranges, however, you will have to build a separate tachometer system for each set of ranges, selecting the values of R301 accordingly.

FORMATTING BY INDIRECT A/D CONVERSION

Direct A/D conversion always includes a process whereby two analog voltage levels are compared. One is the *analog signal* to be digitized; the other is an *error signal*. Whenever the error signal is reduced to zero by a

275

feedback process, the binary number existing at the moment is considered an accurate representation of the analog input level.

The indirect A/D conversion scheme described here yields the same sort of results without any feedback and comparison operations. The general idea is to translate the incoming analog voltage into a frequency and then apply a tachometer technique to digitize the frequency. The overall result is a binary number that represents the input voltage level.

The circuit shown in Fig. 8-14 shows the conversion portions of this scheme. The analog input is applied to the A_{in} terminal and to the control voltage (CV) input of a free-running, 555-type oscillator, Z300. The base frequency of that oscillator is determined by the values of the timing components: C301, R302, R301 and R300. That frequency is altered, however, by the amount of voltage at the CV connection: The higher the voltage at the CV terminal is, the lower the frequency is from the oscillator.

The output of Z300 is thus a frequency inversely proportional to the amount of voltage at the A_{in} input. This is the voltage-to-frequency conversion part of the indirect A/D conversion operation.

The remaining circuitry should look rather familiar. It works very much like a four-bit frequency tachometer of the kind described in the previous section of this chapter. There is one big difference here, however. The LOAD and CLR pulses from Z301-A and Z301-B, respectively, are synchronized to the *voltage-variable frequency* rather than to the output of a fixed frequency oscillator. Furthermore, the counter (Z302) is clocked by a fixed-frequency source, f_s, rather than from some incoming frequency that is to be digitized.

The input points for the signal to be digitized and the necessary fixed-frequency sources are turned around here for one good reason: The frequency from the voltage-to-frequency converter is inversely proportional to the voltage level at A_{in}, which means there is a nonlinear relationship between that voltage and resulting frequency. By switching around the input points for the frequency to be digitized and the reference frequency, this tachometer becomes an *interval-measuring system*. It counts the time intervals between successive pulses from the voltage-to-frequency converter—intervals that are, indeed, proportional to the voltage present at the A_{in} input.

The higher the voltage at A_{in} is, the lower the frequency from Z300 is. And the lower the frequency from Z300 is, the greater the intervals between successive pulses are. And the greater the interval between successive pulses are, the higher the counter (Z302) counts between the occurrence of the LOAD/CLR sequence of pulses. And finally, the higher the counter counts, the larger the value is of the binary number latched at the output of Z303. In short, the higher the voltage is at A_{in}, the larger the binary number is that appears at the four AS outputs of the circuit. That's an *A/D converter* by any definition.

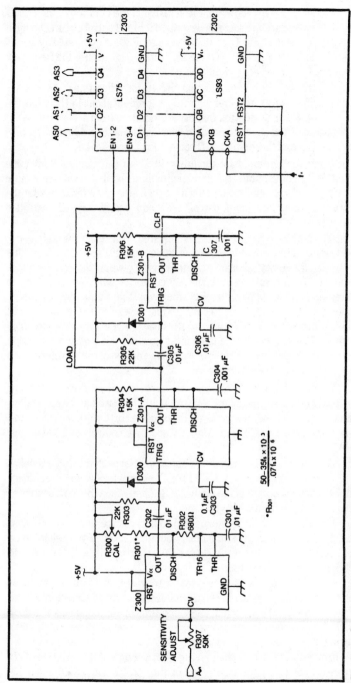

Fig. 8-14. Complete circuit for a four-bit, indirect A/D converter.

277

Using the values shown on the schematic, the duration of the LOAD and CLR pulses are the same as described for the control circuit shown in Fig. 8-11—about 17 μs. The value of resistor R301 is calculated by means of the equation in Fig. 8-14. The term, f_s, in the equation is the frequency of the *fixed-frequency counter source*.

Figure 8-15 shows two different circuits for generating f_s. Circuit A is a simple 555-type astable multivibrator that runs at about 1 kHz, while the one in Fig. 8-15B generates that same frequency from a 1-MHz crystal. Which one of these two frequency standards you use depends on how precise you want the indirect A/D conversion process to be.

If the application is not terribly critical—if you can live with some "slop" in the conversion—the simpler circuit in Fig. 8-15A will serve quite nicely. But if your application calls for a precise conversion, consider the more accurate, but more complicated and costly, crystal-controlled version.

The 1-kHz frequency standard suggested in these figures is applied to the f_s input of the counter section of the circuit of Fig. 8-14. Using this particular frequency standard, the intervals between successive pulses generated by the voltage-to-frequency converter will be counted out in increments of $1/f_s$, or 1ms. So whenever the analog input voltage generates a frequency having a period 9 ms, for example, the counter will count up to the binary version of decimal 9 when the latch circuit is enabled. In other words, the latch picks up a four-bit binary number that represents the number of milliseconds in the period of the waveform from the voltage-to-frequency converter.

You can alter the f_s frequency, of course, to suit your own needs, but the 1-kHz standard suggested here can work for just about any situation. Any scaling that is necessary to get this f_s of 1-kHz to work for you can be done with the SENSITIVITY ADJUST and CAL controls at Z300 in Fig. 8-14.

To calibrate this circuit, apply the lowest-expected voltage level at the A_{in} input and adjust the SENSITIVITY control for the smallest binary number from the output of the latch, Z303. Then adjust the CAL control to get a binary 0000 from the latch.

Next, apply the highest-expected analog voltage to A_{in} and readjust the SENSITIVITY control for a binary output of 1111. If you cannot get a 1111 output, change the setting of the CAL control a little bit until you do get the maximum binary output.

You will have to work those two controls back and forth a little bit as you switch the A_{in} input between the minimum and maximum levels. It's a tricky process, but it is fortunately a one-time operation. The job is done when the minimum input at A_{in} creates a 0000 output from Z303, and a maximum input shows a 1111 output from Z303.

It is possible to construct an eight-bit version of this indirect A/D converter. The counter/latch combination (Z302 and Z303) must be

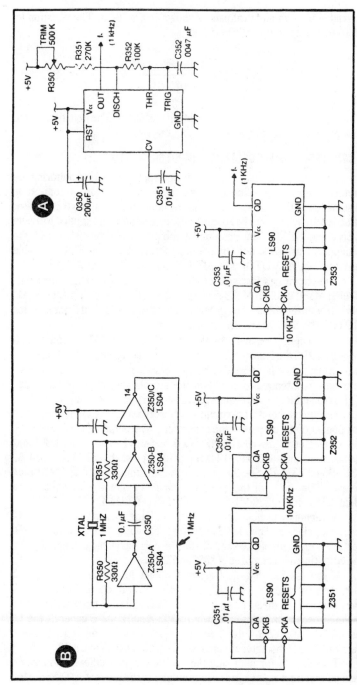

Fig. 8-15. Frequency sources for the indirect A/D converter in Fig. 8-14. (A) 555-type source. (B) Crystal-controlled source.

cascaded with an identical pair as shown in Fig. 8-13. The equation for R301 then has to be slightly modified:

$$R_{301} = \frac{800 - 35f_s \, 4 \times 10^{-3}}{.07f_s \times 10^{-6}}$$

This equation applies only to an eight-bit converter. The equation for the four-bit version is in Fig. 8-14.

A COMPLETE EXAMPLE OF SENSORY INPUT FORMATTING

The circuit shown in Fig. 8-16 is one appropriate for monitoring the motor speed and position of a two-motor manipulator system. It is assumed that the speed of the motors is sensed by a small generator attached to each motor. The outputs of the generators are applied to inputs GEN1 and GEN2. The position of the two moveable frames of the manipulator are picked up by potentiometers R327 and R337: variable resistors labeled MOTOR 1 POSITION and MOTOR 2 POSITION.

The circuit has four sets of four-bit outputs. The output terminals labeled M1S0 through M1S3 represent the actual speed of MOTOR #1, while terminals M2S0 through M2S3 show the same speed parameter for MOTOR #2. See Table 8-2 for the parts list.

The output terminals labeled M1D0 through M1D3 and M2D0 through M2D3 are four-bit binary outputs that represent the position of the two manipulator frames. Everything between the input and output terminals is intended to condition the signals and translate them into TTL-compatible, four-bit binary formats for further processing.

The speed-monitoring circuits both work according to the frequency component of the outputs of the generators. The output of GENERATOR #1, connected to the GEN1 INPUT, is half-wave rectified by D300 and transformed into a clean, TTL-compatible pulse by Z300-A, Z314-A, Z315-B and Z314-B. See the discussion concerning Fig. 7-26 for further details. The speed of the second manipulator motor is handled in an identical fashion by Z300-B, Z314-C, Z315-B and Z314-D.

A frequency tachometer circuit then digitizes each of these motor-frequency signals. The four-bit counter/latch combination for the GEN1 section is made up of Z306 and Z310. The corresponding combination for the GEN2 input is composed of Z307 and Z311.

If you assume that the frequency ranges for the two motor-speed monitoring sections are virtually identical, the two counter/latch circuits can operate from the same source of LATCH and CLR pulses. Z301-A is the main oscillator in this case, while Z302-A and Z302-B generate the 17 μs LATCH and CLR pulses, respectively. This is simply a two-input version of the tachometer circuit described in Figs. 8-11 and 8-12.

The circuits for digitizing the position of two different moveable manipulator frames work in a fashion similar to the indirect A/D converter

Table 8-2. Parts List for the Circuit of Fig. 8-16.

```
Z300—LM3900 quad Norton operational amplifier
Z301, Z302, Z303, Z304, Z305—556 dual timer
Z306, Z307, Z308, Z309—74LS93 4-bit binary counter
Z310, Z311, Z312, Z313—74LS75 quad D latch
Z314—74LS14 hex Schmitt-trigger inverter
Z315—74LS00 quad 2-input NAND gate
D300, D301, D302, D303, D304, D305
D306, D307, D308, D309—any small rectifier diode, such
as 1N4001
R300, R310—4.3 K, ¼W resistor
R301, R307, R311, R317—1 Meg, ¼W resistor
R302, R312—10K, ¼W resistor
R303, R313—250K printed-circuit potentiometer
R304, R305, R314, R315—1.5 Meg, ¼W resistor
R306, R316, R344, R345—10 Meg, ¼W resistor
R308, R309, R318, R319—470 Ohm, ¼W resistor
R320—1 Meg printed-circuit potentiometer
R321—1.8 Meg, ¼W resistor
R322, R329, R339—680 Ohm, ¼W resistor
R323, R325, R330, R332, R340, R342—22K, ¼W resistor
R324, R326, R331, R341, R343—15K, ¼W resistor
R327, R337—1M standard potentiometer (connected to the manipulator's
motion-sensing system)
R328, R338—220K, ¼W resistor
R334—500K printed-circuit potentiometer
R335—270K, ¼W resistor
R336—100K, ¼W resistor
C300, C301, C308, C311, C316, C319, C26, C329—.001 μF capacitor
C302, C303, C305, C313, C323—0.1 μF capacitor
C304, C306, C307, C309, C310, C312, C314, C315
C317, C318, C320, C322, C324, C325, C327, C328—0.01 μF capacitor
C321—.0047 μF capacitor
```

of Fig. 8-14. Rather than going through the extra trouble of generating an analog voltage proportional to the position of the moveable frames, this particular circuit works directly with the change in resistance of the manipulator position sensing potentiometers.

The frequency of the free-running oscillator, Z301-B, is determined by the setting of the MOTOR 1 POSITION resistor, R327. It is the period of this frequency that is measured by the subsequent circuitry: Z303-A, Z303-B, Z308 and Z312. Other than having the oscillator respond directly to the change in resistance at R327, the circuit is identical to the one explained in connection with Fig. 8-14.

The position of the second manipulator section is sensed by MOTOR 2 POSITION, R337, and digitized in the same way by the timers, counter and latch that follow that stage of the operation. The two position-digitizing circuits are run from the same frequency source, Z304-A. The standard frequency in this case is 1 kHz.

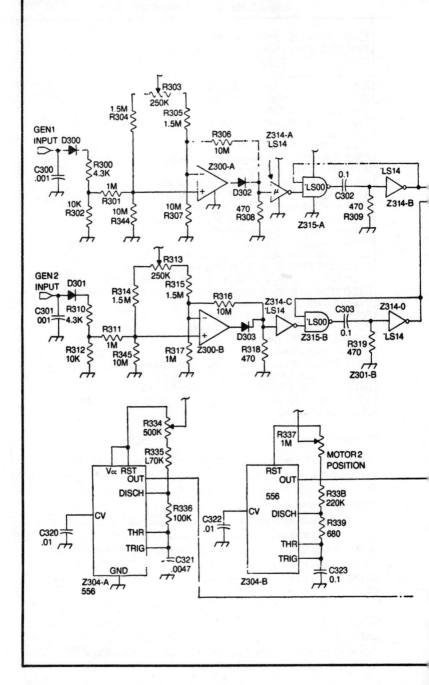

282

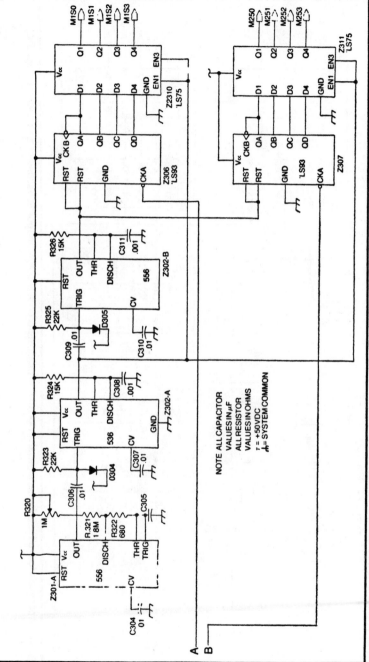

Fig. 8-16. A complete four-bit data formatting system for a pair of manipulator control motors and position sensors.

283

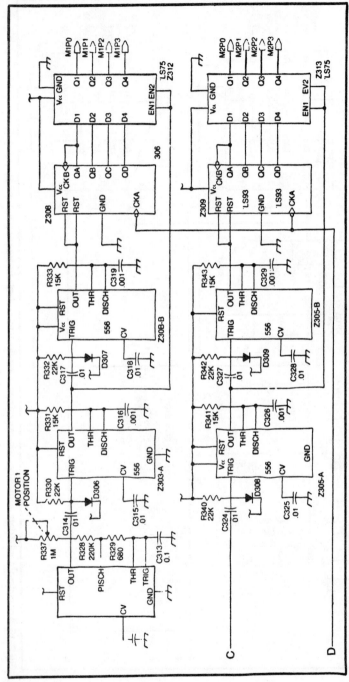

Fig. 8-16. A complete four-bit data formatting system for a pair of manipulator control motors and position sensors (continued from page 283).

This circuit is representative of the level of complexity any parabot/robot experimenter is bound to encounter when trying to build sensory systems with multibit output formats. Considering that it merely conditions sensory input signals for two sections of a single manipulator, one is justified in wondering whether or not robotics is a worthwhile pursuit. A working system built along this line of sophistication requires other such circuits for monitoring drive and steering motors, battery status and a host of other sensory parameters.

One way to simplify the job is to use multibit formatting only when absolutely necessary. After giving the matter some careful and creative thought, you might be surprised how well the system can function with simpler, one-bit data formatting.

But then there is an entirely different approach to the whole matter of data formatting. Microprocessors were not born out of a whim; they were invented to deal with complex data-handling situations. Any experimenter who sees the sensory circuitry getting far more complex than desired should give serious thought to handling the whole matter with a microprocessor-based system. In fact, it is sometimes feasible to run the entire sensory input system from a microprocessor that is separate from the one used for the main processing operations of the machine.

This is not to say that all the circuits described in the three previous chapters become useless when building a microprocessor system. You will find that most of these signal conditioning and data formatting circuits do have a place in microprocessor-based systems. The hardware is simpler in the long run, however. The circuits described in these chapters should be used for fairly simple parabot/robot systems that are too simple to justify the cost of a microprocessor controller.

MICROPROCESSOR DATA FORMATTING

A microprocessor scheme lends itself quite readily to direct A/D conversions of the kind described earlier in this chapter. Thus any group of sensory inputs that can be translated into a digital format by means of that direct A/D conversion process can be handled quite nicely by a microprocessor controller.

Microprocessors, for instance, can be made to count in binary rather easily. The binary count can be applied to a resistor-type D/A converter,

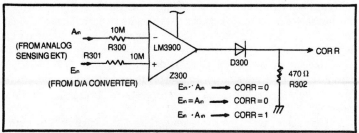

Fig. 8-17. Comparator circuit for microprocessor-controlled A/D conversion.

thus satisfying one of the conditions for performing a direct A/D conversion operation.

What's more, microprocessors can run quite rapidly. And that makes it possible to service a number of different A/D data formatting systems in sequence—all in about the same time it takes to balance one such circuit without the help of a microprocessor.

The circuit shown in Fig. 8-17 is a *one-channel version* of an A/D converter that can be used in conjunction with a microprocessor system. It is somewhat simpler than the direct A/D comparator section described for Fig. 8-8. In a microprocessor system, the A/D conversion process cannot take place continuously, but only when it is instructed to do so. This is why this circuit is so simple. Then, too, there is no need to run the converter in two different directions; it is enough to start E_{in} from zero and count upward until the D/A converter voltage exceeds the analog input voltage. And when that happens, the CORR output shows a sudden change from logic 0 to logic 1.

Figure 8-18 shows how this basic circuit can be used as a module for building a multichannel A/D converter—eight channels with eight-bit (256-level) resolution. This is a pretty good system in anyone's book, and you can see that it is much simpler than the four-channel version of Fig. 8-16. Hardware is undoubtedly saved by using a microprocessor system.

The analog inputs to be digitized are represented in Fig. 8-18 by inputs AI0 through AI7. These are analog voltage levels that have been buffered and scaled by appropriate circuits already described in the earlier chapters. AI0, for example, might be an analog voltage level between 0V and 5V, representing the actual running speed of a motor. AI1 might be an analog voltage level representing the position of a manipulator frame, while AI2 might be an analog voltage representing the charge status of a battery. Any sensory parameter, suitably buffered and scaled as described earlier, can be applied to these AI inputs.

The analog input signals are applied to the inverting inputs of eight separate voltage comparator circuits. Input AI7, for instance, goes to the inverting input of Z300-A. These inputs correspond to the A_{in} signal in Fig. 8-17.

The analog inputs at each comparator are compared with an analog reference voltage level from an eight-bit D/A converter network. That converter is composed of resistors R300 through R315, and its output from the function of R300 and R308 goes to the noninverting inputs of all eight comparator circuits. That connection corresponds to the E_{in} signal in Fig. 8-17.

Take special note of the fact that this eight-channel A/D converter uses just one D/A resistor network. This represents quite a departure from earlier versions that called for a separate D/A converter for each A/D channel.

The resistor-type D/A converter gets its digital input directly from a microprocessor output port. The level of the analog voltage from the D/A

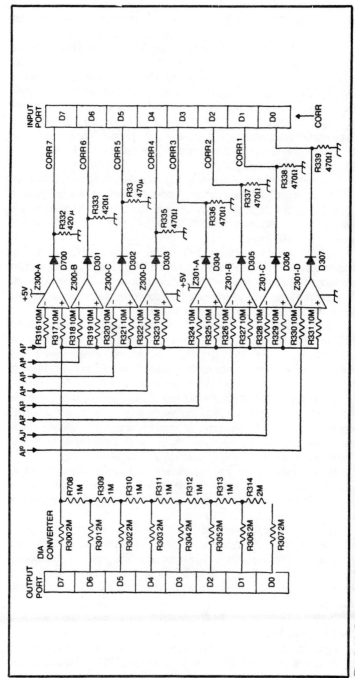

Fig. 8-18. Complete A/D circuitry for a microprocessor-controlled eight-channel, eight-bit data formatting scheme.

287

network is proportional to the value of the binary number from that output port.

But how is it possible to service eight different comparators from a single D/A source of reference voltage? The answer is simple: It isn't. As you will see shortly, the microprocessor system deals with the comparator circuits—the analog input signals—*one at a time*. So even though a single D/D resistor network simultaneously provides analog reference voltages to all eight channels, things are kept in order by having the processor look at just one of the comparator outputs at any given time. The microprocessor senses the status of the comparators—checks the CORR bits—through an eight-bit input port.

Suppose the system is trying to translate the analog voltage at input AI7 into an eight-bit binary format. Under the control of a program or subroutine in the microprocessor system, the output port generates a binary 0, or hexadecimal number 00. The resistive D/A network responds by developing a 0V level.

If this 0V level does not match the analog voltage at AI7, as sensed by comparator Z300-A, CORR7 at input port bit D7 is at logic 0. The program then increments the binary number at the output port to 1, or hexadecimal 01. If the resulting voltage from the D/A converter still does not exceed the voltage at AI7, CORR7 remains at logic 0.

The binary number at the output port is incremented again and again until the voltage from the D/A network exceeds the analog input voltage at AI7. At that moment, CORR7 (bit D7 of the input port) goes to logic 1. And when the microprocessor senses that logic-1 level at input-port bit 7, it knows a match between the analog voltages has occurred.

Here is an important point: Once that CORR7 goes to logic 1, and the microprocessor senses that fact, the binary count present at the output port (input to the D/A converter network) is a digitized version of the analog voltage at AI7. The system can then save that number in memory or process it immediately, depending on how your system is set up.

Bear in mind that this whole scheme works according to the same general principle for direct A/D conversion described earlier in this chapter. The main difference here is that a microprocessor system is replacing all the counters and latches.

After digitizing the signal at AI7, the whole process can begin again, but working with a different analog input and comparators. Maybe you want the system to deal with the AI6 input next. The process is identical to the one just described, only the CORR6 logic level at bit D6 of the input port is the one determining how far the counting operation should progress.

Figure 8-19 describes the necessary microprocessor operations in a flowchart form. The first block in the operation selects the analog input to be digitized, while the second block initializes a microprocessor register/counter at zero. The next step is to OUTPUT the count to the output port, and then INPUT the CORR byte at the input port.

288

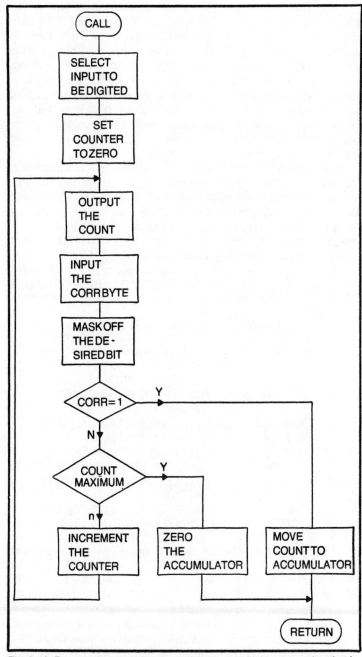

Fig. 8-19. Program flowchart for generating an eight-bit binary number that is proportional to any selected analog voltage level.

The system is dealing with just one of the CORR signals, however, so it is necessary to mask that one from the rest. If the selected CORR bit is equal to 1, the content of the counter is moved to the processor's accumulator where it remains while returning to the main program. If the selected CORR bit shows a logic 0, however, the next step is to see whether or not the counting operation has counted as far as it possibly can. The idea in this case is to take into account the possibility that something has gone wrong and that there is no valid analog voltage level available. Rather than locking the program into an endless loop and tying up the system forever, the scheme has provisions for returning to the main program with the accumulator carrying a zero.

If the counter is not at its maximum point, however, it is incremented. The new value is then sent to the output port. The system loops around, incrementing the counter and monitoring the appropriate CORR bit until one of two things happen: either the CORR bit goes to logic 1 (indicating the current count is a good representation of the value of the analog input voltage) or the counter runs to its maximum level of 11111111 without seeing a good match (indicating something is wrong).

The same flowchart and program can be applied to all eight analog inputs. Only the first operation, SELECT INPUT TO BE DIGITIZED, is different in each case. Here is a sample program written in 8080A/8085 mnemonics:

```
          MVI D, 02H        ;SELECT INPUT AI1
          MVI C, 00H        ;SET THE COUNTER TO ZERO
CMORE:    OUT 0             ;OUTPUT THE CURRENT COUNT
          IN 0              ;INPUT THE CORR BYTE
          ANA D             ;MASK OFF THE DESIRED BIT
          JNZ OKDON         ;CHECK FOR CORR EQUAL 1
          MOV A,C           ;MOVE COUNT TO ACCUMULATOR
          CPI  FFH          ;CHECK FOR MAXIMUM COUNT
          JZ BADON          ;JUMP IF COUNT IS MAXIMUM
          INR C             ;INCREMENT THE COUNTER
          JMP CMORE         ;JUMP BACK TO DO ANOTHER CYCLE
OKDON:    MOV  A,C          ;MOVE CURRENT COUNT TO AC-
                              CUMULATOR
          RET               ;RETURN TO MAIN PROGRAM
BADON:    XRA A             ;ZERO THE ACCUMULATOR
          RET               ;RETURN TO MAIN PROGRAM
```

All eight analog inputs can be digitized one at a time and in any desired order. Again, what is done with the results in the accumulator at the end of each digitizing operation is a matter for future consideration. The main point here is the tremendous reduction in complexity and number of parts used made possible by a microprocessor-oriented data formatting scheme.

If the circuit shown in Fig. 8-18 appears terribly complicated, remember that it is servicing eight sensory inputs and translating each of them into a precision eight-bit format. Such a task would be exceedingly complex without the help of a microprocessor.

Designing Some Internal Controls

9

I wish we could sit down with a cup of coffee and discuss your plans for building a parabot or robot. If we could do that, I would be in a much better position to help you work out the main circuit details. As it is, however, I have no way of knowing what you want your machine to do; I don't know what you need in the way of circuitry. So the best I can do here is paint a rather broad picture and illustrate the basic ideas with a few specific examples.

I have actually been doing that all along. The circuits I have suggested do not represent the only way to do things; in fact, some of the suggestions might not be appropriate at all for the project you are working on. Then, too, you might be able to come up with better circuits for the very examples I have cited. (Incidentally, I would appreciate hearing about such circuits. Just write to me, c/o TAB BOOKS, Inc., Blue Ridge Summit, PA 17214.)

I mention this point here because it is especially relevant to the subject of this chapter. More than anywhere else so far in this book, the matter of designing internal control mechanisms is wide open to diverse procedures and opinions. There might be a dozen good ways to do a particular job here, and perhaps another dozen not-so-good, but workable, approaches.

So I am going to suggest some approaches to planning and designing internal control mechanisms, then cite some specific examples. If you already know a lot about digital logic design, you are certainly free to express your own approaches and design habits. You are probably more comfortable with them, anyway.

This chapter is mainly intended for experimenters who are not very confident about their ability to design custom digital control circuits. If that is your situation, and if you find you still don't have the background necessary for understanding this material, try working your way through my book, *Handbook of Digital IC Applications*. Of course, other texts explain the basics, but this one uses the same general approaches suggested in this chapter.

THE GENERAL LOGIC DESIGN PROCEDURE

The first step in designing a logic control system is to define all the input and output variables, giving them names or labels that clearly distinguish them. If, for example, your system has four microswitches as touch sensors, you might define their states as logic 0=energized and logic 1=deenergized. Then label the signals, such as TS0, TS1, TS2 and TS3.

As far as the output signals are concerned, you might want to control the direction of a motor according to the pattern of input signals—the pattern of touch-sensor openings and closing. Maybe then you can define those motor outputs such that a logic 1 makes the motor run in a forward direction and a logic 0 makes it run in reverse. If you want the motor to stop, you will have to include a second signal for that purpose; maybe a logic 0 makes it run and a logic 1 makes it stop. Thus you have two output signals for the motor, and you can label them MD for motor direction, and MS for motor start/stop. Chances are good that your internal control mechanism will have more inputs and outputs than cited in the example, but the basic idea holds up.

The matter of defining and labeling input and output signals is complicated a bit whenever the control mechanism includes some flip-flops or timing circuits. In that case, the flip-flops and timers can be considered separate input/output devices. You can define and label all the input signals to those sections and then do the same for their outputs.

At the risk of sounding a bit academic, I should say that there are two basic kinds of digital circuits: *combinatorial logic* and *sequential logic* circuits. Combinatorial logic circuits perform the customary types of Boolean logic: AND, OR, NAND, NOR, Exclusive-OR, equality and invert operations. The sequential logic circuits are concerned with oscillators, timers, flip-flops and counters. In short, sequential logic circuits are time-dependent circuits, whereas combinatorial circuits are not.

It is quite easy to deal with combinatorial logic circuits in terms of truth tables and Boolean algebra. Every sequential logic circuit, on the other hand, must be considered in its own fashion, generally eluding precise truth-table and Boolean expression. That is why I suggest dealing with timers and flip-flops (sequential logic circuits) as separate I/0 circuits, although they are actually integral parts of your internal control system.

The second step in the general design procedure is to develop complete truth tables for the combinatorial logic circuits. You have already defined and labeled the inputs and outputs, and the truth table—a *complete* truth table—forces you to look at all possible input and output states. The truth table also provides the keys necessary for selecting the logic devices and working out their necessary interconnections.

The third step is to use the truth tables to rough out a preliminary combinatorial logic design. Coming up with the best possible design is a difficult matter. This is so not because the job is technically difficult, but because there is so much latitude with regard to the selection of devices, the complexity of the circuit and, ultimately, its cost and overall reliability. You will most likely find yourself working over that circuit a number of times, modifying it and respecifying parts until you are satisfied with the result.

The sequential logic *cannot* be designed from a truth table. Instead, you have to think out the sequence of operations you want, defining timing intervals and frequencies for yourself. This isn't a difficult task in principle, but it can be made a bit difficult because a number of different devices might be available for doing the job. Of course, some are going to be more suitable for your particular application than others are, and you will have to explore the marketplace and data books for the best compromise between function, complexity, cost and—don't forget—availability. The fourth step is to finalize the circuit design, gather the necessary parts and assemble them in a breadboard fashion.

Finally, it is time to test the circuit under as many ideal and nonideal conditions as possible. This is where any shortcomings or oversights in your design should become apparent. Rework the circuit and check it again before committing it to its final, circuit-board form.

Be sure to document the system; keep a notebook for all your design procedures and schematics. Such notes will prove vital for future expansion, modification and troubleshooting.

A PURELY COMBINATORIAL DESIGN

The model here is a little parabot that follows a white stripe on the floor. For the sake of convenience, the stripe is a strip of masking tape or ½-inch adhesive tape. In any case, the stripe is white.

As suggested in Fig. 9-1, the parabot has two independent drive wheels, ML and MR. The front wheel is a passive, idle wheel used only to balance the machine. To keep things simple, the drive motors run either at full-forward speed or not at all. There are no provisions for running either motor in reverse.

Elements L1, L2 and L3 are separate LED/phototransistor combinations. In each case, an LED shines a light down toward the floor, and its phototransistor picks up any of that light reflected back from the floor. Presumably, the only time when any significant amount of light is reflected back to the phototransistor is when it is situated over the light-colored stripe on the floor.

The overall objective of the system is to keep the parabot running with assembly L2 situated over the stripe. When the machine moves off

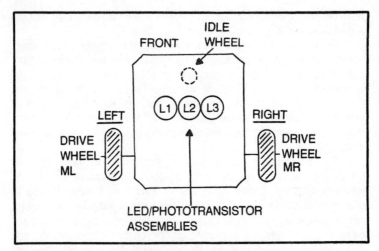

Fig. 9-1. Mechanical configuration for a tape-tracking parabot.

center, L1 or L3 will pick up reflections. It's the job of the internal control logic to stop one of the motors to realign the machine.

To get the design procedure started, suppose any one of the two LED/phototransistor assemblies generates a logic 0 whenever it is situated over the stripe on the floor, and generates a logic 1 level whenever it is off the stripe. Further, suppose either of the motors will run when it gets a logic-0 signal and will stop running whenever it gets a logic-1 signal.

That defines all the input and output states for the system. As far as the labels are concerned, just use those already specified in Fig. 9-1: L1, L2 and L3 for the input signals, and ML and MR for the output signals.

The next step is to set up a complete truth table for the system. A single truth table will do the job because there are no sequential logic operations specified here, although they could be useful for this sort of project.

The truth table should show three input signals under all possible combinations. There are always 2^n possible combinations, where n=the number of inputs. So in this case, there are eight possible logic combinations for the three inputs. Table 9-1 illustrates this point, as well as the very first step in setting up the truth table.

Before specifying the output states for each combination of inputs, it is necessary to specify the spacing of the LED/phototransistor assemblies relative to the width of the stripe on the floor. For the most reliable operation, the LED/phototransistor assemblies should be spaced in such a way that the stripe can fall under any two of them at the same time, but never all three. And there should be no gaps—spaces between the assemblies where the stripe can fall without energizing at least one of them. This is a critical point in the machine design. A more elaborate

Table 9-1. Step 1 in Devising a Combinatorial
Truth Table: Defining All Possible Input Conditions.

L1	L2	L3	ML	MR	MACHINE STATUS	CORRECTION
0	0	0				
0	0	1				
0	1	0				
0	1	1				
1	0	0				
1	0	1				
1	1	0				
1	1	1				

For L inputs: 0 = on the stripe   For M outputs: 1 = stop the motor
                1 = off the stripe   0 = run the motor

control system can overcome that tight specification, but we are trying to keep things simple at the control end of the operation.

The truth table is extended in Table 9-2, showing an interpretation of the position status of the machine under all eight possible combinations of sensory inputs. One of the important features of a complete truth table—one that considers all possible input states—is that it often exposes some operating conditions you might not have considered before.

The first line, for example, shows zeros from all three input assemblies. That means all three are seeing "whiteness." And in the context of the machine cited here, something must be terribly wrong. Perhaps the machine has moved off the stripe and onto a white floor. In any event, something has gone wrong.

Table 9-2. Step 2 in Devising A Combinatorial
Truth Table: Interpreting the Input Conditions.

L1	L2	L3	ML	MR	MACHINE STATUS	CORRECTION
0	0	0			ON WHITE FLOOR	
0	0	1			OFF TO RIGHT	
0	1	0			???	
0	1	1			OFF TO RIGHT	
1	0	0			OFF TO LEFT	
1	0	1			CENTERED	
1	1	0			OFF TO LEFT	
1	1	1			OFF THE STRIPE	

For L inputs: 0 = on the stripe   For M outputs 1 = stop the motor
              1 = off the stripe   0 = run the motor

**Table 9-3. Completing a Combinatorial Truth Table
by Defining the Outputs Under All Possible Input Conditions.**

L1	L2	L3	ML	MR	MACHINE STATUS	CORRECTION
0	0	0	1	1	ON WHITE FLOOR	STOP BOTH MOTORS
0	0	1	1	0	OFF TO RIGHT	STOP ML
0	1	0	1	1	???	STOP BOTH MOTORS
0	1	1	1	0	OFF TO RIGHT	STOP ML
1	0	0	0	1	OFF TO LEFT	STOP MR
1	0	1	0	0	CENTERED	RUN BOTH MOTORS
1	1	0	0	1	OFF TO LEFT	STOP MR
1	1	1	1	1	OFF THE STRIPE	STOP BOTH MOTORS

For L inputs: 0=on the stripe   For M outputs: 1=stop the motor
           1=off the stripe   0=run the motor

In the second line, where L3 is not picking up any light, the machine must be situated a bit to the right. It is even farther off to the right in the fourth line, where L1 is the only assembly seeing the stripe.

The third line in the truth table indicates a rather curious condition. Something is wrong with the setup when the two outside input assemblies can see a stripe, but the middle one cannot. It is difficult to imagine the machine ever encountering such a situation, so there isn't much point in trying to define it. A set of *question marks* suffices.

The bottom half of the truth table deals with situations where the machine is off to the left, centered or completely off the stripe. The CENTERED status, where the middle input assembly is the only one seeing the stripe, is the one the system always attempts to maintain. The status suggested by the last line in the truth table might not have occurred to you at first, but you must deal with it. That's the case where not one of the input assemblies is seeing the stripe; the machine is apparently off the stripe together.

So the eight possible input combinations are interpreted in light of this particular project. The next step is to suggest corrective action in each case and assign the corresponding logic levels to the ML and MR outputs. See the completed truth table in Table 9-3.

Looking over the listing of MACHINE STATUS, there are three undesirable, something-is-wrong conditions: ON WHITE FLOOR, ??? and OFF THE STRIPE. Perhaps the best corrective action in all three instances is to stop the machine altogether. That will give you a chance to pick up the machine and set things right. Also, the motors will stop running whenever you pick up the machine from the first floor. So in those three *default* conditions, specify a corrective action: STOP BOTH MOTOR . This means delivering a logic-1 signal to both motors.

Whenever the machine is OFF TO RIGHT, you can correct the matter by stopping the left motor, ML, as long as the situation persists.

That will leave MR running to correct the offset. That output condition is specified by ML=1, MR=0.

Whenever the machine veers OFF TO LEFT, the corrective action is to STOP MR. Specify outputs ML=0, MR=1.

That leaves just the ideal CENTERED condition. The motor response should be one where both motors are running: ML=0, MR=0.

The truth table is thus complete. You can certainly alter the corrective actions to suit your own system; there are some viable alternatives. Specifying different corrective actions will change the motor-response patterns of ones and zeros, but that does not change the basic process being illustrated here.

The next step is to come up with Boolean logic expressions for ML and MR. You will need two expressions, one for each motor, in terms of one or more of the input variables.

There are two basic steps involved in this matter of preparing Boolean expressions from a truth table. The first is to generate the equations directly from the table, and the second is to simplify the equations as much as possible (the simpler the equations, the simpler the circuits become).

To generate the preliminary Boolean equation for output ML, look down the ML column for logic-1 entries, then write an AND expression for the L inputs that are specified for each case, using a L when the input level is a logic 1, and a L for an input level of logic 0.

There are five conditions where ML=1. The input expressions are:

$$\overline{L1} \cdot \overline{L2} \cdot L3 \qquad \overline{L1} \cdot L2 \cdot L3 \cdot$$
$$\overline{L1} \cdot \overline{L2} \cdot L3 \cdot \qquad L1 \cdot L2 \cdot L3$$
$$\overline{L1} \cdot L2 \cdot \overline{L3}$$

Then OR those expressions to come up with a preliminary equation for ML:

$$ML = \overline{L1} \cdot \overline{L2} \cdot \overline{L3} + \overline{L1} \cdot \overline{L2} \cdot L3 + \overline{L1} \cdot L2 \cdot \overline{L3} + L1 \cdot L2 \cdot \overline{L3} + L1 \cdot L2 \cdot L3$$

This is a pretty terrible equation, but you can apply some Boolean identities to simplify it. If you are weak on this point, you must pick up the necessary knowledge from another source. At any rate, the equation can be simplified to this form:

$$ML = \overline{L1} + L2 \cdot L3 \qquad \qquad \textbf{Equation 9-1}$$

The right motor output variable, MR, also has five input conditions that set it to logic 1:

$$\overline{L1} \cdot \overline{L2} \cdot \overline{L3} \qquad L1 \cdot L2 \cdot \overline{L3}$$
$$\overline{L1} \cdot L2 \cdot \overline{L3} \qquad L1 \ L2 \ L3$$
$$L1 \cdot \overline{L2} \cdot \overline{L3}$$

OR-ing these together to get a complete expression for MR;

$$MR = L1 \cdot L2 \cdot L3 + L1 \cdot L2 \cdot L3 + L1 \cdot L2 \cdot L3 + L1 \cdot L2 \cdot L3 + L1 \cdot L2 \cdot L3$$

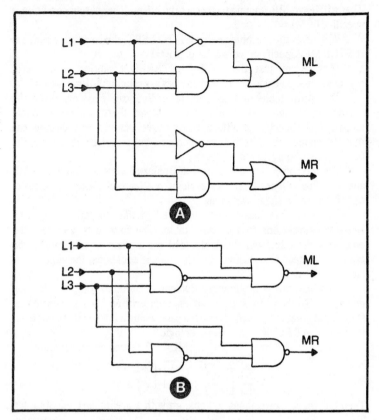

Fig. 9-2. Logic circuitry for the tape-tracking parabot. (A) preliminary design. (B) Finished, practical design.

And that simplifies to this form:

$$MR=\overline{L3}+L1 \cdot L2$$ **Equation 9-2**

With the simplified equations at hand, it is possible to devise a logic circuit that will perform the operations indicated by those equations. Figure 9-2A shows a preliminary design. The only trouble is that it calls for three different kinds of gates and thus three separate IC devices.

Taking advantage of the Boolean-equivalent properties of inverting gates, however, it is possible to do the same job with just four NAND gates and a single quad NAND-gate IC package. You can't get a control circuit much simpler than that. See Fig. 9-2B.

The logic design task is done. All that remains is to breadboard it, apply the patterns of ones and zeros indicated on the original truth table, and monitor the two outputs. The results should match those specified on the table.

Of course you will need some input interfacing circuits for the L signals and some power amplifiers for the motors. But all things considered, the proper approach to designing the logic turns out a very simple circuit.

AN INTERNAL LOGIC DESIGN WITH SOME SEQUENTIAL OPERATIONS

Figure 9-3 shows a top view of a very common sort of parabot/robot machine. It is driven and steered with two independently operated drive/steer motors and wheels, LM and RM. Two idle wheels ensure good machine balance under normal running conditions.

The design problem here is to create a logic scheme for working the machine out of a contact situation—one sensed by wire bumpers FS, RS, BS and LS. Whenever the machine makes contact with an obstacle such that the front bumper switch is energized, it should respond with some sort of reverse motion at the drive/steer motors. The same general idea applies to the other three bumpers and switches and, indeed, any and all possible combinations of two or more bumper switch contacts.

While this situation offers plenty of opportunities for doing some combinatorial logic design, the nature of the thing makes it necessary to work with some *sequential logic* as well. That is the point of this particular example.

Some Preliminary Combinatorial Logic Design

The design operation must begin with some clear-cut definitions of how the bumper switches and drive/steer motors work. Table 9-4A, for instance, shows how either of the drive/steer motors responds to some control signals applied to them.

Table 9-4A assumes you are using some sort of motor drive amplifier and logic (see Chapter 5). In any event, applying a pair of zeros to the

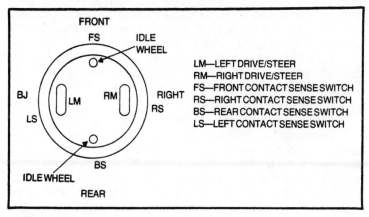

FRONT
FS   IDLE
       WHEEL

BJ
       LM   RM   RIGHT
LS   RS

BS
IDLE WHEEL
REAR

LM—LEFT DRIVE/STEER
RM—RIGHT DRIVE/STEER
FS—FRONT CONTACT SENSE SWITCH
RS—RIGHT CONTACT SENSE SWITCH
BS—REAR CONTACT SENSE SWITCH
LS—LEFT CONTACT SENSE SWITCH

Fig. 9-3. Mechanical configuration for a simple parabot machine.

Motor Signal		Motor Response
MF	MR	
0	0	STOP
0	1	REVERSE
1	0	FORWARD
1	1	STOP

Table 9-4. Tables Defining the Output Conditions for a Simple Contact-Sensing Machine.

Motor Signal				Machine Response
LMF	LMR	RMF	RMR	
0	0	0	0	STOP*
0	0	0	1	REVERSE, LEFT*
0	0	1	0	FORWARD, LEFT*
0	0	1	1	STOP
0	1	0	0	REVERSE, RIGHT*
0	1	0	1	REVERSE*
0	1	1	0	CCW SPIN*
0	1	1	1	REVERSE, RIGHT
1	0	0	0	FORWARD, RIGHT*
1	0	0	1	CW SPIN*
1	0	1	0	FORWARD*
1	0	1	1	FORWARD, RIGHT
1	1	0	0	STOP
1	1	0	1	REVERSE, LEFT
1	1	1	0	FORWARD, LEFT
1	1	1	1	STOP

1 = Energized
0 = Deenergized

B

motor circuit causes the motor to stop turning—so does applying a pair of ones. Setting MR=1 and ML=0, however, causes the motor to turn in reverse, while setting up the opposite state (applying a logic 0 to MR and a logic 1 to ML) causes the motor to turn in its forward direction. The table applies equally well to both of the drive/steer motors.

Table 9-4B extends this basic information, making it applicable to both motors at the same time and relating it to the response of the machine. Whenever the left and right motors are both seeing pairs of zeros, for example, the machine stops. That pattern of control signals to the motors stops the creature dead in its tracks.

The second line in Table 9-4B suggests a situation where the left motor is stopped, but the right motor is turning in a reverse direction. According to the machine-response comments, this makes the machine move in reverse with a right turn.

Table 9-4B summarizes all 16 possible combinations of zeros and ones to the motor control circuit, and it indicates the machine response in each case. There are a lot of redundant responses: four different STOP

Table 9-5. Complete Combinatorial Logic Truth Table for the Contact-Sensing Machine.

Contact Situation				Motor Response				Contact Status	Machine Response
FS	RS	BS	LS	LMF	LMR	RMF	RMR		
0	0	0	0	0	1	0	1	FREE	FORWARD
0	0	0	1	1	0	0	0	LEFT	FORWARD, RIGHT
0	0	1	0	0	1	1	1	REAR	FORWARD
0	0	1	1	1	0	1	1	REAR, LEFT	FORWARD, RIGHT
0	1	0	0	0	0	1	0	RIGHT	FORWARD, LEFT
0	1	0	1	0	1	0	1	RIGHT, LEFT	FORWARD
0	1	1	0	0	0	1	0	RIGHT, REAR	FORWARD, LEFT
0	1	1	1	0	1	0	1	RIGHT, REAR, LEFT	FORWARD
1	0	0	0	0	1	0	1	FRONT	REVERSE
1	0	0	1	0	1	0	0	FRONT, LEFT	REVERSE, RIGHT
1	0	1	0	1	0	0	1	FRONT, REAR	CW SPIN
1	0	1	1	1	0	0	1	FRONT, REAR, LEFT	CW SPIN
1	1	0	0	0	0	0	1	FRONT, RIGHT	REVERSE, LEFT
1	1	0	1	0	1	0	1	FRONT, RIGHT, LEFT	REVERSE
1	1	1	0	0	1	1	0	FRONT, RIGHT, REAR	CCW SPIN
1	1	1	1	0	0	0	1	FULLY TRAPPED	STOP

conditions, two REVERSE LEFT, two FORWARD LEFT, etc. The comments suffixed with an asterisk are those selected to represent the entire family of possible responses. That selection process reduces the number of motor signal inputs from 19 to nine without reducing the actual number of machine responses. So later in the design procedure, a bumper contact situation calling for a REVERSE, LEFT response will be handled by outputting ·0001—the REVERSE, LEFT response marked with an asterisk.

It is now time to look at all 16 possible bumper-switch contact situations and specify an appropriate motor response signal for each one of them. Table 9-5 provides one possible set of machine responses for all 16 possible combinations of contacts with obstacles. Some of the responses are purely arbitrary, and you might be able to come up with some responses that seem more appropriate than those suggested here. Generally speaking, I have chosen to make the machine respond to contact situations by executing a complementary sort of machine response. If, for example, the machine makes contact at its REAR and LEFT bumpers, it responds by moving FORWARD and to the RIGHT. This is a rather straightforward approach, but not necessarily the only good one.

As far as the combinatorial logic design is concerned, we need a "black box" that accepts the four different bumper switch signals (FS, RS, BS and LS) at its inputs and delivers the four motor response signals (LMF, LMR, RMF and RMR) to its outputs. In principle, at least, it is a logic design situation that is not much different from the one illustrated for the tape-tracking machine in the previous section of this chapter.

To this point, you have worked through the first two steps in the basic design procedure: first, defining and labeling terms, and then working out a logic truth table. The next step suggested in the basic procedure is to come up with a set of Boolean logic expressions for each of the four motor response outputs—all in terms of the four bumper switch inputs.

Trying to work out a set of Boolean expressions for this particular truth table, however, turns out to be a ghastly task. Things simply don't work out as nicely as they did for the tape-tracking machine.

If your skill with Boolean algebra is quite good, but you find the equations getting out of hand, that's a sure sign it is time to resort to an entirely different sort of combinatorial logic design. If you are considering using *Karnough maps* ("What are they?" some of you might ask), forget about them; things won't work out much better in the long run.

Wouldn't it be nice if you had a small, permanently programmable memory for doing this combinatorial logic job? Wouldn't it be nice if you could address that memory with the bumper switch input signals and immediately see the appropriate motor responses at the output of such a circuit? You can do that with a *digital multiplexer circuit*. A digital multiplexer can be wired to behave as though it is a ROM (read-only memory).

We haven't the space to go into great detail on this matter here. I can,

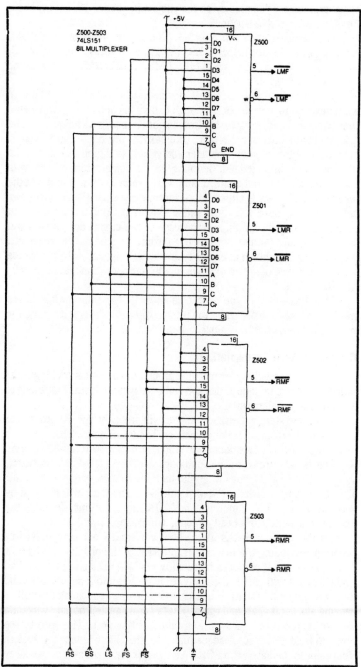

Fig. 9-4. Combinatorial logic circuit for satisfying the truth table of Table 9-5.

however, suggest you read up on the subject in Section 12.2 of my book, *Handbook of Digital IC Applications.*

The result is shown in Fig. 9-4. The circuit, working as a ROM, accepts switch inputs RS, BS, LS, FS and an inverted version of FS, labeled $\overline{\text{FS}}$. The outputs are LMF, LMR, RMF, RMR and inverted versions of all four.

The job is done with just four IC devices: ordinary 74LS151 8:1 digital multiplexers. Anyone stubborn enough to fight his or her way through the combinatorial logic design using Boolean or Karnough techniques will be rewarded with a circuit much more complex that this one—more complex in that it requires a greater number and variety of digital IC devices. So what is the point? Indeed, learning to do combinatorial logic with multiplexers is worth the effort. And what's more, you will find that the responses can be easily altered at a later time if you use the circuit shown in Fig. 9-4. Suppose you aren't satisfied with just one bit in the output response. Using the traditional approach to the circuit design, you will have to start all over and, perhaps even worse, rebuild the entire circuit. Using the circuit shown in Fig. 9-4, however, you need only change the connection to one of the multiplexer inputs—an operation that might take only 10 minutes.

That takes care of the main combinatorial logic design job. If you aren't convinced that is true, you'll have to work through it, using the multiplexer-as-a-ROM technique.

The Need for Some Sequential Logic

Suppose you build your machine around the combinatorial logic just suggested. If you let it go at that and build the system, it won't take long to find a serious problem with it.

When you start up the machine, you will probably have it setting near the middle of a room not making contact with any obstacles. That means all four bumper switches are deenergized and the logic circuit tells the motors to run such that the machine moves along in a FORWARD direction. So far, so good. But what happens when the machine runs headlong into a wall? The front bumper switch, FS, will energize. The logic will respond by generating a REVERSE motion. That's OK—up to a point. Certainly the machine responds to this head-on contact in an appropriate fashion.

As the machine backs away, however, that front bumper switch deenergizes, telling the machine it is FREE of the obstacles. And how does it respond? It responds by running FORWARD right into that wall again. Then it will REVERSE away, FORWARD into it again—back and forth without end. In short, the machine will simply respond to the wall by banging against it again and again. That's hardly an appropriate way for a parabot or robot to deal with a problem of this sort. The cure to the problem is to insert some sequential logic that will prevent the system from going immediately to a FORWARD motor command as it moves away from a contact situation.

304

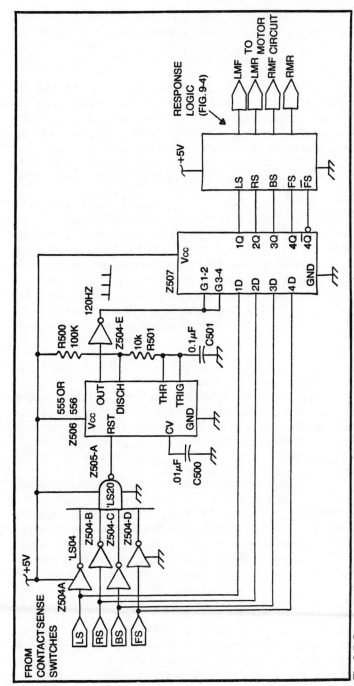

Fig. 9-5. Sequential logic circuit for a machine that will run in a direction dictated by the nature of its most recent contact situation.

305

There are two different ways to handle the problem. One is to work out the sequential logic so that the machine does not change its pattern of motion whenever it is FREE from a contact. The other is to have the machine execute a response for a preset period of time before returning to its FORWARD motion.

In the first case, the machine will respond to a front-only contact by moving in REVERSE until it makes contact with another object. If that second contact happens to be a REAR, LEFT sort of contact, the machine will respond by doing a FORWARD, RIGHT motion and continue executing that motion until it contacts another obstacle.

The machine ends up doing some nonforward motions most of the time. It is just as likely to move across the room with a wide, REVERSE, LEFT motion as it is to move with a straight FORWARD motion.

Table 9-5 still holds, and the combinatorial logic circuit in Fig. 9-4 is still good. The only real difference is that the FREE contact status does not automatically deliver a FORWARD response. Instead, the machine executes the response it called just before the FREE status occurs. If that sounds like a flip-flop, or latching, operation to you, you are on the right track. It is a good example of sequential logic.

But maybe you don't like the idea of having the machine spending so much of its time moving in directions other than straight ahead. In that case, you can choose the second option: having the machine execute its prescribed response for a fixed period of time, then returning to straight FORWARD when that time interval is over. The idea is to give the machine some time to get away from the obstacle that called for the prescribed response in the first place.

Figure 9-5 suggests a solution to the problem of making the machine move away from a contact situation and continue moving in that fashion until it runs into something else. Using this particular scheme, it is rather easy to modify it to suit the second option: making the machine return to its FORWARD motion after being FREE for a prescribed period of time. This option is discussed in connection with Fig. 9-6.

In either case, the system must have some provision for sensing a contact—any sort of contact. And in principle, this can be done with a four-input OR gate. The output of such a gate will be at logic 0 whenever there are not contacts at the bumper switches, but it will rise to logic 1 if any one, or any combination of two or more, bumper contacts are made.

While a four-input OR gate is available in the TTL family of digital devices, it isn't exactly the most popular and easy-to-find device. So we are going to specify a Boolean equivalent: a four-input NAND gate preceded by logic inverters at each input. That setup does the job of a four-input OR gate. See Z505-A and its input inverters in Figs. 9-5 and 9-6.

So far, things have been fairly straightforward, but here is where you have to exercise some caution and careful thinking. The basic idea is to load a latch circuit, Z507, whenever a contact situation occurs. The latch is loaded with the pattern of zeros and ones that represent the nature of the

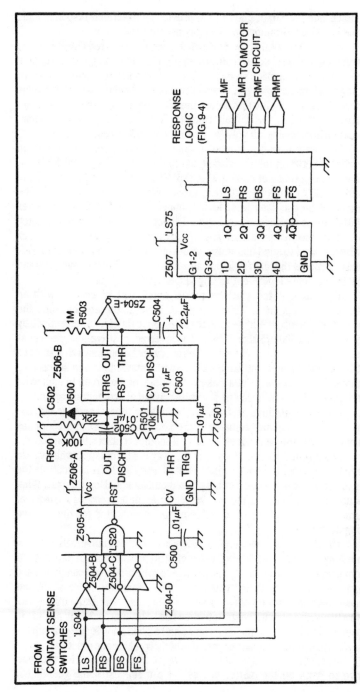

Fig. 9-6. Sequential logic circuit for a machine that returns to FORWARD motion after a prescribed of time.

contact, and its output goes to the previously described response logic circuit to generate an appropriate response from the motors.

If the output of the OR-type sensing circuit were fed directly to the gating input of the latch circuit, the system would respond properly only when the machine made an initial contact with some obstacle. There are going to be times, however, when a response to a contact situation brings up another contact situation before the machine has a chance to get to its FREE state. The OR circuit cannot distinguish one sort of contact situation from another: it only senses that a contact has occurred.

One way around this troublesome state of affairs is to insert an oscillator between the OR circuit and the gate connection on the latch. That is the purpose of Z506 in Fig. 9-5. The oscillator is held off by the output of Z505-A as long as there is no contact at the bumper switches. The output of the oscillator is at logic 0 under that condition, and inverter Z504-E turns that around to a logic-1 level. This level opens the latch circuit to accept and output any bit pattern appearing at the bumper switches. Bear in mind this status occurs only while the machine is FREE. And that means the latch is seeing all zeros from the bumper switches and that the response logic is outputting a FORWARD response.

The moment the machine makes some sort of contact with an obstacle, the output of Z505-A goes to logic 1, thereby enabling the oscillator. The oscillator is set up to run at about 120 Hz, as determined by the values of R500, R501 and C501. The inverted, positive-going pulses at the gating input of the latch circuit thus query the bumper switch status at a 120-Hz rate, loading and latching the switch status between successive pulses. This latch-pulse idea makes it possible for the circuit to update its response codes as the machine works its way out of a contact situation.

Once the machine is free, the oscillator stops running, but the last-executed response remains in the latch. So the machine continues running that last-executed response until another contact situation occurs.

Laying out the circuit for you in this way is a bit unfair in what we are trying to accomplish with this discussion. The main idea is to illustrate the importance of setting up a carefully designed sequential logic circuit. What is not reflected here is the number of different approaches *I tried* that turned out to be either unworkable or too complicated for the relative simplicity of the task.

Unfortunately, there are no widely accepted and universally applicable approaches to designing sequential logic circuits. In the case of combinatorial logic circuits, it is possible to specify all input conditions and their corresponding output conditions, then crank out expressions that make the two work together. That isn't the case with sequential logic. Nearly every sequential logic situation has to be approached in an unique fashion, and it takes some experience and a good knowledge of available devices to come up with a design that is both workable and practical.

I cannot provide that kind of experience and knowledge for you; you have to work on the matter yourself. I can suggest some guidelines, however. First, don't be reluctant to abandon an approach that seems to run into a dead end. The most obvious approaches are not always the best. You often have to loosen up your thinking and apply a bit of creativity and imagination to get things going your way.

Second, it is a good time to look at a different approach if you find yourself having to add more devices again and again. When the circuit for a fairly simple task seems to be growing in complexity and getting out of hand, it is time to back away and try looking at the matter from a different point of view.

Third, try to avoid tricky little schemes that use devices in ways they were never intended to be used. It is certainly possible to use a CMOS inverter as a low-power audio amplifier, but the limits on its performance make it a questionable idea for serious projects. Leave such tricks to the simple project books.

And finally, develop a knack for testing the operation of the circuit on paper. You can eliminate unworkable circuits and avoid the time and expense invested in building prototypes that are destined to fail from the beginning.

Bearing in mind these general guidelines, look over the circuit of Fig. 9-6. This is the circuit that lets the machine execute its response for a prescribed period of time, outputting a FORWARD response when that time interval is over. The circuit is practically identical to the one shown in Fig. 9-5. It uses the same basic approach, but with one additional feature: the timer, Z506-B.

The timer is set up as a retriggerable 3-second timer. Whenever the machine makes some sort of contact with an obstacle, the train of 120-Hz pulses from Z506-A triggers the timer. Since that timer is retriggerable, though, its output looks exactly like that of the oscillator: a series of pulses that operate the latch circuit.

The big difference comes about when the machine frees itself from a contact situation. The oscillator stops running as in the previous circuit, but here the timer continues working out its timing interval for about 3 seconds, a time determined by the value of R503 and C504.

During this time interval, the latch circuit holds the bumper switch pattern it had just prior to the machine getting free of the obstacle. The machine thus responds for 3 seconds as though it is still in that particular contact situation.

At the end of the timing interval, assuming the machine is still FREE, the output of the timer drops to logic 0, the inverter (Z504-E) changes that to a logic 1, and the latch accepts any bit pattern present at the bumper switches. And since the machine is FREE at this point, it figures that the bumper switch pattern is all ones. The response logic generates a motor code for FORWARD, and the machine begins running straight ahead. The next contact situation begins the cycle all over again.

Microprocessors for Parabot/Robot Machines

10

Perhaps the foremost reason for using a microprocessor in a parabot or robot project is to simplify the hardware and, at the same time, give the system a high degree of flexibility that lets you change your mind about things and expand the machine later on. Even the complexity of parabot machines can get out of hand when attempting to monitor a lot of sensory parameters and make a number of output responses.

As discussed in some detail in Chapter 2, a microprocessor cannot be justified solely on the basis of a saving in initial setup costs. While a certain microprocessor IC might sell for less than $20, the cost of supporting memories and buffers can cost 10 times that much. So count on saving time, and not money, by using a microprocessor from the start. The real benefits of lower cost and flexibility generally show up later in the project.

The main purpose of this chapter is to introduce the hardware elements of microprocessor systems in the context of parabot and robot machines. Much of what remains in the book after this chapter deals with software.

The discussions that follow are quite general in some respects but rather specific in others. It is important that you understand some of the criteria for assembling this chapter so that you can get the most from it and not become discouraged or disappointed before you are finished with it.

There is no practical way to describe in detail the finer points of every microprocessor chip in a book such as this one. Such descriptions fill many volumes of technical literature already, and it is up to you to dig into that material as far as necessary to understand what is going on here.

THE BASIC MICROPROCESSOR REQUIREMENTS

The material in this chapter picks up at a point where some microprocessor system—the basic elements—is already built and running. Whether you build such a system on your own and from scratch, or acquire one of the popular one-board computers is up to you. It is assumed you have a basic system that includes a *ROM monitor* and *bootstrap loader*—things absolutely necessary for getting a microprocessor up and running. Furthermore, the basic system should have at least 1K of programmable RAM as well as some provision for programming that RAM: toggle switches, a hexadecimal keypad or a full-blown alphanumeric keyboard.

In short, the assumption is that you have access to a little system that generates a 16-bit address bus, at least an eight-bit bidirectional data bus and some control signals for controlling peripheral devices. That might be a KIM, SYM, Z-80 starter kit, or some other sort of one-board computer. And for the benefit of home computer buffs, a system bus or expansion connector provides the necessary signals for doing any of the projects in this chapter.

Figure 10-1 shows the sort of microprocessor scheme needed to begin work in this chapter. How you go about getting that scheme is up to you. The discussions here are general enough to cover most of these systems that are available today.

You need that 16-bit address bus. Those points on your unit are most often labeled A0 through A15. And you will need access to a birdirectional (two-way) data bus, usually carrying labels such as D0 through D7. A 16-bit data bus (D0 through D15) will work even better in some instances, although all the examples in this book are for the older eight-bit data bus systems.

Finally, you must have access to some control lines. The designations in this case are only partly standard through the industry, and you might come across some variations. In Fig. 10-1, these control lines are labeled $\overline{\text{MEMR}}$, $\overline{\text{MEMW}}$, $\overline{\text{IOR}}$ and $\overline{\text{IOW}}$. In the event you might have to translate these designations to suit a different sort of system, here is a rundown of what they mean:

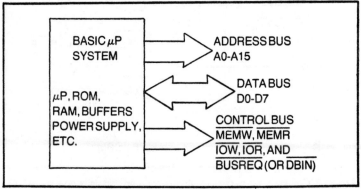

Fig. 10-1. Bus connections required from an available microprocessor system.

$\overline{\text{MEMR}}$—a control line that drops to logic 0 whenever the microprocessor is to accept data from an external memory device or memory-mapped input port.

$\overline{\text{MEMW}}$—a control line that drops to logic 0 whenever the microprocessor is to write data into an external memory device or memory-mapped output port.

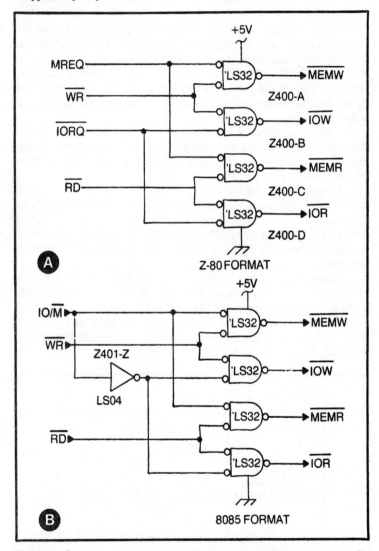

Fig. 10-2. Obtaining the necessary signals for read and write operations. (A) From a system similar to the Zilog Z-80. (B) From a system similar to the Intel 8085.

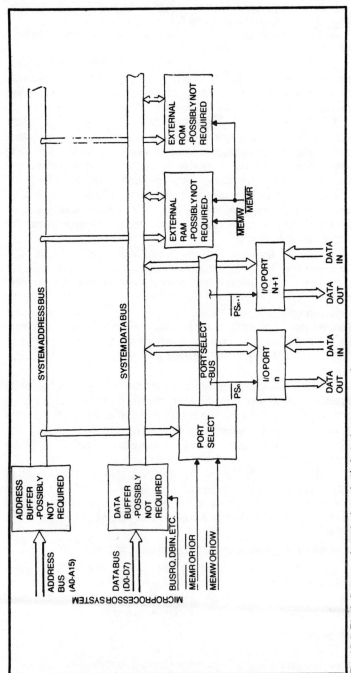

Fig. 10-3. Block diagram of a typical parabot/robot microprocessor bus system.

$\overline{\text{IOR}}$—a control line that drops to logic 0 whenever the microprocessor is to accept data from an input port.

$\overline{\text{IOW}}$—a control line that drops to logic 0 whenever the microprocessor is to deliver data to an output port.

You will also need access to a special control signal that is labeled something such as $\overline{\text{BUSREQ}}$ or $\overline{\text{DBIN}}$, depending on the microprocessor device you are using. This is a more primitive sort of signal that usually has a direct connection to the microprocessor IC device.

Like the $\overline{\text{MEMR}}$ and $\overline{\text{IOR}}$ control signals, this one goes to logic 0 whenever the microprocessor is expecting to see data from the system data bus. The internal timing is different in this case, however. And as far as any add-on, custom circuitry is concerned, the timing is used only for controlling the direction of data flowing through some data-bus buffers.

Some bare-bone microprocessor systems do not develop the control lines to the extent suggested in Fig. 10-1. That being the case, Fig. 10-2 shows how the four main control signals can be developed from the microprocessor IC, itself.

The circuit shown in Fig. 10-2A applies specifically to the Z80 microprocessor, but it is actually representative of several different devices. The $\overline{\text{MREQ}}$ terminal drops to logic 0 whenever the microprocessor is supposed to exchange information with a memory device or memory-mapped I/0 port. $\overline{\text{IORQ}}$, on the other hand, drops to logic 0 whenever the processor is expecting data from an I/0-mapped I/0 port. The $\overline{\text{WR}}$ and $\overline{\text{RD}}$ signals drop to logic 0 whenever the processor is to WRITE or READ information, whether to or from the memory or some I/0 port.

Figure 10-2B is representative of the Intel 8085 system, but it also applies to several others. In this case, the $\overline{\text{IO/M}}$ line is at logic 1 when communicating with an I/0-mapped I/0 port and drops to logic 0 when communicating with a memory device or memory-mapped I/0 port. The WR and RD signals go to logic 0 to indicating a WRITE or READ operation.

Your selection of a basic microprocessor system might include other control lines, but these are the only ones specifically mentioned in this book. Using any other control signals that might be available is up to the experimenter's own understanding of what they do and how they are to be applied.

PLANNING THE CUSTOM MICROPROCESSOR BUS SYSTEM

Given the basic microprocessor circuitry just described, Fig. 10-3 blocks out the custom hardware scheme. This is a very general arrangement that is applicable to just about any sort of parabot/robot experiment, and you might not need some of the sections.

Address and Data Buffers

The purpose of the address and data buffers is to beef up the current-drive capability of those buses. Generally speaking, each one of the bus lines is capable of driving about 10 LS-TTL inputs. Your one-board computer—the basic microprocessor circuit—might have three or four devices already connected to the bus lines, leaving you with six or seven more for the custom portion of the system. That would be quite adequate for most small systems, but you should give serious consideration to adding the buffers under a couple of conditions.

First, you should use the address and data buffers if you have any plans for future expansion of the system. It is easier in the long run to add the buffers now than to worry about fitting them into the circuit later.

And second, you should insert the buffers if you are going to operate the custom portion of the circuit from the expansion port of a commercial

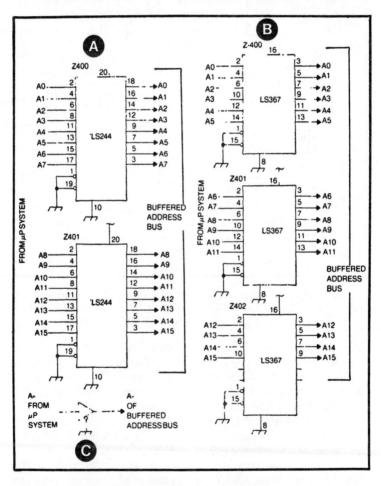

Fig. 10-4. Address buffers. (A) Using the 74LS244 octal buffer. (B) Using the 74LS367 hex buffer. (C) How the buffers are connected for either circuit.

315

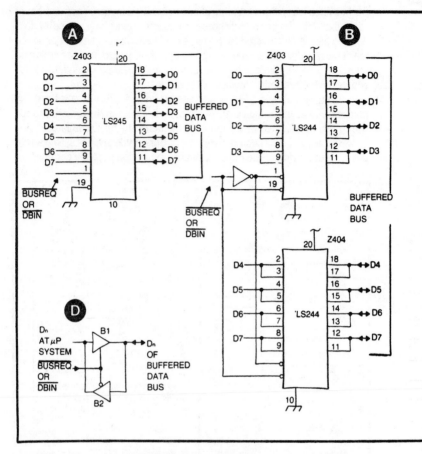

Fig. 10-5. Data bus buffers. (A) Using the 74LS245 eight-bit, bidirectional bus transceiver. (B) Using the 74LS244 octal buffer. (C) Using the 74LS367 hex buffer. (D) How the buffers are connected for any of the three versions.

home computer. Even if the technical manual for the computer lists an adequate drive capability for the address and data buses, it's a good idea to isolate the custom circuitry from a home computer system. To put it bluntly, this buffer-type isolation protects the expensive computer in the event of serious foulups in the circuits you must build yourself.

The diagrams in Fig. 10-4 suggest two types of address bus buffers. In both cases, the devices are simply noninverting, tri-state buffers that are always enabled. See the one-unit example in Fig. 10-4C.

The only difference between the two circuits is the designation for the ICs. The circuit in Fig. 10-4A requires only two IC devices. These are 74LS244 octal buffers, but in many cases, you have to pay for the relative simplicity: The 'LS244 device is usually more expensive and sometimes

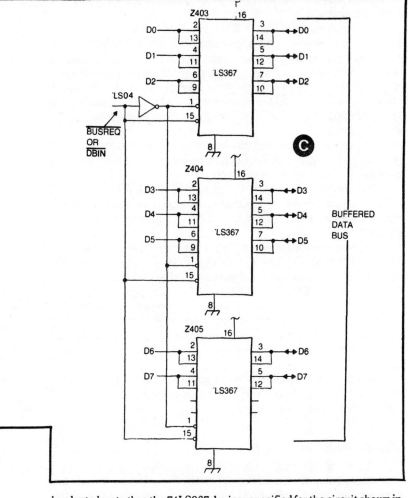

harder to locate than the 74LS367 devices specified for the circuit shown in Fig. 10-4B.

The 'LS367 devices listed in the second version of the circuit are hex buffers. Because there are only six of them in each package, you need three devices to do the same job. The lower cost and better availability might make the extra effort required for building this version somewhat more attractive to you.

Figure 10-5 suggests three versions of an eight-bit data bus buffer. Handling data-bus information is somewhat more involved because the data has to flow in both directions, both from the microprocessor to the custom circuitry and from the custom circuitry to the microprocessor. So there are still 16 buffers involved, but they are enabled in groups of eight.

317

This means you must have some provision for "shifting gears" for input and output data operations.

The circuit shown in Fig. 10-5D illustrates how this bidirectional bus scheme works. Whenever the microporcessor's $\overline{BISREQ}$ or $\overline{DBIN}$ signal goes to ligic 0, the processor ie expecting to receive date from the outside world. That "gear-shifting" signal enables buffer B2 and disables B1 so dat aflows only from custom circuitry to the microprocessor system. At any other time, while $\overline{BUSREQ}$ or $\overline{DBIN}$ is at logic 1, buffer B1 is enabled and B2 is disabled, thus allowing the microprocessor to send data to the external circuitry.

All three of the suggested data-bus buffer circuits do that same job in about the same way. The only difference between the circuits is the particular IC device specified for the job.

The very simple version shown in Fig. 10-5A uses a 74LS245 eight-bit bus transceiver. This is a relatively expensive device that can be hard to locate at times, but it's difficult to beat this version in terms of simplicity.

The second circuit uses two 'LS244 devices for doing the same job. These are the octal buffers already described in connection with the address buffer. The idea is to wire pairs of these buffers in inverse parallel (output to input) and enable one buffer in each pair for input operations and enable the complementary one in each pair for output operations. The purpose of the logic inverter circuit is to separate the two enabling operations. The data buffer suggested shown in Fig. 10-5C is far less expensive than the other two, but of course it takes more fussing around to build it.

All the address and data buffers suggested in these drawings work equally well. It is up to you to select the ones you want to use, trading off cost, time and complexity as you see fit.

Setting Up the Port Select Circuit

The overall microprocessor system communicates with the rest of the parabot/robot system through I/0 ports. Sensory input signals that have been previously formatted into a binary scheme are all applied to input ports. Microprocessor signals intended to make the parabot or robot do something are all applied to output ports.

The number of I/0 ports required for the job depends, of course, on the complexity of your system. The ports described in this book are all eight-bit ports. Each can handle eight bits of input and eight bits of output information. If your own design calls for working with more than eight input and output bits, you will have to count on using more than one I/0 port. Here is where the advantages of a 16-bit processor can pay off.

If you have any ideas that might call for using more than one I/0 port, you must include a port select circuit as indicated in Fig. 10-3. And as indicated in the block diagram of Fig. 10-3, the port-selection task is carried out by looking at a pattern of binary numbers from the system address bus.

There are two distinctly different ways to go about selecting I/O ports. One is commonly known as an I/O-mapped process, and the second is called a memory-mapped I/O selection process. The following discussion deals with both procedures, and it is up to you to select the one you like better. Do not make the decision until you understanding everything involved, because you are going to have to live with the decision through the remainder of your work with the system.

I/O-mapped I/O is conceptually simpler and usually easier to implement. Its primary disadvantage is that it runs more slowly than its memory-mapped counterpart.

Look through the instruction set for your microprocessor, and you will find a pair of instructions having mnemonics that read something like IN *data* and OUT *data*. These instructions are used exclusively for I/O-mapped port operations.

An instruction such as IN Ø55, for instance, tells the microprocessor to accept data from input port No. 5. An OUT Ø5, by the same token, tells the microprocessor to load data into the output port designated No. 5. So assuming you have a port select circuit that is capable of decoding a binary version of the number 5 at the address bus and an I/O port that responds to that selection process, the microprocessor communicates with the outside world at port 5 by simply writing IN Ø5 or OUT Ø5 instructions at the appropriate places in your program.

Whenever your program includes an IN *data* or OUT *data* instruction, the microprocessor places that *data* (a binary version of the port number) onto the lower eight bits of the address bus. Figure 10-6A shows how the binary port number is decoded into a 1-of-8 port-select signals.

Suppose the system is executing an IN Ø5 or OUT Ø5 instruction. That being the case, the lower eight bits of the address bus (bits A0 through A7) show a 5 in binary: 0000 0101. And as far as the circuit in Fig. 10-6A is concerned, output $\overline{PS5}$ drops to logic 0, while the other outputs remain at logic 1. The device, in other words, selects output port No. 5.

The circuit, as shown, responds to IN *data* and OUT *data* program instructions, where *data* is any number (in hexadecimal) from ØØ to Ø7. The I/O-mapped port-select scheme can actually handle 256 different port designations: ØØ through FF. Few systems require that many I/O ports, however, and experimenters tend to use the lower-numbered ports. If you need more than the eight ports that are selectable by the circuit in Fig. 10-6A, consult a good text or databook on the subject to see how two or more of these 'LS138 devices can be cascaded to increase the number of available I/O ports in groups of eight.

Before comparing this I/O-mapped scheme with its memory-mapped counterpart, it is important to note that the $\overline{IOR}$ and $\overline{IOW}$ signals described earlier in this chapter are also closely related to the I/O-mapped scheme. It so happens that the port that is selected by a circuit such as the one shown in Fig. 10-6A must be set for inputting or outputting data-bus information according the IN *data* or OUT *data* instruction that is being executed at the moment.

The I/O port select circuit shown in Fig. 10-6A, you see, merely selects the port that is to be enabled. That circuit has nothing at all to do with telling whether the data should go from the bus to the outside world or from the outside world to the data bus. That is the task of the $\overline{IOR}$ and $\overline{IOW}$ signals.

So two things actually happen when the microprocessor executes an IN *data* or OUT *data* instruction. First, the port number designated by the *data* appears on the lower eight bits of the address bus. The port select circuit then enables the appropriate I/O port. Then that port is set up to send data in a direction determined by $\overline{IOR}$ and $\overline{IOW}$.

If the system is executing an IN *data* instruction, the $\overline{IOR}$ signal drops to logic 0 to *read* information from the I/O port. But if the system is executing an OUT *data* instruction, it is the $\overline{IOW}$ signal that drops to 0, thereby letting the microprocessor *write* data-bus information into the designated I/O port.

The following list of comments summarizes the essential character of I/O-mapped I/O operations:

☐ The operations are programmed by means of IN *data* and OUT *data* instructions

☐ The port number, designated by the *data* portion of the program instruction, is fed to the lower eight bits of the address bus and decoded by a port-select circuit, such as the one shown in Fig. 10-6A

☐ The $\overline{IOR}$ and $\overline{IOW}$ signals from the microprocessor determine the direction of data-flow.

You have access to 256 different I/O-mapped port numbers, between hexadecimal ∅∅ and FF. There is an outside chance that your pre-existing microprocessor board uses some I/O-mapped selection of its own. You can determine whether or not this is the case by consulting the system's technical manual. What you want to avoid is working out a port select circuit that selects ports already used in your basic system. Perhpas your one-board computer uses I/O-mapped ports ∅∅ through ∅7 for some of its own, on-board operations. If that is the case, you will have to alter the pattern of address lines to the port select circuit to decode ports ∅8 through ∅F (connect A3 directly to input G1). Then, of course, your program will access the ports with IN *data* and OUT *data* instructions, where *data* is between ∅8 and ∅F.

The alternative, *memory-mapped I/O*, is far more popular these days. The only problem with it is that it is difficult to describe in general terms. The underlying principle, however, is rather simple: Treat all I/O devices and ports as though they are RAM memories. This might not mean much to a newcomer to the world of microprocessors, but the notion carries some powerful implications.

Every microprocessor system has a particular memory map. Generally speaking, a *memory map* shows the address of all the memory devices and memory-mapped I/O ports. For example, the basic one-board computer probably has memory locations 0000 through 03FF dedicated to a ROM-based monitor and loader. You cannot use that region of memory for any custom memory-mapped operations.

Likewise, your basic system might allocate 0400 through 07FF for on-board program RAM. That's a nice place to put your system program, but you cannot access memory-mapped I/O operations in that region, either.

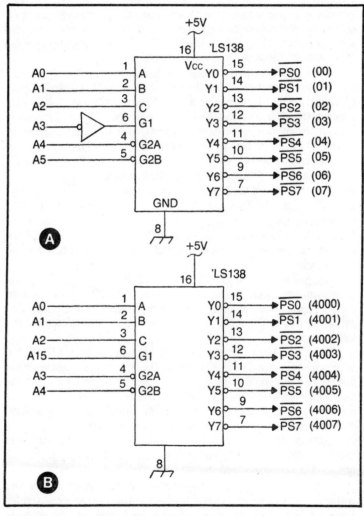

Fig. 10-6. The port select circuit. (A) I/O-mapped I/O version. (B) Memory-mapped version.

Finally, there is probably some address space already set aside for the I/O ports necessary for operating the one-board computer. Some of this space is used for the keypad, some for the readout assembly, cassette tape I/O, ROM burner and so on. You cannot use those addresses for running your custom parabot/robot.

The main point here is to demonstrate the importance of knowing exactly what regions of the memory map of your system are already dedicated to operations for your pre-existing microprocessor system. It is virtually impossible to describe the addresses that are available in a general way; you simply have to determine tht for yourself. Look for spaces in the memory map of your system that carry notes that read something such as UNUSED or USER EXPANSION. That's where you can fit in the memory-mapped port select circuit.

The memory-mapped port select circuit shown in Fig. 10-6B assumes you have access to memory locations 4000 and above. It is specifically set up to decode port-accessing addresses between 4000 and 4007. That means you will be able to access I/O port number 1 with any memory instruction that involves memory location 4001. Using Intel mnemonics, an instruction such as LDA 0140 will then load the eight bits from I/O port 1 into the microprocessor's accumulator; and an STA 0140 instruction will dump the contents of the accumulator into output port 1.

The IN *data* and OUT *data* instructions are not used in a memory-mapped I/O scheme. All communication with the outside world is handled via memory instructions. In a sense, the microprocessor is faked into thinking the I/O ports are actually RAM devices. The task of the memory-mapped port selector is to select a particular port, based on the memory address fed to it.

And just as the IN *data* and OUT *data* instructions to not apply to a memory-mapped I/O scheme, neither do the $\overline{IOR}$ and $\overline{IOW}$ signals from the microprocessor. Instead, the memory-mapped scheme uses the $\overline{MEMR}$ and $\overline{MEMW}$ signals to do the same job: *read* from memory and *write* into memory.

One of the most attractive features of memory-mapped I/O is the wide range of instructions available to it. The *I/O-mapped I/O* scheme has just the two IN and OUT instructions, both of which work only with the microprocessor's accumulator. Treating I/O ports as memory devices, however, it is possible to pass information directly to other registers, including the system stack and bonafide memory locations.

Unfortunately, a complete discussion of I/O-mapped I/O and memory-mapped I/O is beyond the scope of this book. Simply dig out any further information you need from other sources.

This is the time to make the decision regarding which process you will be using, because your choice dictates the nature of the addressing for the port select circuit. You cannot move a step farther until the matter is settled. Even the specific examples cited here and elsewhere in this book might be appropriate for your system, and using them in blind faith can get you into a real mess.

Here is one of those points for many experimenters where it is necessary to do A-plus work in order to pass the course. Good luck!

Setting Up I/O Ports

Any microprocessor system is virtually useless without at least one I/O (input/output) port. It is through such a port that the processor communicates in some meaningful way with the world outside.

I/O ports do not have to be very elaborate, although it is certainly possible to buy some truly sophisticated ICs for doing the job. In keeping with the general philosophy of this book, however, the ports suggested here are rather simple ones that can be applied to just about any sort of I/O operation.

Figure 10-7 is a complete schematic diagram of such an I/O port. Eight bits of digital information from the outside world are applied to the port inputs of Z411. Those input bits are labeled IPn0 through IPn7, where n is the port number selected by a port select circuit.

The input port device is a 74LS244 octal, noninverting buffer that is enabled whenever that particular port is selected by the PSn signal and the system calls for a reading operation by means of the $\overline{\text{IOR}}$ or $\overline{\text{MEMR}}$ signal. So if you have set up a memory-mapped I/O as described in the previous section of this chapter, and you have worked out the port select scheme such that the port is designated port No. 0, the microprocessor will see the input byte on its data bus (connections D0 through D7) whenever the programming calls for a memory-read ($\overline{\text{MEMR}}$) operation for port No. 0.

The output portion of this same port is built around Z410, which is an octal data latching circuit. Upon selecting that particular port and executing a program instruction that calls for a *write* operation, the eight bits on the system data bus are latched into Z410, where they remain available at the output connections (OPn0 through OPn7) until changed by some subsequent output operation to the same port.

The only real different between the input and output port is that bus data is latched into the output port, while in the case of the input port, the input data is available at the data bus only as long as it is needed to complete the operation. Of course, the two kinds of port operations are also different in that one outputs data during a $\overline{\text{IOW}}$ or $\overline{\text{MEMW}}$ operation and the other inputs data during a $\overline{\text{IOR}}$ or $\overline{\text{MEMR}}$ operation. That distinction should be rather obvious by now, however.

And whether you should enable the ports with an IO or MEM signal depends on whether you have chosen to use I/O mapping or a memory-mapped scheme. As mentioned in the previous section dealing with the port select circuit, you have to live with that choice from then on.

To summarize the operation of this I/O port scheme, suppose you have opted for I/O-mapped I/O and you want to call the port *Port 0*. That means the $\overline{\text{PS}n}$ signal to the two post circuits will be designated $\overline{\text{PS0}}$, and it

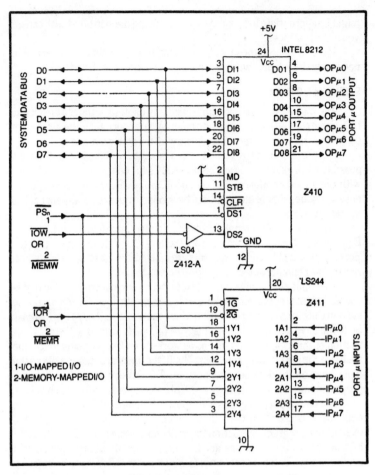

Fig. 10-7. A typical eight-bit I/O port.

will come from your port select circuit—from the line carrying that particular label. Furthermore, you will use the $\overline{\text{IOW}}$ signal to Z410 and the IOR signal to Z411.

Whenever you want the microprocessor to pick up the eight-bit word (or *byte*) available at the Port 0 inputs, all you have to do in the program is an IN 00. Upon executing this instruction, the input byte will be dumped onto the system data bus and into the accumulator register of the microprocessor. What you do with the byte once it is in the accumulator is up to you.

To send a byte from the accumulator register of the microprocessor out to the Port 0 outputs, simply do a program instruction, OUT 00. That will do the job, dumping the contents of the accumulator onto the system data bus and then loading it into the latches in the output port IC. This byte will remain available at the Port 0 outputs until you do another OUT$\emptyset\emptyset$ instruction later in the program.

If you have chosen to use memory-mapped I/O operations, the only difference as far as the I/O circuit is concerned is using $\overline{\text{MEMW}}$ (instead of $\overline{\text{IOW}}$) and $\overline{\text{MEMR}}$ (instead of $\overline{\text{IOR}}$). Then if the port happens to be selected by memory address 4$\emptyset\emptyset\emptyset$—as dictated by the addressing to your port select circuit—you can acquire the byte available at the input port by doing an instruction such as LDA$\emptyset\emptyset4\emptyset$. Likewise, you can output a byte to the output port by doing an STA$\emptyset\emptyset4\emptyset$. Those two instructions use Intel mnemonics, but the basic idea is to load the accumulator with data from locations 4$\emptyset\emptyset\emptyset$or store accumulator data at location 4$\emptyset\emptyset\emptyset$.

It is not absolutely necessary to assign the same port number to both the input and output port devices. You could, for instance, enable the input port with a PS0 (Port 0) signal and enable the output port with a PS1 (Port 1) signal. There is no way the system can work with both the input and output port at the same time, so there is nothing wrong with using the same port-select designation for both of them. Doing so simplifies your programming somewhat.

You will need a circuit such as the one shown in Fig. 10-7 for every I/O port your system requires. Most parabot/robot systems call for more than a single I/O port combination, and the port select circuit of Fig. 10-6 is adequate for handling eight such combinations.

Suppose your design calls for looking at 15 different bits of sensory input information and controlling output devices requiring 23 bits. Because the input and output ports suggested here all have an eight-bit capacity, you will need to sets of input ports (with one bit left over) and three output ports (also with one bit for a spare). You are free to organize this particular port configuration any way you choose, but the most efficient approach is to use two I/O combinations and one output-only circuit—one that uses just one Z410 from the circuit shown in Fig. 10-7. The first pair of I/Os could be enabled by Port-0 operations, the second pair could be enabled by Port-1 operations, and the output-only port could be enabled by a Port-3 operation.

A MINIMUM-SYSTEM EXAMPLE

Figure 10-8 is an example of a small microprocessor I/O scheme. It uses memory-mapped I/O operations, as indicated by the $\overline{\text{MEMW}}$ and $\overline{\text{MEMR}}$ control signals from Z400-A and Z400-B. If you do not intend to use any I/O-mapped I/O operations, there is no need for the two additional logic gates suggested in Fig. 10-2. Simply use the two gates required for generating the control signals appropriate for the type of mapping you intend to use.

And because the data and address buses from the microprocessor system are to be connected to a relatively few IC devices in this case, there is no need for adding data and address buffers. The connections shown here come directly from the microprocessor, itself.

The port select circuit is built around Z401. Only the first two outputs are used, however: $\overline{PS0}$ and $\overline{PS1}$. The address lines to that IC are set up so that PS0 is enabled at memory location 4000(hexadecimal) and $\overline{PS1}$ is enabled by 4001.

The scheme has two separate output ports. Z403 outputs six bits from the data bus, and those outputs might control the speed and direction of a pair of drive/steer motors through some appropriate interfacing circuitry. The second output, Port 1, is at Z405. There are just four bits from this one, and they might be used for carrying out a four-bit A/D conversion operation necessary for monitoring the speed of the drive/steer motors.

The circuit of Fig. 10-8 has just one eight-bit input port that is paired with output Port 0. It is paired in this fashion simply for the sake of convenience, and there is no necessary logical connection between the outputs of Port 0 and those inputs at the same port designation. Because we have used a drive/steer motor example to this point, however, you can consider the input bits as *motor-monitoring bits*.

The point is that the I/O scheme need be only as complicated as the system you are designing. Certainly you will have to include a basic microprocessor system and some suitable sensory/response interfacing, but you have to do much of that work whether or not you need a microprocessor.

Incidentally, the circuit suggested in Fig. 10-8 can be expanded quite easily to accommodate future needs for your system. The port select circuit, for example, already has six additional port-selecting outputs available, and all you would have to do in order to add another I/O system is to build in another set of 8218 output latches and 74LS224 input buffers.

ADDING OUTBOARD MEMORY

Generally, it is simpler to add memory to the on-board microcomputer system, itself, than to add it externally. There are circumstances, however, where it is probably less expensive to do the job externally.

Two of the most useful and least expensive RAM devices these days are the 2102 and the 2114. Both are static RAMs that have proven reliability. Their only disadvantage, compared to their dynamic counterparts, is that they do not have as much data capacity.

Most on-board computers and home computer systems use dynamic RAMs for the sake of getting the greatest possible amount of memory with the fewest number of IC devices. These dynamic RAMs generally cost more than static RAMs do. It is also extremely difficult to use a static RAM properly in a socket intended for a dynamic RAM.

Here is the point of this particular discussion: You can add a great deal of extra RAM at a fairly low cost by using your own external memory setup. Forget about upgrading the memory capabacity of the system by

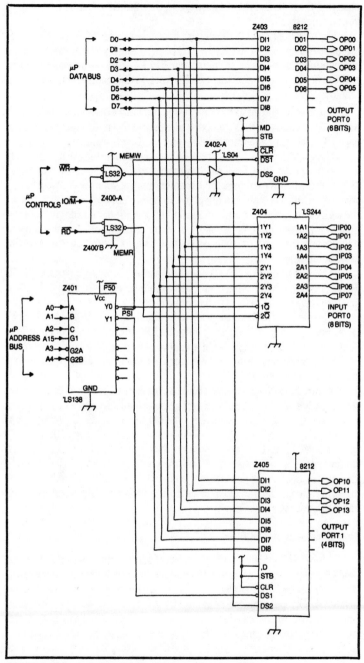

Fig. 10-8. Schematic for a minimum-system bus/port I/O.

changing the dynamic RAMs as suggested by the manufacturer. Outboard your own, instead.

Figures 10-9 and 10-10 suggest how this can be done. Both circuits increase the memory capacity by 1K, but of course you can use the basic arrangement to add as much memory in blocks of 1K, as you choose. The main difference between the two circuits is the memory device each employs.

The circuit shown in Fig. 10-9 uses the very economical 2101 RAM device. It is a 1024x1 RAM, and that means you will need eight of them for each 1K of memory you want to add—one chip for each 1K bit.

The eight-input NAND gate, Z501, and the inverters that precede it are programmed to access this memory between addresses 4C00 and 4FFF. You will have to consult the memory map of your own system to see whether or not you can actually place some memory at that point. If not, the addressing arrangement will have to be modified accordingly. Exactly how that is to be done is beyond the scope of this book. Consult a basic microprocessor text or data manual to find out how to manage the programming yourself.

The circuit of Fig. 10-10 does the same job, using just two 2114L RAM devices. These chips cost three or four times more than the 2102s, but they have a 1024x4 format that lets you use fewer ICs.

The memory in this case is set to be enabled between 4000 and 43FF. The only reason for placing it at a different location in the memory map is to illustrate a second example of memory-select addressing. The RAM SELECT circuit from Fig. 10-9 will work equally well, putting the memory between 4C00 and 4FFF.

While the notion of paying a couple of dollars apiece for 2102 RAM and $10 or $12 apiece for the 2114s might seem to load the choice in favor of the less expensive devices, consider the overall picture. Certainly eight RAMs a $2 apiece figures out to a cost of about $16 for the 2102 version, and two 2114s cost between $20 and $24. But look at the difference in complexity. Consider the amount of time you have to invest in wiring eight devices, as opposed to just two of them. And if you still aren't sold on using the 2114s, consider the extra drain on the power supply when using six more memory ICs.

Port-Addressed Memory

More sophisticated robot designs, especially those using Beta-Class and Gamma-Class, self-programmable memory, call for a memory capacity that exceeds the 64K capacity of 16-bit address microprocessor systems. Or, perhaps for reasons of your own, you want to use some memory that is addressable by more than 16 bits and can handle data words larger than eight or 16 bits. Whenever the reason, the idea is to have some memory available that is outside the usual memory map of your system.

328

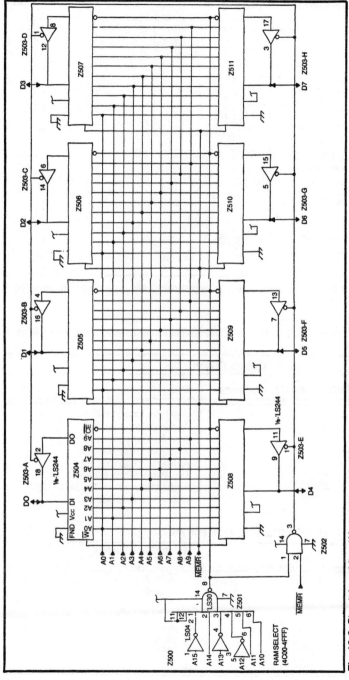

Fig. 10-9. Circuit for adding 1K of outboard RAM using the 2102 static RAM IC.

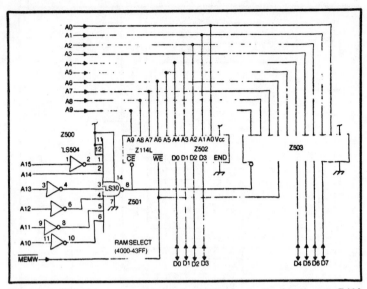

Fig. 10-10. Circuit for adding 1K of outboard RAM using the 2114L static RAM IC.

Table 10-1. Select Codes for the Port-Addressed Memory Scheme in Fig. 10-10.

PORT SELECT ADDRESSES (Z400)		
Addr. (hex)	Port Enabled	Notes
4000	0	Available for general I/O
4001	1	Available for general I/O
4002	2	Available for general I/O
4003	3	Available for general I/O
4004	4	Available for general I/O
4005	5	Low Byte for Port-Addressed Addr.
4006	6	High Byte for Port-Addressed Addr.
4007	7	Port-Addressed Data I/O

PORT-ADDRESSED MEMORY SELECT (Z407)

Select	Memory Block Selected (hex)
PMSO	0000-03FF
PMS1	0400-07FF
PMS2	0800-0BFF
PMS3	0C00-0FFF
PMS4	1000-13FF
PMS5	1400-17FF
PMS6	1800-1BFF
PMS7	1C00-1FFF

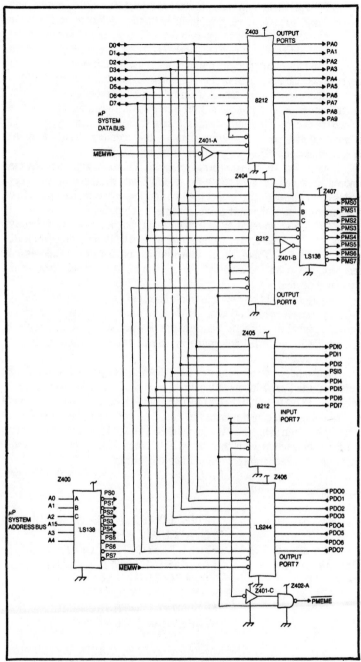

Fig. 10-11. Port selection for port-addressed memory.

331

Adding *port-selected memory* is a rather straightforward operation, and the only trade-off is as lower throughput of memory information. The basic idea is to address some memory from sets of I/O ports and work with the data for the addresses through yet another I/O port. Two output-only ports can be used for setting up the memory address, for example, and an I/O port can take care of the data. This is the scheme illustrated in Fig. 10-11.

In this particular instance, the port-addressed memory uses output ports 5 and 6 for the addressing. Port 5 carries the lower eight bits, while Port 6 takes care of the higher-order bits for the special memory. Port 7 is a bidirectional port, delivering data to the memory during a write operation and pulling in memory data during a read operation.

Z400 in Fig. 10-11 is the *port select circuit.* This one happens to decode microprocessor addresses 4000 through 4007, with the three higher select codes being the ones that select the two address ports and data port. See the general plan for this port-select scheme in Table 10-1A.

Z407 is also an eight-bit selector circuit, but it is playing a slightly different role. This circuit selects one of eight possible groups of 1K memory circuits. Table 10-1B shows which blocks of port-addressed memory are used for each of the memory-select signals from this particular device.

Figure 10-12 shows how this whole business can be connected to a 2K memory composed of 2114L static RAM ICs. You can see that port outputs PA0 through PA9 directly address the RAMs, while memory-select signals $\overline{PMS0}$ and $\overline{PMS1}$ determine which pair of RAMs will be *on-line* at any given instant.

Incidentally, when using the 2114 RAMs, as opposed to the 2102 versions, the Port 7 inputs and outputs in Fig. 10-11 must be connected together (PDI0 to PD00, PDI1 to PD01, etc.). In fact, if you know what you're doing, you can eliminate the Port 7 I/O devices, Z405 and Z406, altogether when using the 2114 memories. The 1024s, however, must be used in conjunction with the formal I/O ports, simply because those memories have separate data-in and data-out terminals.

Here is how a byte of data, say an AF, can be loaded into port-addressable memory location 0571:

```
MVI A, 71      ;MOVE LOWER BYTE OF ADDRESS
                 TO ACCUMULATOR
STA 05 40      ;STORE THAT BYTE AT OUTPUT PORT 5
MVI A, 05      ;MOVE HIGH BYTE OF ADDRESS
                 TO ACCUMULATOR
STA 06 40      ;STORE THAT BYTE AT OUTPUT PORT 6
MVI A, AF      ;MOVE DATA TO THE ACCUMULATOR
STA 07, 40     ;STORE THAT BYTE AT OUTPUT PORT 6
```

There might be simpler ways to handle the job, but this one illustrates the basic process in a straightforward fashion. The idea is to put the low

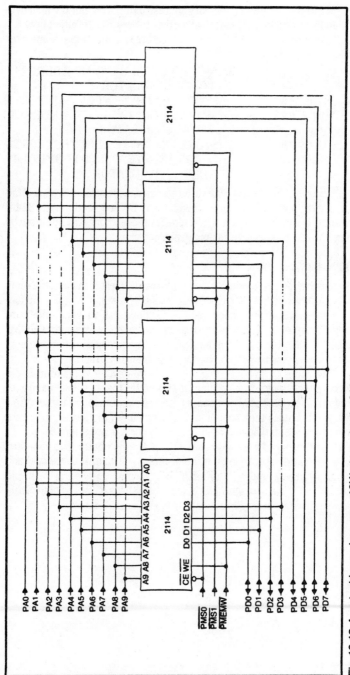

Fig. 10-12. A port-addressed memory of 2K bytes, using the 2114L.

byte of the port-addressed memory address into Port 6, put the high byte of the address into port 7, and move the data to be stored to output Port 7.

Whenever you want to retrieve that stored information at some later time, do this sort of thing:

MVI A, 71	;MOVE LOWER BYTE OF ADDRESS TO ACCUMULATOR
STA Ø5 4Ø	;STORE THAT BYTE AT OUTPUT PORT 5
MVI A,Ø5	;MOVE HIGH BYTE OF ADDRESS TO ACCUMULATOR
STA Ø6 4Ø	;STORE THAT BYTE AT OUTPUT PORT 6
LDA Ø7 4Ø	;LOAD THE BYTE AT INPUT PORT 7 TO THE ACCUMULATOR

Using this sort of port-addressed memory, it is possible to expand the memory capacity of the system indefinitely. Do you need a 64K, 32-bit memory? It will cost you a small fortune for the memory ICs, but it can be done with this approach to memory expansion.

ADDING CUSTOM ROMs

Custom *ROMs* (read-only memories) can be treated in much the same fashion described here for RAMs. It is usually simpler to add a custom ROM to space already allocated for that purpose in your one-board computer system. There is no reason, however, why custom ROMs cannot be added to your own address and data bus and even made port addressable (for whatever reason one might want to do that).

The tricky part about using custom ROM devices is inherent in the device, itself, and has nothing to do with the peculiar nature of parabot/robot systems. It isn't easy for many experimenters to *burn* their own ROM, unless they happen to be using one of the better one-board computer systems that has a built-in ROM burner, or programmer.

Whether or not you use a custom ROM at all is wholly up to you—and so is the matter of working out the program you want it to execute and finding some way to *burn* it in yourself. If it seems I am passing the whole matter to you, you are right. Just bear in mind that a ROM is a nonvolatile memory device that is addressed and read just as a RAM is.

Some Microprocessor-Controlled Parabot Systems

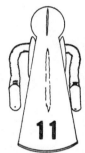

11

The greater the commitment one makes to a microprocessor-controlled system, the greater the number of design options that are available. Earlier in this book—before bringing up the subject of microprocessors—it was possible to make specific recommendations concerning the selection and design of semiconductor circuits. Given a certain task that is to be carried out with conventional semiconductors, there are only a few cost-effective and efficient options available. This makes things especially easy for beginners.

But moving into the world of microprocessors changes the picture altogether. Any given task can be done in a great many ways—all of them having a satisfactory degree of simplicity and effectiveness. The burden of making final design decisions thus rests with the user, and all I can do here for you is point out some possibilities in a general fashion and show some examples. It is up to you to rework the details to suit your own system requirements and programming style.

The three systems described in this chapter illustrate different kinds of control situations that are commonly encountered in parabot machines. Each system deals with a different sort of problem, and while I suggest only one solution, you must bear in mind that there are countless others, some of which might be more appealing to you than my examples are.

In fact, I do not suggest using these examples verbatum, because they can be rather restrictive. You will most likely want to do more with such systems than I show here, and it is your job to modify the ideas and programs as you see fit. Study these examples to the extent your knowledge and patience allow, however, because they will shed some light on problems you might not have anticipated and they might provide some valuable clues for solving problems that might otherwise stump you.

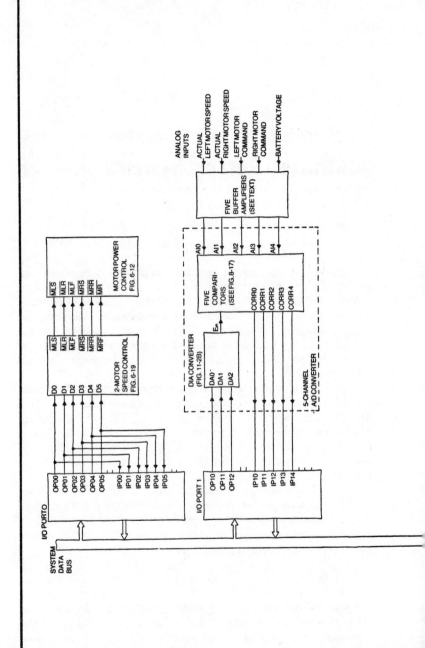

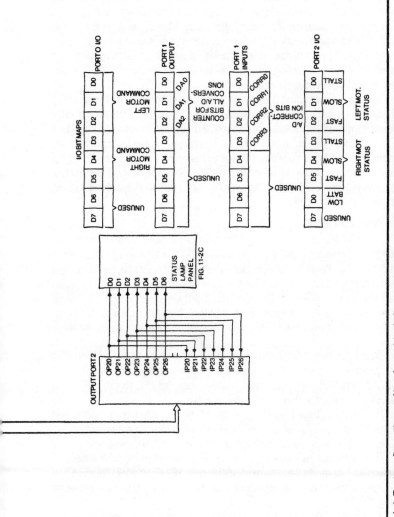

Fig. 11-1. Port configuration and hardware block diagram for the two-motor, drive/steer control system.

337

A COMPLETE TWO-MOTOR DRIVE/STEER CONTROL SYSTEM

The example described in this section gives the user a control panel for setting the speed and direction of two drive/steer motors. Rather than setting the speed and direction parameters by means of toggle switches, as shown in all previous examples, the user can adjust the parameters with a pair of potentiometers—one for each of the motors.

The two controls can be labeled LEFT MOTOR and RIGHT MOTOR. The programming is set up in such a way that both motors are stopped whenever the controls are set to their center positions. Moving either control clockwise from the center position causes its respective motor to run faster in a forward direction. Moving the control counterclockwise from its center position causes the motor to run in reverse, and at a speed proportional to the amount of turn. In short, the speed and direction of the two drive/steer motors is independently adjustable from a corresponding set of user controls.

An LED lamp panel indicates the status of the two motors, telling whether they are running faster than they should be, slower than they should be or stalled. There is no speed-control feedback in this particular example. The system, in other words, has no provisions for automatically correcting any error between the speed and direction you specify at the potentiometer control and the actual speed and direction of the corresponding motor. That sort of servo control is left to the second system example in this chapter.

The Hardware Layout

Figure 11-1 shows the layout for peripheral devices involved in this control example. It is assumed you have access to a standard microprocessor data bus and that you have set up three I/O ports: Port Ø, Portland Port 2. It is further assumed that you have set up the appropriate port select circuitry as described in Chapter 10. That select circuitry may work according to an *I/O-mapped I/O* scheme, but the machine-language program presented later in this section assumes you are using *memory-mapped I/O*.

Referring to the diagram in Fig. 11-1, you can see that Port Ø handles the speed and direction commands for the two motors. Six of the eight available bits at this port are used in this example: P00 through P05. Do whatever you want with the two remaining bits.

The output bits from Port Ø go to two places: to a four-speed two-direction drive/steer control circuit (such as the one described in Fig. 6-19) and back to the corresponding input portion of that same port.

The logic function table for the drive/steer control circuit appear in Table 6-4. You might want to mark its place in the book now, because you will have to refer to it a number of times while studying the programming example.

Ultimately, the program is going to generate the six speed/direction control bits to be applied to this drive/steer control circuit. The output of that control circuit then provides signals for controlling the speed and direction of the motors through the motor power control section, a circuit that can be identical to the one described in connection with Fig. 6-12. Because the theory of operation for these two circuits is already presented in some detail in Chapter 6, there is no point in going over it again.

The output bits at Port Ø are also connected back to the input bits of that same port. The idea here is to give the program an ability to fetch the current motor direction/speed command without having to save it in some special place in the RAM of the system.

Port 1 is dedicated to A/D operations. The A/D circuit is a five-channel version of the one described in Fig. 8-17. The analog input signals include voltage levels from the two user-operated potentiometers, a pair of motor-speed sensors (one for each motor) and a battery voltage sensor.

Figure 11-2A shows the sources for the LEFT MOTOR and RIGHT MOTOR commands. These are the potentiometers that are used for manually setting the speed and direction commands for the drive/steer motors. The voltages from the wiper arms can be isolated from the system by running them through a simple buffer circuit, such as the one shown back in Fig. 7-19. That particular amplifier should be set up for unity gain in this case (making R_i equal to R_f, in other words). Alternately, you might want the advantages of a scaling amplifier of the sort shown in Fig. 7-28. That circuit was originally described as one for scaling the output of a continuous manipulator position sensor, but it works equally well as a scaling amplifier for this input control situation.

Or if you are trying to make the system as simple as possible, you can connect the wiper arms of the two controls directly to the AI2 and AI3 inputs of the comparitor circuits, bypassing the buffer/amplifier section altogether. That decision is up to you.

The scheme described here calls for monitoring the actual speed status of the two motors. For the sake of simplicity, the actual direction status of the motors is ignored, assuming the motor will at least turn in the correct direction.

Any motor speed sensing circuit that generates a voltage proportional to running speed will serve here. Perhaps you like the electromechanical motor speed sensing scheme. That being the case, the buffer/amplifier section should be set up along the lines suggested in Fig. 7-25. You will need two of these sensing setups, with one for each of the two motors. Their outputs are to be connected to inputs AI0 and AI1 of the A/D comparitor section.

The last of the five analog inputs is a voltage representing the charge status of an on-board battery. The buffer/amplifier circuit in this case should be constructed according to the circuit shown in Fig. 7-29. The output of that section goes to the AI4 input of the A/D converter.

The CORR bits are labeled CORR0 through CORR4, indicating a correspondence between those bits and the analog inputs. CORR0, for example, is the correction bit for AI0, the actual LEFT MOTOR SPEED status, while CORR4 is the correction bit for the analog BATTERY VOLTAGE signal. In short, the A/D section has five separate analog inputs and five corresponding CORR-bit outputs. Those CORR bits go to corresponding input bits for Port 1. Any processor action calling for reading the data at Port 1 will thus pick up all five CORR bits from the A/D converter section of this system.

The analog error signal, E_{in}, that is vital to the operation of the A/D converter comes from a resistor-type, three-bit D/A converter. See Fig. 11-2B. The microprocessor software takes care of the three-bit counting

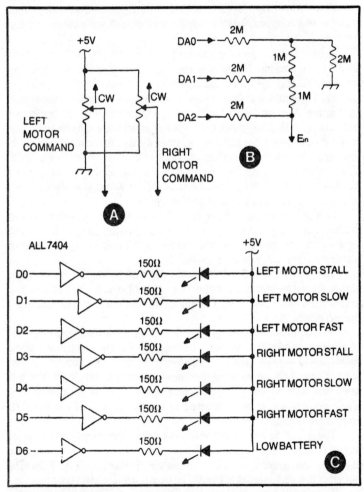

Fig. 11-2. Special circuits for the control system. (A) User's control inputs. (B) Three-bit D/A converter. (C) LED buffer/inverters and lamp circuit.

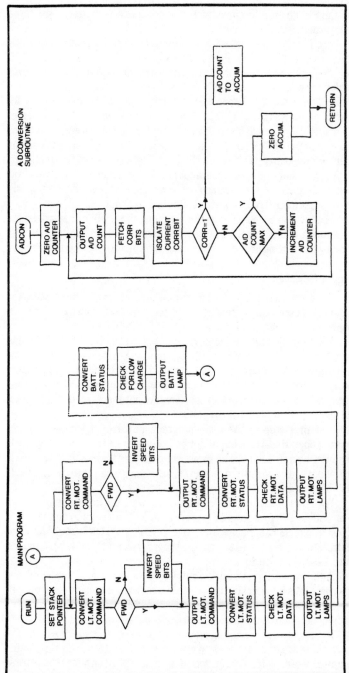

Fig. 11-3. Flowcharts for the two-motor, drive/steer control system.

operation, generating the input bits for the D/A converter. And whenever it is time to do an A/D conversion, the counter bits appear at the three lower-order bits of Port 1 output.

Port Ø is thus responsible for getting control commands to the drive/steer motors, while Port 1 is totally dedicated to carrying out A/D operations necessary for formatting command inputs, actual motor status and battery voltage level. Port 2 is wholly dedicated to operating the status lamps. There are five lamps that are interfaced to the microprocessor system, as shown in Fig. 11-2C. These lamps light up to indicate whether either motor is running faster than the input command direction, running slower (but not stalled) or completely stalled. The seventh lamp goes on whenever the battery voltage reaches a critically low level determined by the programming. Any software output, or write, command involving Port 2 will thus affect the status lamps, while an input, or read, instruction will provide information regarding the status of those lamps.

The bit map accompanying the diagram of Fig. 11-1 summarized the assignments of bits for all port inputs and outputs. Such a map is especially helpful when setting up the machine-language microprocessor program.

Program Flowchart

The flowchart in Fig. 11-3 shows how the software for this system can be set up. There are two main parts to the flowchart: a chart of the *main program* and a separate one for doing the *A/D conversions*.

The flowchart for the main program can, itself, be broken down into three parts: one for dealing with the left motor operations, one for dealing with right-motor operations and a third for checking out the battery voltage. And finally, the motor operations can be divided into two phases: first check for the command signal from the motor control potentiometer and then check the actual motor status.

The first step in the main program, however, specifies a stack-pointing operation. This might not be necessary when some ready-made computer boards that have an established stack of their own. If you aren't sure about your own system, however, set a stack location to be on the safe side. The program here calls for two bytes of stack memory.

With that housekeeping task out of the way, the flowchart gets down to business. The first main step is to do an A/D conversion on the LEFT MOTOR command signal from the LEFT MOTOR speed/direction control. In a manner of speaking, the program wants to know what the left motor is supposed to do.

After formatting that analog voltage into a three-bit binary word, the flowchart calls for checking the word for a forward-direction command. If, indeed, the control is specifying some speed in a forward direction, no further operations are to be performed on the command word.

If the control is specifying motion in a reverse direction, however, the chart shows that it is necessary to invert the two speed bits in the command word. This has to be done if the motor is to run faster in a reverse direction as the control is rotated courterclockwise of its center, stop command position.

Once the command word is properly formatted, it is output to the motor speed and direction control circuitry. Presumably, the left motor will respond as prescribed by that command word.

So the system now knows what you want the left motor to do, and it has begun executing the command. Now is the time to check the left motor status to see if it is, indeed, doing the job correctly.

This phase of the job begins by doing an A/D conversion on the signal from the left motor speed sensor. When that conversion is done, the system has a two-bit number representing the actual motor speed.

That actual motor speed status is then compared with the command word generated earlier in the program. If there are any discrepencies, if the motor is stopped when it shouldn't be, or if the A/D conversion overflows for any reason, the system output logic levels to light the appropriate status lamps.

Remember that this system has no provisions for automatically taking care of such descrepancies; it simply lights some LEDs to *indicate* the nature of the discrepancies when they exist. The system featured in the following section of this chapter includes some of these automatic, servo-type controls.

The next phase of operations concerns the same set of steps as applied to the right motor. From a flowcharting point of view, the operations are identical, so there is no need to dwell on them again.

Finally, the system takes a look at the battery level. It first converts the battery voltage into a two-bit binary number that is proportional to the voltage, compares that number with some user-specified criteria for a low-battery condition, and then either sets or resets a BATTERY LOW lamp accordingly. When the battery-testing operations are done, the main program cycles back to the point where it begins checking out the left motor again.

The program runs in an endless loop that can be summarized this way:

☐ Do an A/D conversion on the LEFT MOTOR command.

☐ Format the LEFT MOTOR command.

☐ Output the LEFT MOTOR command.

☐ Do an A/D conversion on the actual LEFT MOTOR speed.

☐ Compare the actual LEFT MOTOR speed with the LEFT MOTOR command.

☐ Indicate any discrepencies as LEFT MOTOR SLOW, LEFT MOTOR FAST or LEFT MOTOR STALL.

☐ Do an A/D conversion on the RIGHT MOTOR command.

☐ Format the RIGHT MOTOR command.

☐ Output the RIGHT MOTOR command.

☐ Do an A/D conversion on the actual RIGHT MOTOR speed.

☐ Compare the actual RIGHT MOTOR speed with the RIGHT MOTOR command.

☐ Indicate any discrepencies as RIGHT MOTOR SLOW, RIGHT MOTOR FAST or RIGHT MOTOR STALL.

☐ Do an A/D conversion on the battery voltage.

☐ Check the voltage agains some criteria for a low-battery condition.

☐ Light the LOW BATTERY lamp if the battery voltage is low.

☐ Loop back to the first step.

Programming the Operations

Like all other major program listings in this book, the examples that follow show only assembler line numbers (for reference purposes) and the source-program mnemonics. There is little point in showing the machine-language object codes because the addresses of the instructions will vary so much from one kind of microprocessor system to another.

If you should ever decide to use one of these programs verbatum, you will have to assemble it yourself, either by hand or with the help of a computer and suitable assembler program. And if none of this makes any sense to you at all, you can be quite sure it is time to find a good book on assembly language programming for the microprocessor device you plan to use.

There are two different listings here: one using Intel 8080A/8085 mnemonics and the other using Z80 mnemonics. Study the one that suits your own understanding better, using the flowchart in Fig. 11-1 as a guide.

The stack pointer is set in line 1000 at whatever TOPS address is suitable for your system. Generally, it is placed at the highest available RAM address.

Line 1030 begins the process of converting the LEFT MOTOR command from an analog to a digital format. At that point, register B is first loaded with a code that puts a logic 1 in bit position 2, only. For the purpose of A/D conversion, this particular code signals a conversion on the CORR2 bit—the correction bit for the LEFT MOTOR input command.

Line 1050 calls the A/D conversion subroutine, ADCON. The finer features of that subroutine will be discussed later in this section. What is important now is that the operations return to line 1070 of the main program with the accumulator carrying a three-bit binary number representing the LEFT MOTOR speed and direction command. The actual operation in line 1070 saves that command in register D while some other operations are performed on the version that remains in the accumulator.

According to the flowchart, the system must now decide whether the command word calls for forward or reverse direction of motion. So 1070 isolates that direction bit. If it is a logic 1, it means that the command is calling for a forward direction of motion, and operations jump down to line 1140. But if the direction bit is a zero, the motor is supposed to run in reverse, and the program must invert the speed bits to get the speed control set up to provide increasing, reverse-direction motion as the speed/direction control is moved counterclockwise from center.

The speed bits are inverted by first fetching them from register D (where they were stored back in line 1070), then doing an Exclusive-OR with binary ones in the two speed-bit positions. See line 1120. That operation inverts only the speed bits, leaving the direction bit unchanged. The command word, thus modified, is again saved in register D by the instruction in line 1130.

Whether the command word is inverted or not, the next step (line 1140) is to fetch an unadulterated version of what the motors are supposed to be doing from Port Ø. The LEFT MOTOR direction and speed bits are cleared by AND-ing them with a hexadecimal F8; the direction and speed bits for RIGHT MOTOR operations are left unchanged. Then in line 1180, the new command word, previously saved in register D, is inserted into the command word in the accumulator. Then the revised command word is finally output to Port Ø. See line 1200.

Finally, the motors have a chance to respond to the new command from the LEFT MOTOR control. And then it is time to see what the LEFT MOTOR is actually doing in response to these commands.

At line 1220, the system is set up for doing an A/D conversion on CORRØ, the LEFT MOTOR speed status input. The appropriate code is placed into the B register before the program calls the A/D conversion subroutine, ADCON.

After executing ADCON, control returns to the instruction in line 1250. As before, the accumulator carries a digitized version of an analog input, but this time it represents the actual LEFT MOTOR speed in a two-bit format. The motor speed sensor suggested for this project cannot confirm the direction of the motor, so there is no third direction-indicating bit.

At any rate, the digitized version of the actual LEFT MOTOR speed is saved in register B at line 1250. The program then clears the LEFT MOTOR status lamps by fetching the lamp status word (line 1260), doing a logical AND operation to zero the three lower-order bits (line 1270) and then outputting the result to the Port 2, status lamp output (line 1280).

With the LEFT MOTOR status lamps thus cleared, the program fetches the current motor command again from Port 0, isolates the LEFT MOTOR bits, and compares them with the actual LEFT MOTOR speed bits previously saved in register B. This is where the system begins checking for problems—errors between what the LEFT motor is being *told to do* with what it is *actually doing*.

In line 1330, if there is no difference between the command and actual LEFT MOTOR status, operations jump to RTCYC, where operations begin for the RIGHT MOTOR system. If the actual motor speed happens to be greater than the command, however, the jump-if-carry instruction in line 1340 transfers operations to LMFAS, which is a set of instructions that begins at line 1480 that turns on the LEFT MOTOR FAST status lamp.

If the motor isn't running at just the right speed (as detected by line 1330) and it isn't running too fast (as detected by line 1340), it is either running too slowly or it is stalled. The next set of instructions determine which is the case.

The actual speed of the motor, previously saved in register B, is put into the accumulator again by line 1350. It is then compared with zero to see whether or not the motor is stalled. If, indeed, the motor is stopped, line 1380 sends operations down to an LMSTA line—one beginning a set of instructions that light up the LEFT MOTOR STALL lamp. But if the comparison in line 1360 turns up some actual speed other than zero, the logical conclusion is that the motor is running too slowly. Lines 1390 through 1470 thus light up the LEFT MOTOR SLOW lamp and transfer operations down to RTCYC, where checks on the RIGHT MOTOR begin.

This entire set of LEFT MOTOR checks can be summarized as follows:

☐ The program first compares the speed command and actual speed to determine if the motor is running at just the right speed or too fast. If the motor is running at the right speed, operations procede to RTCYC to begin running the same sort of checks on the RIGHT MOTOR. But if the motor is running too fast, the LEFT MOTOR FAST lamp is lighted, and *then* the program begins the RIGHT MOTOR checks.

☐ The program then compares the actual LEFT MOTOR speed with zero. If the LEFT MOTOR speed is *not* zero, it must be running slower than the command suggests, and the system responds by lighting the LEFT MOTOR SLOW lamp. Otherwise, the motor must be stalled, and the system lights the LEFT MOTOR STALL lamp.

Operations for checking out the RIGHT MOTOR follow the same general pattern in program lines 1700 through 2370. The only difference is that the program works with different sets of bits for converting the RIGHT MOTOR command, converting the RIGHT MOTOR status, comparing the two, and setting the status of the RIGHT MOTOR status lamps.

The final section of the main program checks the battery voltage. Line 2420 sets the B register for converting the battery voltage to a two-bit binary format, using the correction bit, CORR4, from the A/D conversion port, Port 1. Line 2440 calls ADCON to do the actual A/D conversion on the battery voltage.

The result is compared with hexadecimal 2 in line 2460. That operation compares the actual battery voltage for 75-percent charge. Lines 2480 and 2490 are sensitive to the cases where the battery voltage is at 75-percent charge or less, respectively; and if either of these conditions is satisfied, control goes to BATLOW, which is a set of instructions that light up the LOW BATTERY lamp.

Otherwise, the program clears the LOW BATTERY lamp and jumps all the way back to the beginning of the main program. See the jump to LTCYC in line 2560.

The remainder of the listing deals with the A/D conversion subroutine, ADCON. Beginning at line 2700, the program checks the operation code that is always placed into the B register before the main program calls this subroutine. If it is a hexadecimal 4 or 8, the D register specifies a maximum A/D count of 3. That is the maximum A/D count for the two-bit motor-status conversions. Otherwise, the program uses a maximum A/D count of 7 for the remaining A/D conversions. Beyond that point, the listing follows the flowchart for the A/D conversion subroutine.

8080A/8085 Program Listing

```
2-MOTOR, 4-SPEED PARABOT CONTROL SYSTEM

8080A/8085 PROGRAM LISTING


1000 START:      LXI SP,TOPS       ;SET STACK
1010 ;BEGINNING OF LEFT MOTOR OPERATIONS
1020 ;FIRST, CONVERT LEFT MOTOR COMMAND
1030 LTCYC:      MVI B,04H         ;SET FOR LT.
1040                               ;MOT. COMM.
1050             CALL ADCON        ;DO A/D CON-
1060                               ;VERSION
1070             MOV D,A           ;SAVE COMM.
1080             ANI 4H            ;ISOLATE DI-
1090                               ;RECTION BIT
1100             JNZ NEXT1         ;JMP IF FWD
1110             MOV A,D           ;GET COMM.
1120             XRI 3H            ;INV SPEED
1130             MOV D,A           ;SAVE COMM.
1140 NEXT1:      LDA PORT0         ;FETCH MOT.
1150                               ;STATUS
1160             ANI 0F8H          ;CLR LT. MOT.
1170                               ;BITS
1180             ORA D             ;INSERT NEW
1190                               ;COMMAND
1200             STA PORT0         ;OUTPUT COMM.
1210 ;NEXT, CONVERT LEFT MOTOR STATUS
1220             MVI B,1H          ;SET LT. MOT.
1230                               ;STATUS
1240             CALL ADCON        ;CONVERT
1250             MOV B,A           ;SAVE
1260             LDA PORT2         ;GET LAMPS
1270             ANI 08FH          ;RESET LT.
1280             STA PORT2         ;MOT. LAMPS
1290             LDA PORT0         ;FETCH COMM.
1300             ANI 3H            ;ISOLATE LT.
```

347

```
1310                              ;MOT. BITS
1320              CMP B           ;COMPARE
1330              JZ RTCYC        ;JMP IF SAME
1340              JC LMFAS        ;JMP IF FAST
1350              MOV A,B         ;GET STATUS
1360              CPI 0H          ;CHK STALL
1370                              ;
1380              JZ LMSTA        ;JUMP OF SO
1390              LDA PORT2       ;GET LAMP
1400                              ;STATUS
1410              ANI 0F8H        ;RESET LT.
1420                              ;MOT. LAMPS
1430              ORI 2H          ;SET SLOW
1440                              ;LAMP
1450              STA PORT2       ;OUTPUT TO
1460                              ;LAMPS
1470              JMP RTCYC       ;DO RT. MOT.
1480 LMFAS:       LDA  PORT2      ;GET LAMP
1490                              ;STATUS
1500              ANI 0F8H        ;RESET LT.
1510                              ;MOT. LAMPS
1520              ORI 4H          ;SET FAST
1530                              ;LAMP BIT
1540              STA PORT2       ;OUTPUT TO
1550                              ;LAMPS
1560              JMP RTCYC       ;DO RT. MOT.
1570 LMSTA:       LDA PORT2       ;GET LAMP
1580                              ;STATUS
1590              ANI 0F8H        ;RESET LT.
1600                              ;MOT. LAMPS
1610              ORI 1H          ;SET STALL
1620                              ;LAMP BIT
1630              STA PORT2       ;OUTPUT TO
1640                              ;LAMPS
1650 ;
1660 ;
1670 ;
1680 ;BEGINNING OF RIGHT MOTOR OPERATIONS
1690 ;FIRST, CONVERT THE MOTOR COMMAND
1700 RTCYC:       MVI B,8H        ;SET FOR RT.
1710                              ;MOT. COMM
1720              CALL ADCON      ;DO A/D CON-
1730                              ;VERSION
1740              MOV D,A         ;SAVE COMM
1750              ANI 4H          ;ISOLATE DI-
1760                              ;RECTION BIT
1770              JNZ NEXT2       ;JMP IF FWD
1780              MOV A,D         ;GET COMM
1790              XRI 3H          ;INV SPEED
1800              ADD A           ;SHIFT LT.
```

```
1810                ADD A           ;SHIFT LT.
1820                ADD A           ;SHIFT LT.
1830                MOV D,A         ;SAVE COMM
1840 NEXT2:         LDA PORT0       ;FETCH MOT.
1850                                ;STATUS
1860                ANI 0C3H        ;CLR RT. MOT.
1870                                ;BITS
1880                ORA D           ;INSERT NEW
1890                                ;COMMAND
1900                STA PORT0       ;OUTPUT COMM.
1910 ;NEXT, CONVERT RIGHT MOTOR SPEED STATUS
1920                MVI B,2H        ;SET RT. MOT.
1930                                ;STATUS
1940                CALL ADCON      ;DO A/D CON-
1950                                ;VERSION
1960                ADD A           ;SHIFT LT.
1970                ADD A           ;SHIFT LT.
1980                ADD A           ;SHIFT LT.
1990                MOV B,A         ;SAVE
2000                LDA PORT2       ;GET LAMP
2010                ANI 0C7H        ;RESET RT.
2020                STA PORT2       ;MOT. LAMPS
2030                LDA PORT0       ;FETCH COMM.
2040                ANI 38H         ;ISOLATE RT.
2050                                ;MOT. BITS
2060                CMP B           ;COMPARE
2070                JZ BATCHK       ;JMP IF SAME
2080                JC RMFAS        ;JMP IF FAST
2090                MOV B,A         ;GET STAT
2100                CPI 0H          ;CHK STALL
2110                                ;
2120                JZ RMSTA        ;JMP IF SO
2130                LDA PORT2       ;GET LAMP
2140                                ;STATUS
2150                ANI 0C7H        ;RESET RT.
2160                                ;MOT. LAMPS
2170                ORI 10H         ;SET SLOW
2180                                ;LAMP
2190                STA PORT2       ;OUTPUT TO
2200                                ;LAMPS
2210                JMP BATCHK      ;DO BATT.
2220 RMFAS:         LDA PORT2       ;GET LAMP
2230                                ;STATUS
2240                ANI 0C7H        ;RESET RT.
2250                                ;MOT. LAMPS
2260                ORI 20H         ;SET FAST
2270                                ;LAMP BIT
2280                STA PORT2       ;OUTPUT TO
2290                                ;LAMPS
2300                JMP BATCHK      ;DO BATT.
```

```
2310 RMSTA:      LDA PORT2       ;GET LAMP
2320                             ;STATUS
2330             ANI 0C7H        ;RESET RT.
2340                             ;MOT. LAMPS
2350             ORI 8H          ;SET STALL
2360                             ;LAMP BIT
2370             STA PORT2       ;OUTPUT TO
2380                             ;LAMPS
2390 ;
2400 ;
2410 ;CHECK THE BATTERY VOLTAGE
2420 BATCHK:     MVI B,10H       ;SET FOR
2430                             ;BATT. CHK
2440             CALL ADCON      ;DO A/D CON-
2450                             ;VERSION
2460             CPI 6H          ;LOOK FOR 3/4
2470                             ;CHARGE
2480             JZ BATLOW       ;JUMP IF LOW
2490             JC BATLOW       ;
2500             LDA PORT2       ;GET LAMP
2510                             ;STATUS
2520             ANI 0BFH        ;RESET BATT.
2530                             ;LOW BIT
2540             STA PORT2       ;OUTPUT TO
2550                             ;LAMPS
2560             JMP LTCYC       ;START ALL
2570                             ;OVER AGAIN
2580 BATLOW:     LDA PORT2       ;GET LAMP
2590                             ;STATUS
2600             ORI 40H         ;SET LOW
2610                             ;BATT. BIT
2620             STA PORT2       ;OUTPUT TO
2630                             ;LAMPS
2640             JMP LTCYC       ;START ALL
2650                             ;OVER AGAIN
2660 ;
2670 ;END OF MAIN PROGRAM
2680 ;
2690 ;BEGINNING OF ADCON SUBROUTINE
2700 ADCON:      MOV A,B         ;CHK OP
2710             CPI 4H          ;CHK CORR2
2720             JZ BIT2         ;JMP IF SO
2730             CPI 8H          ;CHK CORR3
2740             JZ BIT2         ;JMP IF SO
2750             MOV D,7H        ;SET FOR 3-
2760                             ;BIT A/D
2770             JMP GOAD        ;
2780 BIT2:       MOV D,3H        ;SET FOR 2-
2790                             ;BIT A/D
2800 GOAD:       MOV C,0H        ;ZERO THE A/D
```

```
2810                              ;COUNTTER
2820 RECYC:       MOV A,C         ;GET A/D CNT
2830             STA PORT1        ;OUTPUT A/D
2840                              ;COUNT
2850             LDA PORT1        ;GET CORR BITS
2860             ANA B            ;ISOLATE CORR
2870             JNZ ADOK         ;JMP IF A/D
2880                              ;IS DONE
2890             MOV A,C          ;GET A/D CNT
2900             CMP C            ;IS IT MAX
2910             JZ ADINV         ;JMP IF SO
2920             INR C            ;INCREMENT
2930                              ;A/D CNT
2940             JMP RCYC         ;DO ANOTHER
2950                              ;A/D CYCLE
2960 ADOK:       MOV A,C          ;A/D CNT TO
2970                              ;ACCUMULATOR
2980             RET              ;RETURN TO
2990                              ;MAIN PROG.
3000 ADINV       XRA A            ;ZERO THE
3010                              ;ACCUMULATOR
3020             RET              ;RETURN TO
3030                              ;MAIN PROG.
3040 ;
3050 ;
3060 ;END OF LISTING
3070 ;USER MUST DEFINE THESE
3080      ;ORG
3090      ;TOPS
3100      ;PORT0
3110      ;PORT1
3120      ;PORT2
```

Z80 Program Listing

2-MOTOR, 4-SPEED PARABOT CONTROL SYSTEM

Z80 PROGRAM LISTING

```
1000 START:      LD SP,TOPS       ;SET STACK
1010 ;BEGINNING OF LEFT MOTOR OPERATIONS
1020 ;FIRST, CONVERT LEFT MOTOR COMMAND
1030 LTCYC:      LD B,04H         ;SET FOR LT.
1040                              ;MOT. COMM.
1050             CALL ADCON       ;DO A/D CON-
1060                              ;VERSION
1070             LD D,A           ;SAVE COMM.
1080             AND 4H           ;ISOLATE DI-
1090                              ;RECTION BIT
1100             JP Z,NEXT1       ;JMP IF FWD
```

```
1110            LD A,D          ;GET COMM.
1120            XOR 3H          ;INV SPEED
1130            LD D,A          ;SAVE COMM.
1140  NEXT1:    LD A,(PORT0)    ;FETCH MOT.
1150                            ;STATUS
1160            AND 0F8H        ;CLR LT. MOT.
1170                            ;BITS
1180            OR D            ;INSERT NEW
1190                            ;COMMAND
1200            LD (PORT0),A    ;OUTPUT COMM.
1210  ;NEXT, CONVERT LEFT MOTOR STATUS
1220            LD B,1H         ;SET LT. MOT.
1230                            ;STATUS
1240            CALL ADCON      ;CONVERT
1250            LD B,A          ;SAVE
1260            LD A,(PORT2)    ;GET LAMPS
1270            AND 0F8H        ;RESET LT.
1280            LD (PORT2),A    ;MOT. LAMPS
1290            LD A,(PORT0)    ;FETCH COMM.
1300            AND 3H          ;ISOLATE LT.
1310                            ;MOT. BITS
1320            CP B            ;CMPARE
1330            JP Z,RTCYC      ;JMP IF SAME
1340            JP C,LMFAS      ;JMP IF FAST
1350            LD A,B          ;GET STATUS
1360            CP 0H           ;CHK STALL
1370                            ;
1380            JP Z,LMSTA      ;JUMP OF SO
1390            LD A,(PORT2)    ;GET LAMP
1400                            ;STATUS
1410            AND 0F8H        ;RESET LT.
1420                            ;MOT. LAMPS
1430            OR 2H           ;SET SLOW
1440                            ;LAMP
1450            LD (PORT2),A    ;OUTPUT TO
1460                            ;LAMPS
1470            JP RTCYC        ;DO RT. MOT.
1480  LMFAS:    LD A,(PORT2)    ;GET LAMP
1490                            ;STATUS
1500            AND 0F8H        ;RESET LT.
1510                            ;MOT. LAMPS
1520            OR 4H           ;SET FAST
1530                            ;LAMP BIT
1540            LD (PORT2),A    ;OUTPUT TO
1550                            ;LAMPS
1560            JP RTCYC        ;DO RT. MOT.
1570  LMSTA:    LD A,(PORT2)    ;GET LAMP
1580                            ;STATUS
1590            AND 0F8H        ;RESET LT.
1600                            ;MOT. LAMPS
```

```
1610              OR 1H           ;SET STALL
1620                              ;LAMP BIT
1630              LD (PORT2),A    ;OUTPUT TO
1640                              ;LAMPS
1650  ;
1660  ;
1670  ;
1680  ;BEGINNING OF RIGHT MOTOR OPERATIONS
1690  ;FIRST, CONVERT THE MOTOR COMMAND
1700  RTCYC:     LD B,8H          ;SET FOR RT.
1710                              ;MOT. COMM
1720              CALL ADCON      ;DO A/D CON-
1730                              ;VERSION
1740              LD D,A          ;SAVE COMM
1750              AND 4H          ;ISOLATE DI-
1760                              ;RECTION BIT
1770              JP NZ,NEXT2     ;JMP IF FWD
1780              LD A,D          ;GET COMM
1790              XOR 3H          ;INV SPEED
1800              ADD A           ;SHIFT LT.
1810              ADD A           ;SHIFT LT.
1820              ADD A           ;SHIFT LT.
1830              LD D,A          ;SAVE COMM
1840  NEXT2:     LD A,(PORT0)     ;FETCH MOT.
1850                              ;STATUS
1860              AND 0C3H        ;CLR RT. MOT.
1870                              ;BITS
1880              OR D            ;INSERT NEW
1890                              ;COMMAND
1900              LD (PORT0),A    ;OUTPUT COMM.
1910  ;NEXT, CONVERT RIGHT MOTOR SPEED STATUS
1920              LD B,2H          ;SET RT. MOT.
1930                              ;STATUS
1940              CALL ADCON      ;DO A/D CON-
1950                              ;VERSION
1960              ADD A           ;SHIFT LT.
1970              ADD A           ;SHIFT LT.
1980              ADD A           ;SHIFT LT.
1990              LD B,A          ;SAVE
2000              LD A,(PORT2)    ;GET LAMPS
2010              AND 0C7H        ;RESET RT.
2020              LD (PORT2),A    ;MOT. LAMPS
2030              LD A,(PORT0)    ;FETCH COMM.
2040              AND 38H         ;ISOLATE RT.
2050                              ;MOT. BITS
2060              CP B            ;COMPARE
2070              JP Z,BATCHK     ;JMP IF SAME
2080              JP C,RMFAS      ;JMP IF FAST
2090              LD B,A          ;GET STAT
2100              CP 0H           ;CHK STALL
2110                              ;
```

353

```
2120              JP Z,RMSTA       ;JMP IF SO
2130              LD A,(PORT2)     ;GET LAMP
2140                               ;STATUS
2150              AND 0C7H         ;RESET RT.
2160                               ;MOT. LAMPS
2170              OR 10H           ;SET SLOW
2180                               ;LAMP
2190              LD (PORT2),A     ;OUTPUT TO
2200                               ;LAMPS
2210              JP BATCHK        ;DO BATT.
2220 RMFAS:       LD A,(PORT2)     ;GET LAMP
2230                               ;STATUS
2240              AND 0C7H         ;RESET RT.
2250                               ;MOT. LAMPS
2260              OR 20H           ;SET FAST
2270                               ;LAMP BIT
2280              LD (PORT2),A     ;OUTPUT TO
2290                               ;LAMPS
2300              JP BATCHK        ;DO BATT.
2310 RMSTA:       LD A,(PORT2)     ;GET LAMP
2320                               ;STATUS
2330              AND 0C7H         ;RESET RT.
2340                               ;MOT. LAMPS
2350              OR 8H            ;SET STALL
2360                               ;LAMP BIT
2370              LD (PORT2),A     ;OUTPUT TO
2380                               ;LAMPS
2390 ;
2400 ;
2410 ;CHECK THE BATTERY VOLTAGE
2420 BATCHK:      LD B,10H         ;SET FOR
2430                               ;BATT. CHK
2440              CALL ADCON       ;DO A/D CON-
2450                               ;VERSION
2460              CP 6H            ;LOOK FOR 3/4
2470                               ;CHARGE
2480              JP Z,BATLOW      ;JUMP IF LOW
2490              JP C,BATLOW      ;
2500              LD A,(PORT2)     ;GET LAMP
2510                               ;STATUS
2520              AND 0BFH         ;RESET BATT.
2530                               ;LOW BIT
2540              LD ,(PORT2),A     ;OUTPUT TO
2550                               ;LAMPS
2560              JP LTCYC         ;START ALL
2570                               ;OVER AGAIN
2580 BATLOW:      LD A,(PORT2)     ;GET LAMP
2590                               ;STATUS
2600              OR 40H           ;SET LOW
2610                               ;BATT. BIT
```

```
2620              LD (PORT2),A      ;OUTPUT TO
2630                                ;LAMPS
2640              JP LTCYC          ;START ALL
2650                                ;OVER AGAIN
2660 ;
2670 ;END OF MAIN PROGRAM
2680 ;
2690 ;BEGINNING OF ADCON SUBROUTINE
2700 ADCON:       LD A,B            ;CHK OP
2710              CP 4H             ;CHK CORR2
2720              JP Z,BIT2         ;JMP IF SO
2730              CP 8H             ;CHK CORR3
2740              JP Z,BIT2         ;JMP IF SO
2750              CP 10H            ;CHK CORR4
2760              JP Z,BIT2         ;JMP IF SO
2770              LD D,7H           ;SET FOR 3-
2780                                ;BIT A/D
2790              JP GOAD           ;
2800 BIT2:        LD D,3H           ;2-BIT CNT
2810 GOAD:        LD C,0H           ;ZERO THE A/D
2820                                ;COUNTER
2830 RECYC:       LD A,C            ;GET A/D CNT
2840              LD (PORT1),A      ;OUTPUT A/D
2850                                ;COUNT
2860              AND B             ;ISOLATE CORR
2870              JP Z,ADOK         ;JMP IF A/D
2880                                ;IS DONE
2890              LD A,C            ;GET A/D CNT
2900              CP C              ;IS IT MAX
2910              JP A,ADINV        ;JMP IF SO
2920              INC C             ;INCREMENT
2930                                ;A/D COUNT
2940              JP RCYC           ;DO ANOTHER
2950                                ;A/D CYCLE
2960 ADOK:        LD A,C            ;A/D CNT TO
2970                                ;ACCUMULATOR
2980              RET               ;RETURN TO
2990                                ;MAIN PROG
3000 ADINV:       XOR A             ;ZERO THE
3010                                ;ACCUMULATOR
3020              RET               ;RETURN TO
3030                                ;MAIN PROG
3040 ;
3050 ;
3060 ;END OF LISTING
3070 ;USER MUST DEFINE THESE
3080      ;ORG
3090      ;TOPS
3100      ;PORT0
3110      ;PORT1
3120      ;PORT2
```

355

A PARABOT-CLASS, THREE-AXIS MANIPULATOR CONTROL

Aside from being an example of how to go about putting together a microprocessor-controlled manipulator, the circuits and programs in this section demonstrate the application of servo control to parabot/robot situations. The system uses three different motors to move three independent sections of a rather complicated manipulator. Because this is a parabot machine, the human operator is responsible for dictating the relative positions of all three axes of motion. This is done by turning a set of three position-determining potentiometers.

The servo feature enters the picture at the point where the system attempts to bring the actual position status of the manipulator in line of the corresponding commands from the human operator. The greater the amount of error—or discrepancy between where the manipulator sections are and where they are supposed to be—the faster the motors run. As the error approaches zero for a particular axis of motion, the motor slows. And when the error is reduced to zero (give or take some slight A/D conversion error), the motor comes to a stop.

The human operator thus specifies the positions of the three axes in a direct fashion but has no direct control over the speed of response to changes in position. The program, built around a classic servosystem model, takes care of the speed and direction commands.

Hardware Layout

Figure 11-4 is a block diagram of the port interfacing for this bus-oriented, three-axis manipulator control system. There are six ports that can be conveniently divided into three groups of two. Ports Ø and 1 care of the parameters for manipulator Axis 1, Ports 2 and 3 handle the parameters for Axis 2, and Ports 4 and 5 take care of Axis 3.

Generally speaking, the port configurations are identical for all three manipulator axes. The only difference is that Port 1 includes five additional outputs that are used for the A/D counting operations.

Ports Ø, 2 and 3 have six outputs. The three lower-order bits set the speed and direction of their respective motors, and the three higher-order bits feed motor status information to the LED panel. The three inputs that are common to Ports 1, 3 and 5 pick up the CORR bits required for doing three separate A/D conversions for each axis of motion.

Figure 11-5 carries the process a step farther, showing the output interface of the system. The nine speed and direction bits from Ports Ø, 2 and 3 go to a four-speed, two-direction motor control circuit—one that can be identical to the one described in Fig. 6-18. Of course, this system calls for a triple version of that single-motor circuit.

The active-low signals from each of the three sections of that motor control circuit go to a motor power amplifier. A triple version of the circuit shown in Fig. 6-12 will work quite nicely here.

As indicated in Fig. 11-5, the motor setup has six different mechanical outputs: a speed and position sensor for each of the three axes. As in the

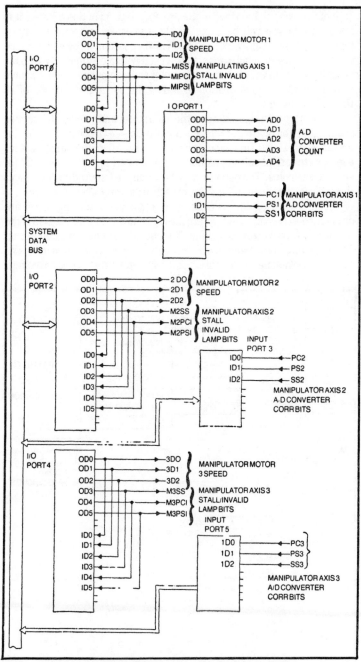

Fig. 11-4. Port configuration for the three-axis manipulator control.

357

case of the system described in the previous section of this chapter, it is assumed that you will be using an electromechanical (generator-type) motor speed-sensing scheme. And it is further assumed you will want the position of each manipulator section sensed with a continuously variable potentiometer. In any case, the main motor and manipulator section should provide mechanical information regarding the speed and position of all three manipulator axes.

The mechanical outputs from the main motor and manipulator section are ganged to an internal status sensing section — a circuit blocked out for you in Fig. 11-6. This block contains a set of buffer amplifiers as appropriate for speed and position sensing. See Figs. 7-25 and 7-27 for specific examples. The outputs of that section are buffered analog signals that indicate the speed and position of each of the three axes of motion.

The raw analog signals from the internal status sensing circuit are directed to A/D comparitors where they are converted with the help of the five-bit D/A converter into six of the nine CORR bits indicated in Fig. 11-4. The three remaining CORR bits for the system come from the user's position controls: a set of three potentiometers that output a voltage

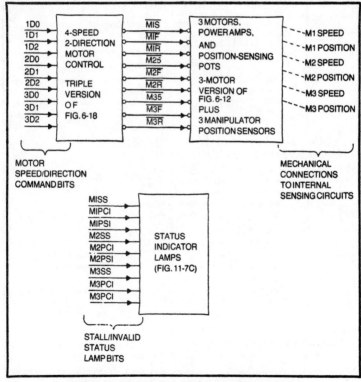

Fig. 11-5. Output hardware for the three-axis manipulator control.

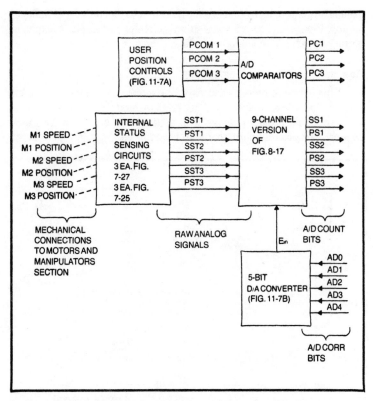

Fig. 11-6. Input hardware for the three-axis manipulator control.

proportional to the desired position of each of the three manipulator sections.

You can get a fairly complete picture of the system by comparing Figs. 11-4, 11-5 and 11-6. If you have done your homework thus far, you should have no trouble filling in the details. The only section not mentioned thus far is the LED lamp panel.

That panel, blocked out in Fig. 11-5, indicates a stalled motor or an invalid A/D conversion for position command and position status signals. MISS, for example, lights a MOTOR 1 STALLED lamp whenever that particular motor is stalled for any reason. The M1PCI signal lights a lamp that might be labeled AXIS 1 COMMAND INVALID, indicating a bad A/D conversion from your control for Axis 1. Finally, the M1PSI signal can light a lamp labeled AXIS 1 POSITION INVALID to indicate a bad A/D conversion for the position sensor attached to Axis 1 of the manipulator.

With luck, you will not see the INVALID lamps lighting very often; they won't light if the axis is properly calibrated and adjusted. You can expect to see the STALL lamps lighting up rather routinely as your manipulator attempts to squeeze things or run to its operational limits of motion. Figure 11-7 shows some specific circuits that might be helpful in getting a good understanding of what is happening here.

Flowchart Analysis of the Programming

This is one of those instances where a main program is quite a bit simpler than the subroutine it calls. As shown on the flowchart in Fig. 11-8, the main program begins with setting the position of the stack pointer. Immediately after that, the program establishes a permanent

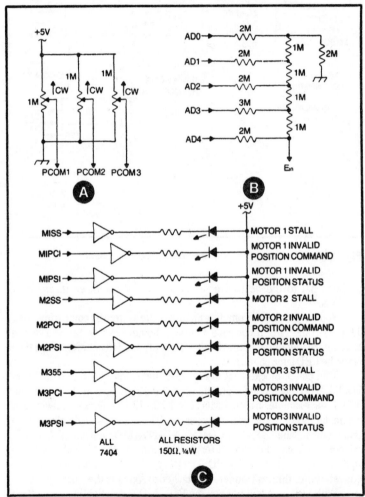

Fig. 11-7. Special circuits for the manipulator control. (A) User's position controls. (B) Five-bit D/A converter. (C) LED indicators and buffer inverters.

address for the Port-1 A/D counter output. With those little tasks out of the way, the main program begins running an endless loop of operations.

The first major step is to point to Port 1 for the purpose of doing some A/D conversions on the parameters for Axis 1. The MSET subroutine then does the actual work. The details of that subroutine will be described shortly.

After the program has dealt with the parameters for M1 (motor 1 or Axis 1 operations), however, it sets up a pointer for the second axis of motion and deals with that. Finally the main program points to Port 5 and calls MSET to take care of the parameters for motor 3, or the 3rd manipulator axis. Then the whole process begins all over with the first motor and axis.

As far as the MSET subroutine is concerned, it can be divided into five basic operations:

□ Convert the user-specified position command.

□ Convert the actual manipulator position status.

□ Compare the position command and status to generate an error signal (actual position minus specified position).

□ Generate a motor speed and direction command based on the magnitude and direction of position error.

□ Convert the actual motor speed status and compare it with the specified speed command to see whether or not the motor is running as it is supposed to be.

The main program calls this MSET sequence for each of the three axes in turn. The result is either a smoothly operating system, or some lights will indicate a problem with motor stalling or bad A/D conversions somewhere along the line.

The first block in the MSET subroutine converts the user-specified position signal into a five-bit binary format. This gives the manipulator a nice resolution of 32 discernable positions. The conditional operation checks the position command (PC) word for a possible overflow, or invalid conditions. If, indeed, the PC word is invalid, the program sets the PCI lamp and ultimately tells the user that something went wrong with the A/D conversion for that control. But assuming the conversion was a good one, the program resets that lamp and saves the position-command word for future reference.

The next step is to check the actual manipulator position, converting its analog signal into a five-bit digital format. As in the case of the position control word, it is checked for a possible problem with making the A/D conversion. And if the conversion is a bad one for some reason, operations turn on the PSI (INVALID POSITION COMMAND) lamp and jump down the program to turn off the motor.

It is important to do these validity checks on the position command and status words. The manipulator is apt to run through some terrible spasms if anything goes wrong with either of those conversion operations.

The MSET subroutine is set up to turn off the motor and light the appropriate INVALID lamp when such a situation arises.

Getting back to the more favorable condition (where neither A/D conversion is invalid), the system calculates the direction and magnitude of position error. If there is no error, the Ø ERROR conditional causes the motor to stop running. Why shouldn't it stop if the manipulator is positioned exactly where you say it is supposed to be?

Of course, there will be positional error whenever you specify a change in position. The next conditional operation figures out whether the actual position is ahead or behind the specified position, and by how much.

If it so happens that the actual position is ahead of where you want it, the system sets up for doing a motor operation in a reverse direction. Otherwise, it sets up for moving the motor in its forward direction. Eventually, the program works out a three-bit word indicating how fast and in what direction the motor should turn, and it sends that speed/direction command to the appropriate motor — the one being serviced at the time.

Finally, the program must compare its own speed/direction command with the actual *speed* status of the motor. As in the case of the system described in the first section of this chapter, the system has no provisions for checking on the actual *direction* of turn of the motor. For all practical purposes, one can safely assume the motor will turn in the right direction, even though the speed might be wrong.

The system checks the speed command (SC) against the actual speed status (SS). If they are the same, everything is going according to plan, and the program turns off the STALL lamp. If there is a discrepancy however, it turns on the appropriate STALL lamp. It is up to the operator to take whatever action is desired for dealing with that stall condition.

Programming the Operations

The program listings for this system follow the flowchart rather closely, but there are a couple of special procedures that must be spelled out carefully. You can use either the 8080A/8085 or Z-80 listings. The line numbers are the same in both instances, and they both operate according to the same flowchart (Fig. 11-8) and general port block diagram (Fig. 11-4).

Bear in mind that the line numbers are for *reference* purposes and are not an integral part of the program. You will have to assign memory locations to the critical terms appearing at the end of the listings and, of course, it is up to you to assemble the source code yourself, making it compatible with the memory map for your own microprocessor system.

The program begins at line 1000, setting the stack pointer. In this particular listing, you will only need two bytes of stack space in ROM, so you won't have trouble coming up with a suitable address value for TOPS.

Line 1010 then sets the H,L pair to the address of Port 1. With this, you will be able to access the A/D counter outputs at any time by simply loading the contents of the A/D counter to the memory address specified by the H,L pair.

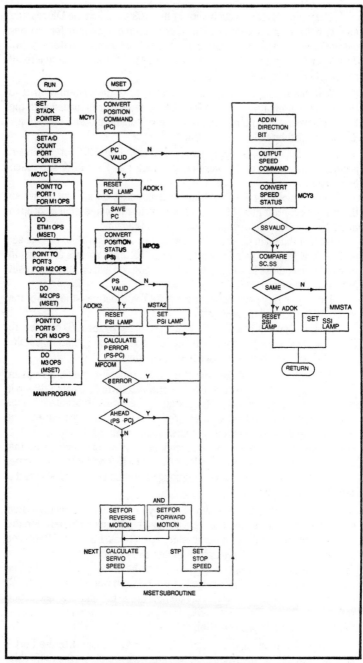

Fig. 11-8. Flowchart for the three-axis manipulator control system.

363

The main program cycle begins at line 1030, loading the DE register pair with the address of Port Ø. This is setting up the program for carrying out operations pertaining to motor 1, or Axis 1 of the manipulator system. That Port Ø address is then carried to the MSET subroutine where all motor-1 operations are carried out.

With the motor-1 operations thus completed, the system returns to the main program at line 1050. The D,E register pair is loaded with the Port 2 address at that point, setting up the system for motor-2 operations. The next line then calls MSET again.

When the motor-2 operations are done, line 1070 of the main programs sets things up for working with motor 3. Then MSET is called once again.

Upon completing the operations for that motor, the system returns to the main program at line 1090. This line returns the whole affair back up to line 1030, and the whole process begins all over again.

All that the main program (between lines 1030 and 1090) does is set up the D, E register pair for working with the three motors and axes in turn. It is an endless looping operation that goes on and on until you reset or turn off the system.

That's the easy part. Now for the MSET subroutine—the part of the program that does all the *real work*.

MSET begins at line 1130 by setting register C, the A/D counter, to zero. That is always the first step in the A/D conversion operation, but the next instruction calls for some special explanation.

As you have seen in the case of the main program, the D, E register pair carries an address relating to the motor or axis to be treated. When it is time to work with motor 1, for instance, the MSET subroutine begins with the address for Port Ø in the register pair. But according to the flowchart and block diagram, the first task of MSET is to do an A/D conversion on the position command setting of the user, and that information is not available from Port Ø. The CORR bit for that job comes from Port 1, instead. So the incrementing instruction in line 1150 sets this port pointer up to Port 1.

As you go through this subroutine, you will find instructions that move the port pointer (the D,E register pair) back and forth between the two ports involved in the data processing operations. So line 1150 sets the port pointer to one port number higher than specified by the calling program. If the main program calls Port Ø at the outset, for instance, this instruction moves the port pointer to Port 1. When working with motor 2, the subroutine begins with the port pointer at Port 2, and this instruction sets it up to Port 3. The general idea is to point to the port that looks at the CORR bits for the A/D conversion.

Line 1190 then loads the contents of the A/D counter to the Port 1 address specified by the H,L pair. Then the system picks up the CORR bits for that motor, or axis, by means of the loading instruction in line 1190.

Line 1210 isolates the position command (PC) correction bit, thus setting up the A/D operation for the user's position command. If the result is a 1 bit, the A/D conversion must be done, and operations move down to ADOK1 in line 1320. But if the conversion is not completed, as indicated by a 0 bit at line 1230, it is time to see whether or not the A/D counter is overflowing and working with an invalid conversion.

To do this, line 1040 moves the maximum A/D count (a five-bit count in this case) to the accumulator and then compares it with the actual A/D count in register C. If these two numbers are equal, the A/D count has reached its maximum point without finding a good A/D conversion value. That situation defines an invalid conversion for the position command parameter. Whenever this happens, the conditional statement in line 1270 carries operations down to MSTA1 in line 1410. In short, that set of instructions lights up the INVALID POSITION COMMAND lamp and ultimately stops the motor.

But assuming the A/D count has not yet reached its maximum point and there has not yet been a good A/D conversion matchup, the program increments the A/D counter (line 1280) and uses a jump instruction to repeat the conversion operation. The actual conversion takes place between MCY1 and ADOK1.

If the conversion has gone well, the instructions for ADOK1 first set the port pointer back to its original place, fetches the current motor and lamp status, sets the bit for the INVALID POSITION COMMAND lamp to 0, and outputs the result. The overall effect is that the INVALID POSITION COMMAND lamp is turned off without affecting the status of any of the other bits at that port.

Here is an important step: The instruction in line 1380 saves the binary value of the good A/D conversion in register B. It will remain there until it is needed for comparing it with the actual position of the manipulator axis.

After saving the digitized value in register B, ADOK1 concludes by doing a jump to MPOS — a set of instructions that check the actual position.

MSTA1 is called whenever there is an invalid A/D conversion — an overflow of the A/D counter in register C. The operations are practically identical to those in ADOK1; except that the INVALID POSITION COMMAND lamp is turned on, the digitized value is *not* saved for future reference, and operations jump to STP (to stop the motor), rather than to MPOS.

There is no point, you see, in attempting to correct the position of the manipulator when the position command, itself, is faulty. It is up to you, the operator of this parabot scheme, to correct an invalid position command by setting the control to a position that creates valid A/D conversions.

In summary, this first phase of MSET attempts to create a five-bit binary number representing the manipulator position you specify at that potentiometer control of the motor. If everything is progressing as it should, the INVALID POSITION COMMAND lamp is turned off and the five-bit binary version of the position command is saved for future reference in register B. But in the event of a A/D overflow, the INVALID POSITION COMMAND lamp is turned on and the motor stops running.

The second phase of the operation is to do an A/D conversion on the actual position of the manipulator axis that is being treated by the subroutine. Those operations begin at line 1510.

Many of the instructions follow the same sequence shown for the first phase. A/D operations are much the same in any siutuation. The A/D counter, for instance, is first set to zero at line 1510, the port pointer is set up to the port carrying the motor's correction bits at line 1570, and the A/D conversion is underway.

Line 1580 isolates the correction bit for the position status (the actual manipulator position) and then goes through a series of conditional operations related to attempting an A/D conversion of that parameter. There is little need for going into a lot of detail at this point, because the operations are so similar to those just described for the position command conversion.

ADOK2 begins a sequence of instructions used whenever the position status conversion is a good one. The INVALID POSITION STATUS lamp is turned off, and the five-bit binary version of the position status is saved by default in register C.

MSTA2 comes into the picture only when there is an invalid A/D conversion of the manipulators actual position status. It turns on the INVALID POSITION STATUS lamp and jumps to a point in the program where the motor is stopped.

If both A/D conversions to this point have been good ones, the program enters the third phase, MPCOM. The actual position command then rests in register B, and the actual position status is in register C. It is time to compare them in order to find out the nature of any discrepancy between them.

The position command word is put into the accumulator by the instruction in line 1880, and the actual position status is subtracted from it by the arithmetic operation in line 1890. If there is no difference between these two position parameters, as detected by the conditional instruction in line 1920, the program jumps down to the point where the motor is stopped. Finding a zero error in this fashion is the goal of every position-changing operation in the system.

But there is bound to be some error whenever you are adjusting the position control potentiometer, and the next step is to determine whether the actual manipulator position is ahead of or behind the point you specify at the control. If it so happens that actual position status is greater than the position command, the manipulator is *ahead* of your specified position. That situation is detected by line 1940. In effect, that instruction says,

"Jump to AHD if the binary value of the actual position is greater than the command position."

If line 1960 is ever executed, it means the actual position of the manipulator is *behind* the position specifed at the control. That instruction sets up a forward-motion bit in register B. If the manipulator is ahead, though, line 1980 is executed instead to set a reverse-motion bit into register B.

In either case, register B is carrying a direction-of-correction bit by the time the program reaches NEXT in line 1990. The next step is to figure out how fast the motor should turn in order to correct the difference in a smooth and efficient fashion. This is where the classic servo control enters the picture.

Line 1990 first checks the magnitude of the error to see if it is larger than hexadecimal 15. That would amount to a rather large error, and line 2000 calls for jumping to a set of instructions that sets up a relatively fast motor response.

If, however, the error is less than 15, line 2020 checks for an error greater than 0A hexadecimal. If the error is greater than that particular value, line 2030 forces the program down to a point where the motor is set for a medium speed.

If the error is less than 0A, however, the error is relatively small, and line 2050 sets a slow response bit into the accumulator. The direction-of-correction bit, saved in register B by line 1960 or 1980, is then added to the correction speed to create a complete motor command.

Operations for MED and FAST are practically identical. Only the magnitude of the correction speed is different. In any case, by the time the program reaches SPED in line 2201, the accumulator is carrying a complete motor command word — one having both direction and speed elements. It is a three-bit command, incidentally, with the highest of the three carrying the direction information and the two lower-order bits carrying the correction speed: FAST, SLOW, MEDIUM or STOP. This marks the end of the servo-response instructions and the beginning of the next phases: outputting the motor command and then checking the motor speed status.

Line 2210 saves the motor speed command for future reference. The program then executes a series of instructions intended to output the new motor speed/direction command without affecting the status of the other bits at that port. That gets us down to line 2280.

In line 2280, the direction bit is eliminated, leaving only the two-bit speed command. This number is saved in register B by the instruction in line 2300. It is then time to do an A/D conversion aimed at determining the actual motor speed. The conversion takes place between lines 2310 and 2500. Except for the CORR bit it deals with, the operation is identical to the one already described for digitizing position command and position status.

367

ADOK3 is executed whenever the actual motor status conversion is a good one. It first compares the actual speed status with the speed command in line 2520. If they are the same — the ideal situation — the instructions in lines 2560 through 2610 simply turn off the MOTOR STALL lamp. The routine concludes by returning to the main program, thus ending the MSET subroutine.

But if the motor is stalled or there is a serious discrepancy between the motor speed command and actual speed status, the program runs the instructions for MMSTA. These instructions turn on the MOTOR STALL lamp before returning operations to the main program.

The system has no provisions for automatically correcting the motor speed. Only the manipulator position parameters are servo controlled.

8080A/8085 Program Listing

PARABOT-CLASS, 3-AXIS MANIPULATOR
CONTROL

8080A/8085 PROGRAM LISTING

```
1000 START:    LXI SP,TOPS    ;SET STACK
1010           LXI H,PORT1    ;POINT TO A/D
1020                          ;PORT 1
1030 MCYC:     LXI D,PORT0    ;SET M1 OPS
1040           CALL MSET      ;DO M1 OPS
1050           LXI D,PORT2    ;SET M2 OPS
1060           CALL MSET      ;DO M2 OPS
1070           LXI D,PORT4    ;SET M3 OPS
1080           CALL MSET      ;DO M3 OPS
1090           JMP MCYC       ;DO IT AGAIN
1100 ;END OF MAINLINE PROGRAM
1110 ;
1120 ;BEGINNING OF MSET SUBROUTINE
1130 MSET:     MVI C,0H       ;ZERO A/D
1140                          ;COUNTER
1150           INX D          ;POINT TO
1160                          ;PC PORT
1170 MCY1:     MOV M,C        ;OUTPUT A/D
1180                          ;COUNT
1190           LADX D         ;FETCH CORR
1200                          ;BITS
1210           ANI 1H         ;ISOLATE PC
1220                          ;CORR BIT
1230           JNZ ADOK1      ;JMP IF DONE
1240           MVI A,1FH      ;SET MAX A/D
1250                          ;COUNT
1260           CMP C          ;SAME AS A/D
1270           JZ MSTA1       ;JMP IF SO
1280           INR C          ;INCREMENT
1290                          ;A/D COUNT
1300           JMP MCY1       ;CONTINUE A/D
```

```
1310                                      ;CONVERSION
1320 ADOK1:      DCX D                    ;POINT TO STAT.
1330                                      ;PORT
1340             LDAX D                   ;GET STATUS
1350             ANI 0EFH                 ;RESET PC IN-
1360                                      ;VALID BIT
1370             STAX D                   ;OUTPUT STATUS
1380             MOV B,C                  ;SAVE STATUS
1390             JMP MPOS                 ;GOTO POS
1400                                      ;CHECKS
1410 MSTA1:      DCX D                    ;POINT TO STAT.
1420                                      ;PORT
1430             LDAX D                   ;GET STATUS
1440             ORI 10H                  ;SET PC IN-
1450                                      ;VALID BIT
1460             STAX D                   ;OUTPUT STATUS
1470             JMP STP                  ;STOP THE MOT.
1480 ;END OF POSITION COMMAND A/D CONVERSION
1490 ;
1500 ;CONVERT ACTUAL POSITION
1510 MPOS:       MVI C,0H                 ;ZERO A/D
1520                                      ;COUNTER
1530             INX D                    ;POINT TO PS
1540                                      ;PORT
1550 MCY2:       MOV M,C                  ;OUTPUT A/D
1560                                      ;COUNT
1570             LDAX D                   ;GET STATUS
1580             ANI 2H                   ;ISOLATE PS
1590                                      ;CORR BIT
1600             JNZ ADOK2                ;JMP IF A/D
1610                                      ;IS DONE
1620             MVI A,1FH                ;SET MAX A/D
1630             CMP C                    ;SAME AS A/D
1640                                      ;COUNT
1650             JZ MSTA2                 ;JMP IF SO
1660             INR C                    ;INCREMENT A/D
1670                                      ;COUNTER
1680             JMP MCY2                 ;CONTINUE A/D
1690                                      ;CONVERSION
1700 ADOK2:      DCX D                    ;POINT TO
1710                                      ;STAT PORT
1720             LDAX D                   ;GET STATUS
1730             AND 0DFH                 ;RESET PS IN-
1740                                      ;VALID BITS
1750             STAX D                   ;OUTPUT STAT
1760             JMP MPCOM                ;JMP TO FIGURE
1770                                      ;SPEED COMM
1780 MSTA2:      DCX D                    ;POINT TO
1790                                      ;STATUS PORT
1800             LDAX D                   ;GET STATUS
1810             ORI 20H                  ;SET PS IN-
```

369

```
1820                                    ;VALID BITS
1830              STAX D                 ;OUTPUT STAT
1840              JMP STP                ;STOP MOTOR
1850 ;END OF MOTOR POSITITION STATUS CONVERT
1860 ;
1870 ;FIGURE MOTOR SPEED COMMAND
1880 MPCOM:   MOV A,B                    ;GET PC BITS
1890          SUB C                      ;SUBTRACT TO
1900                                     ;GET POSITION
1910                                     ;ERROR
1920          JZ STP                     ;STOP MOTOR
1930                                     ;IF NO ERR
1940          JC AHD                     ;JMP IF POS
1950                                     ;IS AHEAD
1960          MVI B,4H                   ;SET FWD DIR
1970          JMP NEXT                   ;
1980 AHD:     MVI B,0H                   ;SET REV DIR
1990 NEXT:    CPI 15H                    ;ERROR BIG
2000          JNC FAST                   ;IF SO, RUN
2010                                     ;FAST
2020          CPI 0AH                    ;ERROR MED
2030          JNC  MED                   ;IF SO, RUN
2040                                     ;MEDIUM
2050          MVI A,1H                   ;SET SLOW
2060          ADD B                      ;INSERT DIR.
2070                                     ;BIT
2080          JMP SPED                   ;OUTPUT NEW
2090                                     ;SPEED
2100 MED:     MVI A,2H                   ;SET MEDIUM
2110          ADD B                      ;INSERT DIR.
2120                                     ;BIT
2130          JMP SPED                   ;OUTPUT NEW
2140                                     ;SPEED
2150 FAST:    MVI A,3H                   ;SET FAST
2160          ADD B                      ;INSERT DIR.
2170                                     ;BIT
2180          JMP SPED                   ;OUTPUT NEW
2190                                     ;SPEED
2200 STP:     XRA A                      ;SET STOP
2210 SPED:    MOV B,A                    ;SAVE SPEED
2220          LDAX D                     ;GET STATUS
2230          ANI 0F8H                   ;CLEAR SPEED
2240                                     ;BITS
2250          ORA B                      ;SET NEW SPD
2260                                     ;BITS
2270          STAX D                     ;OUTPUT STAT
2280          ANI 3H                     ;GET RID OF
2290                                     ;DIR BIT
2300          MOV B,A                    ;SAVE SPEED
2310          MVI C,0H                   ;ZERO A/D
```

```
2320                            ;COUNTER
2330              INX  D        ;POINT TO SS
2340                            ;PORT
2350   MCY3:     MOV  M,C       ;OUTPUT A/D
2360                            ;COUNT
2370              LDAX D         ;GET CORR
2380                            ;BITS
2390              AND  4H        ;ISOLATE SS
2400                            ;CORR BIT
2410              JNZ  ADOK3     ;JMP IF A/D
2420                            ;IS DONE
2430              MVI  A,3H      ;SET MAX
2440                            ;A/D COUNT
2450              CMP  C         ;SAME AS A/D
2460                            ;COUNT
2470              JZ   MMSTA     ;JMP IF SO
2480              INR  C         ;INCREMENT
2490                            ;A/D COUNT
2500              JMP  MCY3      ;CONTINUE
2510                            ;A/D CONV.
2520   ADOK3:    CMP  B         ;COMPARE SS
2530                            ;AND SC
2540              JNZ  MMSTA     ;JMP IF NOT
2550                            ;THE SAME
2560              DCX  D         ;POINT TO
2570                            ;STAT PORT
2580              LDAX D         ;GET STAT
2590              ANI  07FH      ;RESET STALL
2600                            ;BIT
2610              STAX D         ;OUTPUT STAT
2620              RET            ;RETURN TO
2630                            ;MAIN PROG
2640   MMSTA:    DCX  D         ;POINT TO
2650                            ;STAT PORT
2660              LDAX D         ;GET STAT
2670              ORI  8H        ;SET STALL
2680                            ;BIT
2690              STAX D         ;OUTPUT STAT
2700              RET            ;RETURN TO
2710                            ;MAIN PROG
2720   ;
2730   ;END OF MSET SUBROUTINE
2740   ;
2750   ;
2760   ;USER MUST DEFINE THESE
2770        ;START
2780        ;TOPS
2790        ;PORT0
2800        ;PORT1
2810        ;PORT2
2820        ;PORT4
```

PARABOT-CLASS, 3-AXIS MANIPULATOR
CONTROL

Z80 PROGRAM LISTING

```
1000 START:     LD SP,TOPS      ;SET STACK
1010            LD HL,PORT0     ;POINT TO A/D
1020                            ;PORT 1
1030 MCYC:      LD DE,PORT0     ;SET M1 OPS
1040            CALL MSET       ;DO M1 OPS
1050            LD DE,PORT2     ;SET M2 OPS
1060            CALL MSET       ;DO M2 OPS
1070            LD DE,PORT4     ;SET M3 OPS
1080            CALL MSET       ;DO M3 OPS
1090            JP MCYC         ;DO IT AGAIN
1100 ;END OF MAINLINE PROGRAM
1110 ;
1120 ;BEGINNING OF MSET SUBROUTINE
1130 MSET:      LD C,0H         ;ZERO A/D
1140                            ;COUNTER
1150            INC DE          ;POINT TO
1160                            ;PC PORT
1170 MCY1:      LD (HL),C       ;OUTPUT A/D
1180                            ;COUNT
1190            LD A,(DE)       ;FETCH CORR
1200                            ;BITS
1210            AND 1H          ;ISOLATE PC
1220                            ;CORR BIT
1230            JP NZ,ADOK1     ;JMP IF DONE
1240            LD A,1FH        ;SET MAX A/D
1250                            ;COUNT
1260            CP C            ;SAME AS A/D
1270            JP Z,MSTA1      ;JMP IF SO
1280            INC C           ;INCREMENT
1290                            ;A/D COUNT
1300            JP MCY1         ;CONTINUE A/D
1310                            ;CONVERSION
1320 ADOK1:     DEC DE          ;POINT TO STAT.
1330                            ;PORT
1340            LD A,(DE)       ;GET STATUS
1350            AND 0EFH        ;RESET PC IN-
1360                            ;VALID BIT
1370            LD (DE),A       ;OUTPUT STATUS
1380            LD B,C          ;SAVE STATUS
1390            JP MPOS         ;GOTO POS
1400                            ;CHECKS
1410 MSTA1:     DEC DE          ;POINT TO STAT.
1420                            ;PORT
1430            LD A,(DE)       ;GET STATUS
```

```
1440            OR 10H        ;SET PC IN-
1450                          ;VALID BIT
1460            LD (DE),A      ;OUTPUT STATUS
1470            JP STP         ;STOP THE MOT.
1480 ;END OF POSITION COMMAND A/D CONVERSION
1490 ;
1500 ;CONVERT ACTUAL POSITION
1510 MPOS:     LD C,0H        ;ZERO A/D
1520                          ;COUNTER
1530            INC DE         ;POINT TO PS
1540                          ;PORT
1550 MCY2:     LD (HL),C      ;OUTPUT A/D
1560                          ;COUNT
1570            LD A,DE        ;GET STATUS
1580            AND 2H         ;ISOLATE PS
1590                          ;CORR BIT
1600            JP NZ,ADOK2    ;JMP IF A/D
1610                          ;IS DONE
1620            LD A,1FH       ;SET MAX A/D
1630            CP C           ;SAME AS A/D
1640                          ;COUNT
1650            JP Z,1MSTA2    ;JMP IF SO
1660            INC C          ;INCREMENT A/D
1670                          ;COUNTER
1680            JP MCY2        ;CONTINUE A/D
1690                          ;CONVERSION
1700 ADOK2:    DEC DE         ;POINT TO
171                           ;STAT PORT
1720            LD A,(DE)      ;GET STATUS
1730            AND 0DFH       ;RESET PS IN-
1740                          ;VALID BITS
1750            LD (DE),A      ;OUTPUT STAT
1760            JP MPCOM       ;JMP TO FIGURE
1770                          ;SPEED COMM
1780 MSTA2:    DEC DE         ;POINT TO
1790                          ;STATUS PORT
1800            LD A,(DE)      ;GET STATUS
1810            OR 20H         ;SET PS IN-
1820                          ;VALID BITS
1830            LD (DE),A      ;OUTPUT STAT
1840            JP STP         ;STOP MOTOR
1850 ;END OF MOTOR POSITITION STATUS CONVERT
1860 ;
1870 ;FIGURE MOTOR SPEED COMMAND
1880 MPCOM:    LD A,B         ;GET PC BITS
1890            SUB C          ;SUBTRACT TO
1900                          ;GET POSITION
1910                          ;ERROR
1920            JP Z,STP       ;STOP MOTOR
1930                          ;IF NO ERR
```

```
1940              JP C,AHD        ;JMP IF POS
1950                              ;IS AHEAD
1960              LD B,4H         ;SET FWD DIR
1970              JP NEXT         ;
1980 AHD:         LD B,0H         ;SET REV DIR
1990 NEXT:        CP 15H          ;ERROR BIG
2000              JP NC,FAST      ;IF SO, RUN
2010                              ;FAST
2020              CP 0AH          ;ERROR MED
2030              JP NC, MED      ;IF SO, RUN
2040                              ;MEDIUM
2050              LD A,1H         ;SET SLOW
2060              ADD B           ;INSERT DIR.
2070                              ;BIT
2080              JP SPED         ;OUTPUT NEW
2090                              ;SPEED
2100 MED:         LD A,2H         ;SET MEDIUM
2110              ADD B           ;INSERT DIR.
2120                              ;BIT
2130              JP SPED         ;OUTPUT NEW
2140                              ;SPEED
2150 FAST:        LD A,3H         ;SET FAST
2160              ADD B           ;INSERT DIR.
2170                              ;BIT
2180              JP SPED         ;OUTPUT NEW
2190                              ;SPEED
2200 STP:         XOR A           ;SET STOP
2210 SPED:        LD B,A          ;SAVE SPEED
2220              LD  A,(DE)      ;GET STATUS
2230              AND 0F8H        ;CLEAR SPEED
2240                              ;BITS
2250              OR B            ;SET NEW SPD
2260                              ;BITS
2270              LD (DE),A       ;OUTPUT STAT
2280              AND 3H          ;GET RID OF
2290                              ;DIR BIT
2300              LD B,A          ;SAVE SPEED
2310              LD C,0H         ;ZERO A/D
2320                              ;COUNTER
2330              INC DE          ;POINT TO SS
2340                              ;PORT
2350 MCY3:        LD (HL),C       ;OUTPUT A/D
2360                              ;COUNT
2370              LD A,DE         ;GET CORR
2380                              ;BITS
2390              AND 4H          ;ISOLATE SS
2400                              ;CORR BIT
2410              JP NZ,ADOK3     ;JMP IF A/D
2420                              ;IS DONE
2430              LD A,3H         ;SET MAX
```

```
2440                              ;A/D COUNT
2450              CP C            ;SAME AS A/D
2460                              ;COUNT
2470              JP Z,MMSTA      ;JMP IF SO
2480              INC C           ;INCREMENT
2490                              ;A/D COUNT
2500              JP MCY3         ;CONTINUE
2510                              ;A/D CONV.
2520   ADOK3:     CP B            ;COMPARE SS
2530                              ;AND SC
2540              JP NZ,MMSTA     ;JMP IF NOT
2550                              ;THE SAME
2560              DEC DE          ;POINT TO
2570                              ;STAT PORT
2580              LD A,(DE)D      ;GET STAT
2590              AND 07FH        ;RESET STALL
2600                              ;BIT
2610              LD (DE),A       ;OUTPUT STAT
2620              RET             ;RETURN TO
2630                              ;MAIN PROG
2640   MMSTA:     DEC DE          ;POINT TO
2650                              ;STAT PORT
2660              LD A,DE         ;GET STAT
2670              OR 8H           ;SET STALL
2680                              ;BIT
2690              LD (DE),A       ;OUTPUT STAT
2700              RET             ;RETURN TO
2710                              ;MAIN PROG
2720   ;
2730   ;END OF MSET SUBROUTINE
2740   ;
2750   ;
2760   ;USER MUST DEFINE THESE
2770       ;START
2780       ;TOPS
2790       ;PORT0
2800       ;PORT1
2810       ;PORT2
2820       ;PORT4
```

A SIMPLE PARABOT WITH PRESCRIBED RESPONSES

The machine described in this section is something quite different from the two previous examples in this chapter. In this case, the human operator has *no* direct control over the behavior of the machine.

The machine "wants" to run forward at full speed all the time, but of course that is not possible. Whenever the path of the machine is obstructed by an immovable object, it pulls a user-prescribed response out of its memory and executes that response for several seconds.

If, at the end of the response-trying time, the machine is no longer stalled against the object (or any other, for that matter), it resumes its fast, forward mode of motion. But if it so happens that the machine is still stalled at the end of the response-trying period, it pulls an alternate response from memory and tries that one.

While the human operator does not direct the motion of the machine while it is running around the floor, the operator has prescribed the responses during the *programming phase* of the project. These responses are *not* truly intelligent responses. They have been figured out in advance by the human operator, and that defines the machine as a *parabot*, as opposed to a true robot. The point of presenting this particular sort of machine in this chapter is to demonstrate how modes of machine behavior can be prescribed by a programmer and entered into memory in such a way that the machine can get to those responses when they are needed.

The General Hardware Scheme

Figure 11-9 is a block diagram of this fairly simple system. There is just one I/O port, Port Ø. The four lower-I/O bits are responsible for setting the direction of motion of the machine. The motion format is rather similar to that specified in Table 5-6. The fifth I/O bit, labeled STA, sounds an audible alarm whenever the machine encounters an immovable object and one or both of its drive/steer motors is stalled.

Input-only bits at positions ID6 and ID7 of the I/O port carry single-bit logic levels that indicate whether or not the right and left motors are turning. In this case, those lines are at a logic-1 level whenever the designated motor is *not* turning.

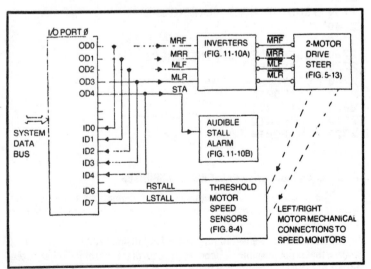

Fig. 11-9. Hardware and port layout for the simple parabot.

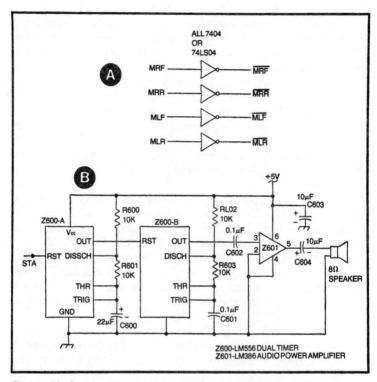

Fig. 11-10. Special circuits for the simple parabot system. (A) Motor logic inverters. (B) Audible stall alarm.

So the machine moves around the floor, operating from motion codes generated by the microprocessor programming and available from the four lower-order bits of the I/O port. Whenever the machine is stuck against an object in the room, an audible alarm sounds until the machine is free and running forward again. Two motor-speed monitors watch over the speed of the drive/steer motors and generate logic-1 levels whenever their motors are stopped. Whether a stopped-motor condition is actually spelling out a stall condition is up to the program to determine.

Figure 11-10 shows some of the special circuits required here. The purpose of the motion-command inverters is to put the motion commands from the I/O port into an active-low format required for the two-motor, drive/steer power amplifier shown in Fig. 5-13.

The stall alarm circuit shown in Fig. 11-10B generates a beeping sound whenever STA goes to logic 1. It continues making that sound until STA drops to logic 0 again.

The whole example is pointless, however, without getting into the matter of prescribing motion codes under a variety of possible object-contact situations. This matter is summarized in Table 11-1.

Table 11-1A shows six different motion codes in a hexadecimal format. These are the four-bit codes from the four lower-order bits of the I/O port. Whenever the machine is executing a 1 code, for instance, it is supposed to be running in a forward direction with a left-turn component. While executing a motion code of hexadecimal two, however, the machine tries to run in reverse with a left-turn component. These six motion codes comprise the total heirarchy of motions of this machine.

Table 11-1B suggests how the machine should respond whenever it is stalled while executing one of six possible motion codes. The first column lists the motion codes the machine is executing at the time a contact or stall condition occurs, while the second column suggests the corresponding response.

For example suppose that the machine is running forward when it runs into something. The motion code being executed at that time is a hexadecimal 5—FORWARD. The prescribed response to a contact while running FORWARD is a hexadecimal 2—REVERSE, LEFT TURN. That should get the machine away from the contact situation.

But suppose it doesn't. Suppose doing a REVERSE, LEFT TURN fails to remedy the contact situation. The response to a contact while executing that motion (hexadecimal 2) is to try a hexadecimal 1—FORWARD, LEFT TURN.

According to Table 11-1B, every possible motion code has a motion-code response associated with it. It is up to the programmer to specify the response codes for every possible contact situation. The

Table 11-1. Motion Codes for the Simple Parabot System. (A) Definition of the Six Available Motion Codes. (B) Motion Code Responses Called by a Contact While a Different Motion Code is Being Executed.

Motion Code (hex)	A	Machine Response
1		FORWARD, LEFT TURN
2		REVERSE, LEFT TURN
4		FORWARD, RIGHT TURN
5		FORWARD
8		REVERSE, RIGHT TURN
A		REVERSE

Motion Code at Stall	B	Programmed Motion Code Response
1		A
2		1
4		8
5		2
8		5
A		4

responses suggested here are pretty good ones, but you can certainly alter them to suit your own taste and ideas about how the machine should respond.

Flowchart Analysis of the Program

The flowchart for this relatively simple system is in Fig. 11-11. It begins by loading the prescribed responses into memory and making them accessible to the machine. That is a one-time operation. If you are satisfied with the results, those responses can be preserved in a custom ROM. For our immediate purposes, assume you must load the response memory at some location in RAM space.

Once the responses are in memory, the next step is to get the machine running across the floor in its typical FORWARD motion. That is accomplished by outputting the FORWARD motion code, 05H.

The next step is to check for a stall condition—one indicating contact with some sort of obstacle. If there is no stall, as picked up by the STALL conditional, the system simply reloads the FORWARD motion code. But if a stall condition exists, as determined by comparing the two stall bits with the motion code being executed at the time, the system looks to its memory for the prescribed response. It then outputs that prescribed response and, incidentally, begins sounding the stall alarm. It tries the prescribed response for several seconds, as determined by the length of the TIME DELAY operation. Then it checks to see if a stall condition still exists.

The machine continues running through the STALL loop while fetching prescribed responses, outputting those responses and trying them through the time delay interval until the stall situation is cleared up. When there is no longer a STALL condition, the system loops back up to the point where it outputs the FORWARD motion code and silences the alarm. It remains in that loop until the next stall condition arises.

How the Program Works

The mnemonics for the program are shown in 8080A/8085 and Z-80 formats in the next two sections of this chapter. Follow through either of them, comparing the ideas in the discussion with the flowchart and system block diagram.

The first phase of the program sets the starting point for the response memory and then loads the set of prescribed responses. All this occupies lines 1000 through 1170.

The stating point for the memory is loaded into the H,L pair at line 1010. It is up to you to select the address value for MEM; pick some place in RAM that will not disturbed by other programming operations.

The prescribed responses are selected on the basis of the motion code being executed at the moment contact occurs. There is no 0 motion code, so line 1020 moves the memory pointer up one address to MEM plus

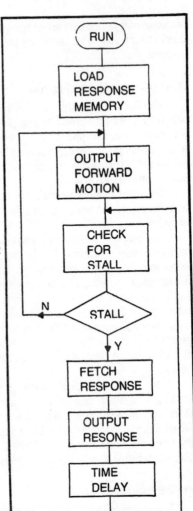

Fig. 11-11. Flow chart for the parabot with prescribed memory response.

1. This location is where the response for dealing with a stall condition (while executing motion-code 1) should be loaded. The response is a hexadecimal A (see Table 11-1B), so that A is placed into the appropriate memory address by line 1030 of the listing. Actually, a 1A is put into that place. The higher-order byte is responsible for turning on the audible alarm. So really all the prescribed responses stored in this section of memory are preceded by a 1, thus turning on the alarm whenever the system calls up a response for a stall condition.

Through the remainder of the memory-loading operation, the response-memory pointer is incremented to a place determined by the value of the motion code being executed. Then the appropriate response is loaded.

Line 1210 makes the H,L register pair point to the Port Ø, memory-mapped address. Line 1230 then outputs the motion code necessary for making the machine run FORWARD and silencing the audible stall alarm.

While some hardware for detecting stall conditions was described earlier in this book, that particular task is carried out by the software in this case. The stall-checking operations include lines 1270 through 1600.

Rather than going into a lot of tedious detail concerning the theory of operation, it is sufficient to say that the STCHK instructions solve the following Boolean expression:

$$STALL=(MRF \oplus MRR)RSTALL \oplus (MLF \oplus MLR)LSTALL$$

A STALL condition exists, for instance, if the right motor is supposed to be turning (MRF $\oplus$ MRR=1) *and* that motor is *not* turning (RSTALL=1). A STALL can also occur if the left motor is supposed to be turning, but it is not.

If the motor is not stalled, as picked up by the instruction in line 1600, the program loops back to FWD in line 1230. If the stall condition exists, however, the program selects the prescribed response from the response memory. See lines 1640 through 1740.

The prescribed response is output to the motor control circuitry by line 1740, and the next step is to give the machine some time to try it out. This is the job of the time delay instructions, lines 1770 through 1840.

Incidentally, if you build this system and find the time delay for your particular machine is not long enough, you can stretch the delay time by making the byte loaded into register B at line 1770 a larger hexadecimal number. Of course, you can get shorter delays by reducing that value. Line 1870 marks the end of the time delay, and the program does an unconditional jump back to STCHK.

8080A/8085 Program Listing

```
PARABOT WITH PRESCRIBED MEMORY
8080A/8085 PROGRAM LISTING


1000 START:        ;LOAD THE RESPONSES
1010               LXI H,MEM        ;POINT TO MEM
1020               INX H            ;POINT TO 1
1030               MVI M,1AH        ;RESP. FOR 1
1040               INX H            ;POINT TO 2
1050               MVI M,11H        ;RESP. FOR 2
1060               INX H            ;
1070               INX H            ;POINT TO 4
1080               MVI M,18H        ;RESP. FOR 4
1090               INX H            ;POINT TO 5
```

```
1100            MVI M,12H       ;RESP. FOR 5
1110            INX H           ;
1120            INX H           ;
1130            INX H           ;POINT TO 8
1140            MVI M,15H       ;RESP. FOR 8
1150            INX H           ;
1160            INX H           ;POINT TO A
1170            MVI M,14H       ;RESP. FOR A
1180                            ;
1190 ;END OF RESPONSE MEMORY LOADING
1200 ;
1210            LXI H,PORT0     ;POINT TO
1220                            ;I/O PORT
1230 FWD:       MVI M,5H        ;OUTPUT FWD
1240                            ;AND SILENCE
1250                            ;THE ALARM
1260 ;BEGIN STALL CHECKS
1270 STCHK:     MOV A,M         ;GET STAT
1280            ANI 1H          ;ISOLATE MRF
1290            MOV B,A         ;SAVE IN B0
1300            MOV A,M         ;GET STAT
1310            ANI 2H          ;ISOLATE MRR
1320            RAR             ;MRR TO A0
1330            XRA B           ;EX-OR WITH
1340                            ;MRF
1350            MOV B,A         ;SAVE IN B0
1360            MOV A,M         ;GET STAT
1370            RAL             ;RSTA TO A7
1380            RAL             ;RSTA TO CY
1390            RAL             ;RSTA TO A0
1400            ANA B           ;AND WITH B0
1410            MOV B,A         ;SAVE IN B0
1420            MOV A,M         ;GET STAT
1430            ANI 4H          ;ISOLATE MLF
1440            MOV C,A         ;SAVE IN C2
1450            MOV A,M         ;GET STAT
1460            ANI 8H          ;ISOLATE MLR
1470            RAR             ;MLR TO A2
1480            XRA C           ;EX-OR WITH
1490                            ;MLF
1500            MOV C,A         ;SAVE IN C2
1510            MOV A,M         ;GET STAT
1520            RAL             ;LSTA TO CY
1530            RAL             ;LSTA TO A0
1540            RAL             ;LSTA TO A1
1550            RAL             ;LSTA TO A2
1560            ANA C           ;AND WITH C2
1570            RRC             ;RESULT TO A1
1580            RRC             ;RESULT TO A0
1590            ORA B           ;OR WITH B0
```

382

```
1600            JZ FWD           ;JMP IF NOT
1610                             ;STALLED
1620 ;END OF CHECK FOR STALL
1630 ;FETCH APPROPRIATE RESPONSE FROM MEMORY
1640            MOV A,M          ;GET STAT
1650            LXI H,MEM        ;POINT TO MEM
1660            ANI 0FH          ;ISOLATE MOTION
1670                             ;COMMAND
1680            MOV B,A          ;SAVE IN B
1690            DAD B            ;ADD TO MEM
1700                             ;POINTER
1710            MOV A,M          ;GET RESPONSE
1720            LXI H,PORTO      ;POINT TO
1730                             ;I/O PORT
1740            MOV M,A          ;OUTPUT RESP
1750 ;RESPONSE NOW BEING TRIED
1760 ;DO A TIME DELAY
1770            MVI B,0FH        ;SET HIGH
1780                             ;BYTE
1790 SETC:      MVI C,0FFH       ;SET LOW BYTE
1800 CNTC:      DCR C            ;COUNT DOWN
1810            JNZ CNTC         ;COUNT AGAIN
1820                             ;IF NOT ZERO
1830            DCR B            ;COUNT DOWN
1840            JNZ SETC         ;START LOW
1850                             ;BYTE COUNT
1860                             ;AGAIN
1870            JMP STCHK        ;CHECK FOR
1880                             ;STALL
1890 ;END OF PROGRAM LISTING
1900 ;
1910 ;
1920 ;USER MUST DEFINE THESE
1930     ;START
1940     ;MEM
1950     ;PORTO
```

Z80 Program Listing

PARABOT WITH PRESCRIBED MEMORY
Z80 PROGRAM LISTING

```
1000 START:     ;LOAD THE RESPONSES
1010            LD HL,MEM        ;POINT TO MEM
1020            INC HL           ;POINT TO 1
1030            LD (HL),1AH      ;RESP. FOR 1
1040            INC HL           ;POINT TO 2
1050            LD (HL),11H      ;RESP. FOR 2
1060            INC HL           ;
1070            INC HL           ;POINT TO 4
```

```
1080              LD (HL),18H      ;RESP. FOR 4
1090              INC HL           ;POINT TO 5
1100              LD (HL),12H      ;RESP. FOR 5
1110              INC HL           ;
1120              INC HL           ;
1130              INC HL           ;POINT TO 8
1140              LD (HL),15H      ;RESP. FOR 8
1150              INC HL           ;
1160              INC HL           ;POINT TO A
1170              LD (HL),14H      ;RESP. FOR A
1180                               ;
1190 ;END OF RESPONSE MEMORY LOADING
1200 ;
1210              LD HL,PORT0      ;POINT TO
1220                               ;I/O PORT
1230 FWD:         LD (HL),5H       ;OUTPUT FWD
1240                               ;AND SILENCE
1250                               ;THE ALARM
1260 ;BEGIN STALL CHECKS
1270 STCHK:       LD A,(HL)        ;GET STAT
1280              AND 1H           ;ISOLATE MRF
1290              LD B,A           ;SAVE IN B0
1300              LD A,(HL)        ;GET STAT
1310              AND 2H           ;ISOLATE MRR
1320              RRC A            ;MRR TO A0
1330              XOR B            ;EX-OR WITH
1340                               ;MRF
1350              LD B,A           ;SAVE IN B0
1360              LD A,(HL)        ;GET STAT
1370              RL A             ;RSTA TO A7
1380              RL A             ;RSTA TO CY
1390              RL A             ;RSTA TO A0
1400              AND B            ;AND WITH B0
1410              LD B,A           ;SAVE IN B0
1420              LD A,(HL)        ;GET STAT
1430              AND 4H           ;ISOLATE MLF
1440              LD C,A           ;SAVE IN C2
1450              LD A,(HL)        ;GET STAT
1460              AND 8H           ;ISOLATE MLR
1470              RRC A            ;MLR TO A2
1480              XOR C            ;EX-OR WITH
1490                               ;MLF
1500              LD C,A           ;SAVE IN C2
1510              LD A,(HL)        ;GET STAT
1520              RLA              ;LSTA TO CY
1530              RL A             ;LSTA TO A0
1540              RL A             ;LSTA TO A1
1550              RL A             ;LSTA TO A2
1560              AND C            ;AND WITH C2
1570              RRC A            ;RESULT TO A1
```

```
1580            RRC A               ;RESULT TO A0
1590            OR B                ;OR WITH B0
1600            JP Z,FWD            ;JMP IF NOT
1610                                ;STALLED
1620 ;END OF CHECK FOR STALL
1630 ;FETCH APPROPRIATE RESPONSE FROM MEMORY
1640            LD A,(HL)           ;GET STAT
1650            LXI H,MEM           ;POINT TO MEM
1660            ANI 0FH             ;ISOLATE MOTION
1670                                ;COMMAND
1680            LD B,A              ;SAVE IN B
1690            ADD HL,BC           ;ADD TO MEM
1700                                ;POINTER
1710            LD A,(HL)           ;GET RESPONSE
1720            LD HL,PORT0         ;POINT TO
1730                                ;I/O PORT
1740            LD (HL),A           ;OUTPUT RESP
1750 ;RESPONSE NOW BEING TRIED
1760 ;DO A TIME DELAY
1770            LD B,0FH            ;SET HIGH
1780                                ;BYTE
1790 SETC:      LD C,0FFH           ;SET LOW BYTE
1800 CNTC:      DEC C               ;COUNT DOWN
1810            JP NZ,CNTC          ;COUNT AGAIN
1820                                ;IF NOT ZERO
1830            DEC B               ;COUNT DOWN
1840            JP NZ,SETC          ;START LOW
1850                                ;BYTE COUNT
1860                                ;AGAIN
1870            JP STCHK            ;CHECK FOR
1880                                ;STALL
1890 ;END OF PROGRAM LISTING
1900 ;
1910 ;
1920 ;USER MUST DEFINE THESE
1930     ;START
1940     ;MEM
1950     ;PORT0
```

An Independent Parabot With Nest Search

12

My first full-scale robot project, through 1974-1975, evolved a little creature I called Buster. Buster is the central figure in my book, TAB book No. 841, *Build Your Own Working Robot.* While microprocessors were widely available at that time, the cost was rather high and few experimenters had the foggiest notion about how to use them. So Buster was built around a purely 7400-series TTL technology.

The creature described in this chapter is an updated version of old Buster. The machine attempts to run across the floor in a full-speed, straight-ahead fashion. And when it gets hung up by an obstacle in its path, the machine consults its response memory for motion codes that promise to get it away from the obstacle. The format in this case is identical to the one described in the final section of Chapter 11.

In addition to blundering around a room, this machine also monitors its main battery level and begins seeking out a nest when the battery reaches a prescribed low-level point. The next, of course, houses a battery charger and a bright light the machine uses for finding it.

Upon noting a low-battery condition, the machine goes into a major search pattern, moving in a wide circular pattern and looking for the bright light attached to the nest. When the machine "sees" that light, it does a fast forward motion, moving directly toward the light and nest. Of course, some obstacles can get in the way, and the response-to-stall mechanism must take priority over the need to recharge the batteries.

Once the machine gets away from that stall situation, it begins looking for the nest again. If the machine happens to be running toward the nest, but suddenly loses sight of the light, it goes into a minor search pattern, moving to the left a bit and then to the right. The idea is to find the light again without running the risk of really getting lost by doing the major search pattern.

All of this is taking place when the machine notes a low battery level. When the battery is fairly well charged, the machine ignores the nest light.

Whenever the machine is in its nest-searching mode of operation, we can only hope it locates and connects to the battery charger before the battery becomes completely exhausted. But assuming the machine does finally make a good connection with the next, it begins monitoring the battery recharge current. It remains at the nest—soaking up fresh battery power—until the recharge current falls below a prescribed level. Then the machine treats the nest as any other obstacle in its path and moves away to resume its normal, blundering activity.

In the perspective of robotics, this is a rather high-level *parabot* machine. It is capable of caring for its own needs and surviving rather well *without* human intervention. The only feature that knocks it out of the realm of real robots is the fact that all of its responses to the environment are *prescribed* in its memory. It makes no decisions independent of the program specified earlier by the programmer who gave birth to the creature.

In keeping with the general character of my earlier Buster machine, this system allows a human operator to intervene at any time, taking over complete control in a manual fashion. The system thus includes a manual control panel that you can enable or disable as you see fit.

There are two reasons for presenting this particular system here. First, it is an example of a complete and independent machine. Second, it demonstrates the operation of a moderately sophisticated goal-seeking routine—seeking out and tracking down the nest whenever the main battery voltage drops to a critically low level.

And, incidentally, the system includes an example of how to interface with *optical speed sensors*. All the systems demonstrated in Chapter 11 used the *electromechanical* speed-sensing scheme.

THE PORT AND PERIPHERAL HARDWARE LAYOUT

Figure 12-1 shows the ports and peripherial circuitry for this independent parabot system. Port 0 has only four output lines that are used for setting up the motion codes of the machine. The motion format, built around a two-motor drive/steer system, is identical to the system described in Fig. 11-9. If you have already studied that example, you will have no trouble seeing how the drive/steer mechanism for this one works.

All eight inputs to Port 0 are used here. The four lower-order bits simply pick up the current motor-control code. The next two bits, IP04 and IP05, are connected to the signal outputs of an optical speed-sensing circuit. Whenever a motor is turning, the signal at the corresponding input port of Port 0 is a series of TTL-compatible pulses that have a frequency proportional to the running speed of the motor. The circuits generating the MLSP and MRSP pulses are identical to those described in connection with the circuit in Fig. 7-28. Two of them are needed here, of course—one for each motor.

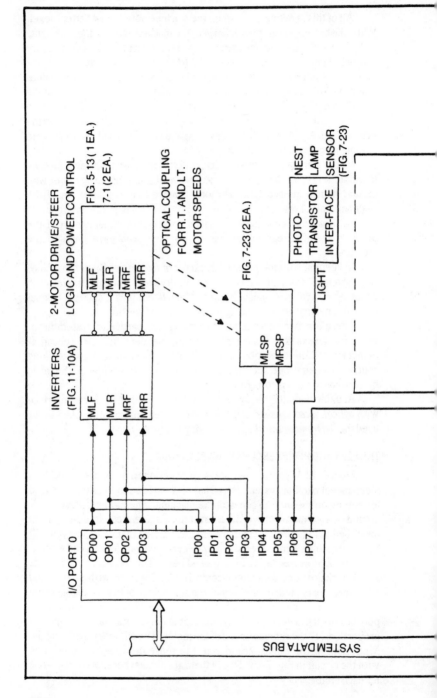

388

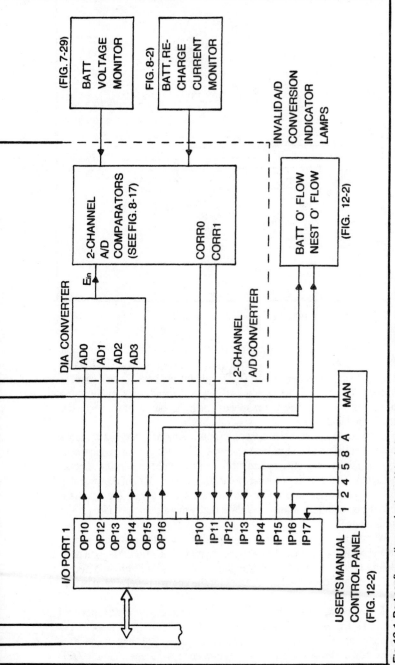

Fig. 12-1. Port configuration and external block diagram for the nest-seeking parabot.

389

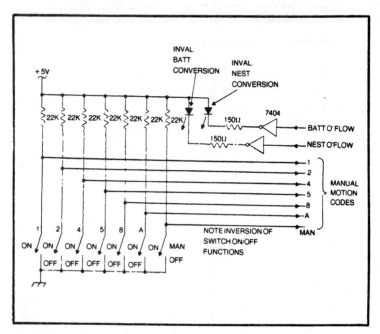

Fig. 12-2. Special circuits for the nest-seeking parabot.

The machine must have some provisions for detecting the bright light on the nest assembly. This light-sensing circuit generates a LIGHT signal that is applied to bit IP06 of Port 0. The phototransistor, in this case, is mounted to the front of the machine. Its signal is conditioned by the same circuit used for monitoring motor speed. The light source, however, is the light attached to the nest.

I/O Port 1 handles A/D conversion operations and provides the input point for manual control signals. As far as A/D conversions are concerned, the system does two of them: one for the battery voltage sensing scheme and another for the battery recharge current sensor. Both of those analog inputs are converted into a four-bit binary format, giving the results a resolution of 16 different levels. This is really more than adequate for the system at hand.

Whenever you want to get manual control over the machine, simply set a MAN switch on your control panel to its ON position. You will have complete control over the activity of the machine by working six different motion code switches. See the circuits for the control panel shown in Fig. 12-2.

The labels on those control circuits reflect the motion codes specified in Table 11-1A. While that table does not show a STOP motion code, it is possible to manually stop the machine in this case. One way to specify a STOP code is by setting the six motion-code switches to their OFF positions. Another way is to set more than one motion-code switch to their ON position. No switch does anything, however, if the MAN switch is set to its OFF position.

Before it causes you any difficulties, please note that the ON/OFF designations for the switches are reversed. This is not an error in the book. Whenever a switch is in the position labeled OFF, it should be closed to output a logic-0 level. And whenever a switch is in the position labeled ON, it should be open to let its pull-up resistor output a logic-1 level.

The two lamps included on this control panel are used only for spotting some alignment problems associated with the two A/D conversions. These are actually A/D conversion overflow indicators. If either lamp lights up, its corresponding A/D conversion is not finding a valid binary value for the analog level being converted.

MAIN PROGRAM FLOWCHART

Figure 12-3 is the overall flowchart for the system. The first set of operations are performed only when the program is started. The first block sets the stack pointer for the program and locates the beginning of the response memory. The second operation simply loads the prescribed responses to normal contact situations.

The actual operating routine begins by outputting a FORWARD motion code that begins the machine running across the floor in a straight-ahead fashion. Once that is underway, the program checks the MAN switch to see whether or not you want to take over manual control. If so, the system goes off to a subroutine that lets you do that. There are a couple of interesting wrinkles built into the manual control routine, but you will have to wait a while before finding out what they are.

If the system does not see that you want to run it manually, or when your manual operations are done, the system then checks for a stall condition. This is the only provision available for sensing contact with an immovable object. The STALL conditional in the flowchart goes through a set of operations that detect a stall for either drive/steer motor.

If, indeed, the system picks up a stall condition, it goes to a STALL RESPONSE routine that fetches an appropriate sort of response from the response memory. The program does not leave that routine until a successful stall-curing motion code is found and tested.

Finally, the system checks its own battery level. If it is low, the machine executes a complicated series of steps aimed at finding the nest and battery charger, making a good connection with the battery charger, sitting at the charger until the battery is charged and moving away.

The program then loops all the way back to the point where it outputs the normal, fast-forward motion code, and things begin all over again. Ideally, the system runs the loop indefinitely, keeping things fairly stirred up in your workshop.

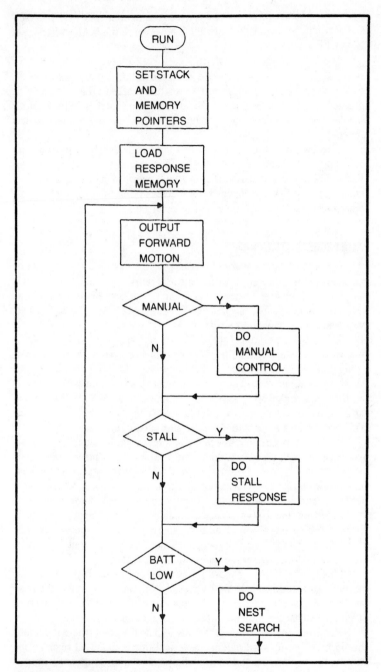

Fig. 12-3. Main program flowchart.

392

MAIN PROGRAM LISTING

Here is the listing for the main program operations. This is all written in Intel (8080A/8085) mnemonics. If you are using a Z80 system, you shouldn't have much trouble making the conversion yourself.

```
1000 ;BEGINNING OF MAIN PROGRAM
1010 START:                      ;
1020 ;SET THE POINTERS
1030          LXI SP,TOPS        ;SET STACK
1040          LXI H,MEM          ;SET BEGINNING
1050                             ;OF MEMORY
1060 ;LOAD THE RESPONSE MEMORY
1070          INX H              ;POINT TO 1
1080          MVI M,0AH          ;LOAD 1
1090          INX H              ;POINT TO 2
1100          MVI M,1H           ;LOAD 2
1110          INX H              ;
1120          INX H              ;POINT TO 4
1130          MVI M,8H           ;LOAD 4
1140          INX H              ;POINT TO 5
1150          MVI M,2H           ;LOAD 5
1160          INX H              ;
1170          INX H              ;
1180          INX H              ;POINT TO 8
1190          MVI M,5H           ;LOAD 8
1200          INX H              ;
1210          INX H              ;POINT TO A
1220          MVI M,4H           ;LOAD A
1230 ;DO STRAIGHT FORWARD
1240 FWD:     LXI H,PORTO        ;POINT TO
1250                             ;PORT 0
1260          MVI M,5H           ;OUTPUT FOR-
1270                             ;WARD
1280 ;CHECK FOR MANUAL CONTROL
1290          MVI A,M            ;GET STATUS
1300          ANI 80H            ;ISO. MAN BIT
1310          JZ STRT            ;JMP IF NOT
1320 ;ELSE GOTO MANUAL CONTROL
1330          CALL MAN           ;
1340 ;CHECK FOR A STALL CONDITION
1350 STRT:    CALL STCHK         ;
1360          JZ BTRT            ;JMP IF NOT
1370 ;ELSE CALL STALL ROUTINE
1380          CALL STALL         ;
1390 ;CHECK THE BATTERY VOLTAGE
1400 BTRT:    CALL BATLOW        ;
1410          JC FWD             ;JMP IF OK
1420 ;ELSE BEGIN SEARCHING FOR NEST
1430          CALL SRCH          ;
1440 ;THEN START ALL OVER AGAIN
```

```
1450            JMP FWD
1460    ;
1470    ;END OF MAIN PROGRAM LISTING
1480    ;USER MUST DEFINE THESE
1490        ;START
1500        ;MEM
1510        ;PORT0
1520    ;ASSEMBLER WILL DEFINE THESE
1530        ;FWD
1540        ;STRT
1550        ;BTRT
1560    ;
1570    ;
```

The beginning of the main program is specified by the START label in line 1010. You will have to assign that address as appropriate for your own microprocessor system.

The instruction in line 1030 sets the address of the top of the stack. The address for TOPS is, again, something you must determine for your system. And the same is true for the location of the response memory that is loaded into the HL pair in line 1040. Bear in mind that the stack will grow downward, while the response memory will be loaded upward; make sure the two will not crash into one another.

Lines 1070 through 1220 are devoted to loading appropriate responses into the response memory. The rationale behind these operations has already been described in connection with the final system example in Chapter 11. By the time the system gets down to line 1230, it is ready to go.

Line 1240 shows that the HL pair is used as a pointer for Port 0 operations. This is a memory-mapped scheme, and by setting the HL pair to the Port 0 address, it is possible to swap information around by means of single-byte instructions.

In fact, this idea begins paying off in the very next instruction (line 1260). That instruction outputs the FORWARD motion code in a rather efficient fashion.

With the pointers now set, the response memory fully loaded and the machine running FORWARD, the next step is to check for possible manual takeover of the operations. The instruction in line 1290 picks up the input information from Port 0. The MAN switch bit is included in that byte, and it is isolated by the instruction in line 1300. If MAN is set (indicating your desire to take over manual control) a 1 bit remains in the accumulator and the conditional instruction in line 1310 is *not* satisfied. If that's so, the program goes by default to line 1330 and calls the manual-operation subroutine that gives you manual control over the system.

If the MAN bit isolation instruction in line 1300 turns up an accumulator full of zeros, you do not want manual control. The conditional in line 1310 is then satisfied. The program thus jumps down to STRT in line 1350. Even if you do elect to run the system manually for a while, the program returns to that STRT in line 1350 when you are done. So whether or not you elect a manual control over the system, the program eventually runs line 1350. And that line begins a series of instructions aimed at testing for a possible motor-stalling condition—contact with an immovable object in the environment.

The instruction in line 1350 calls for a stall-checking subroutine labeled STCHK. That routine, as you will see later in this chapter, returns to the main program at line 1360 carrying a logic-1 bit in the Z-flag register if one or both motors is stalled. Otherwise, it returns with a Z-flag value of 0.

The conditional instruction in line 1360 tests the Z-flag value. If a stall condition exists, that conditional instruction is not satisfied and the program defaults to line 1380. This CALL instruction calls a subroutine (labeled STALL) that fetches appropriate stall responses from the response memory. And the program doesn't return until the current stall condition is remedied.

If there is no stall condition, as detected by the conditional in line 1360, or if a stall condition has been cleared up by means of the STALL subroutine, operations pick up at line 1400.

The instruction in line 1400 calls a battery voltage-checking subroutine labeled BATLOW. It is essentially an A/D conversion operation of the analog voltage level from the battery. That subroutine returns to line 1410 carrying a CY-flag bit of 0 if the battery is low. Otherwise, it returns with a CY-flag bit of 1.

So the conditional instruction in line 1410 restarts the operational part of the main program if it finds the CY flag at logic 1 (indicating a good battery level). But if the BATLOW subroutine returns with the CY bit set to 0, the program goes to line 1430. That line calls the nest-searching subroutine, SRCH. The system continues executing this subroutine until the battery is fully charged.

With the battery thus ready for full operation again, SRCH returns to line 1450, and the unconditional jump in that line returns the whole program back up to FWD (line 1240) to resume normal running.

MANUAL OPERATION SUBROUTINE

Figure 12-4 is the flowchart for this relatively simple subroutine. It should be simple, because the user is doing all the thinking for the machine.

This subroutine is called from several different places, including the main program. In each case, however, it is called only when the user sets the MAN switch on the control panel to its ON position. A software test of that switch condition, in fact, is the signal that calls this routine.

The subroutine begins by setting some relevant pointers for the I/O ports involved in the manual control of the machine. Then the program fetches the motion-code command from the user's control panel, debouncing the switches as necessary.

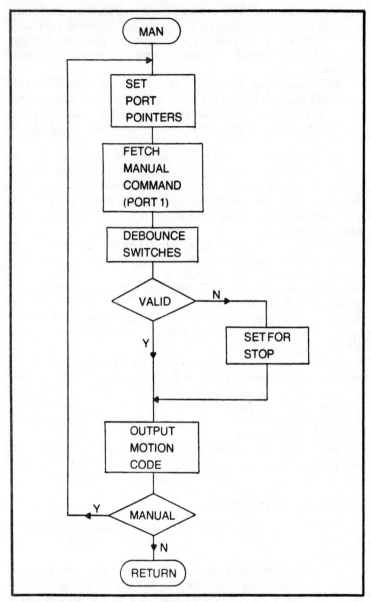

Fig. 12-4. Manual control subroutine flowchart.

It is necessary to debounce the switch contacts. Otherwise, the microprocessor system might well respond to the rapid vibration of the switch contacts as they close, thereby causing the subroutine to execute some potentially weird motor-code responses.

Earlier versions of switch debouncing circuits used a hardware procedure, effectively delaying the output of a switch command until about 10 ms had passed. That delay gives the switch contacts 10 ms to settle down. This is the first instance in this book of *software-controlled switch debouncing*. You will find further details on this matter when studying the actual program listing.

So after the system has fetched the manually generated motion-code command and debounced the switch settings, it checks for a VALID motion code. Recall that the system is using a family of six different motion codes, based on the function table of Table 11-1A. If the user attempts to enter more than one motion code at a time (by turning on more than one motion-code switch), the result is considered *not* valid. You can see from the flowchart that the program responds by setting up the control for stopping the motion of the machine.

Another way to get a *not* VALID response is by having none of the motion-code switches turned on. That, too, sets the system for doing a STOP.

In any event, the motion code is output to the motor control system by the function block labeled OUTPUT MOTION CODE. It is at that point in the scheme that the machine does what it is told to do—provided, of course, the user is entering valid commands.

After outputting the motion-code command, the system checks to see whether or not the MAN switch is still turned ON. If so, the subroutine is repeated from the start. The system, in other words, remains under manual control as long as the MAN switch is set. When the user turns off the MAN switch, however, that condition is picked up at the MANUAL conditional, and the program responds by returning to the program that called it in the first place.

Here is the listing for the MAN subroutine, represented in Intel mnemonics:

```
3000  ;BEGINNING OF MANUAL SUBROUTINE
3010  MAN:                        ;
3020  ;SET THE PORT POINTERS
3030            LXI H,PORT0       ;HL TO PORT0
3040            LXI D,PORT1       ;DE TO PORT1
3050  ;GET MANUAL COMMAND AND DEBOUNCE THE
3060  ;SWITCHES
3070            LDAX D            ;GET COMM
3080            ANI 0FCH          ;ISOLATE THE
3090                              ;COMM BITS
3100  DEBS:     MOV B,A           ;SAVE IN B
3110            MVI C,0AFH        ;SET DEBOUNCE
3120                              ;TIMING
3130  CNTC:     DEC C             ;
3140            JNZ CNTC          ;
3150      ;TIMING DONE
3160            LDAX D            ;GET COMM
```

```
3170              ANI 0FCH        ;ISOLATE THE
3180                              ;COMM BITS
3190              CMP B           ;COMP WITH
3200                              ;BEFORE
3210              JNZ DEBS        ;REPEAT IF
3220                              ;BOUNCING
3230 ;DEBOUNCE DONE -- BEGIN VALID CHECK
3240              MVI C,0H        ;ZERO BIT
3250                              ;COUNTER
3260              MVI D,6H        ;SET CYCLE
3270                              ;COUNTER
3280 CYC1:        RAL             ;
3290              CC BCNT         ;COUNT A
3300                              ;1 BIT IN CY
3310              DEC D           ;
3320              JNZ CYC1        ;JMP IF NOT
3330                              ;DONE
3340              JMP NEXT1       ;
3350 BCNT:        INR C           ;
3360              RET             ;
3370 NEXT1:       MVI A,1H        ;
3380              CMP C           ;
3390              JZ OKMAN        ;JMP IF GOOD
3400                              ;MOVE COMM
3410              XRA A           ;ZERO ACC.
3420              JMP OUTMAN      ;
3430 OKMAN:       MOV A,B         ;GET COMM
3440 ;OUTPUT THE COMMAND
3450 OUTMAN:      MOV M,A         ;
3460 ;CHECK FOR SYSTEM STILL MANUAL
3470              MOV A,M         ;GET COMM
3480              ANI 80H         ;ISOLATE MAN
3490                              ;BIT
3500              JZ MAN          ;REPEAT MAN
3510                              ;IF MAN BIT
3520              RET             ;ELSE RETURN
3530                              ;TO CALLING
3540                              ;PROGRAM
3550 ;END OF MANUAL CONTROL SUBROUTINE
3560 ;
3570 ;USER MUST DEFINE THESE
3580      ;PORT0
3590      ;PORT1
3600      ;MAN
3610 ;ASSEMBLER WILL DEFINE THESE
3620      ;CNTC
3630      ;CYC1
3640      ;BCNT
3650      ;NEXT1
3660      ;OKMAN
```

The first step in the subroutine is to set pointers for Ports 0 and 1. Both are used through the execution of the subroutine, so it is handy to have their memory-mapped addresses available in register pairs HL and DE.

In line 3070, the program fetches switch panel data from Port 1. The instruction in line 3080 then isolates the motion-code bits. The result is that the six motion-code bits are in the accumulator.

Then the debouncing routine begins. First, the motion-code bits are saved in register B for future reference. Then a debounce counter (register C) is initialized. See line 3110. The instructions in lines 3130 and 3140 count down the debounce counter, effectively creating a delay that depends on the basic operating frequency of your system. The ideal time here is about 10 ms. You might have to alter the value of the initial count in line 3110 to suit the timing of your own system.

At any rate, the debounce time delay is concluded by the time the program reaches line 3160. The instruction on that line fetches the control panel information and isolates the motion-code bits once again. The comparing instruction on line 3190 compares this version of the motion-code with the one saved earlier in register B.

If the two versions are the same, the switches must not be bouncing, and the command appearing in the accumulator must be a reliable version of what the user wants. If that's true, the conditional statement in line 3210 is *not* satisfied, and the program defaults to the next phase of the MAN subroutine. But if the new version of the panel command is different from the one saved in B before the time delay was executed, the switch contacts must be bouncing, and it is necessary to take another look at the switches. The whole operation, including the time delay, is run again from label DEBS in line 3100. The system remains in this loop until two consecutive inspections of the control panel turn up identical versions of the switch settings. That's how the debouncing feature works here.

The next major phase of the MAN subroutine picks up at line 3240. The purpose here is really to determine whether or not the system should execute a STOP motion code that fills the Port 0 output with zeros.

Bear in mind that the current switch status is residing in the accumulator at this time. The basic idea behind this valid check is to count the number of ones in that word, and the job is accomplished by rotating the accumulator to the left and through the CY flag bit. As the switch bits are thus rotated, register C will count the number of ones as they pass through the CY flag bit location.

So the ones counter is initialized at zero by the instruction in line 3240. Since it is not necessary to check all eight bits—just the six higher-order ones—a second counter (register D) is set to a value of 6. Now the thing is ready to go.

The first bit is rotated by CY by line 3280. If it happens to be a 1 (indicating that the corresponding switch is ON), the CC instruction in line 3290 detects that fact and seconds the program down to a two-instruction subroutine labeled BCNT. At that point, the ones counter is incremented. Execution then resumes at line 3310.

The instruction in line 3310 decrements the rotation counter. Then line 3320 checks to see whether or not all six bits have been rotated through carry. If not, the program jumps back up to CYC1 to rotate the next bit into CY. Otherwise, the rotation job is done, and register C carries a number equal to the number of 1 bits in the motion code from the control panel.

The instructions in lines 3370 and 3380 compare the number of ones carried in the C register with a 1. If the number in the C register is indeed equal to 1, as determined by the conditional statement in line 3380, the program jumps down to OKMAN in line 3430. That happens whenever the user has set just one of the motion code switches to its ON position.

But if the number of ones in the motion-code command is not equal to 1, the program goes to the accumulator-zeroing instruction in line 3410. If the number in the ones counter is not equal to 1, one of two things must have happened. Either no switch is ON, or more than one is ON. In either case, the machine is supposed to stop; that's the purpose of the XRA A, accumulator-zeroing instruction.

So the program arrives at OKMAN in line 3430 only if one, and only one, command switch is turned ON. At that point, the content of the B register—a register saving the motion code command from the earlier debouncing operation—is output to the appropriate motion-code word to Port 0. And that's precisely what the instruction in that line does. Finally, after all this debouncing and bit counting, the MAN subroutine outputs the user's designated motion code, and the machine responds accordingly.

All that remains to be done in the subroutine is to see whether or not it is supposed to continue in this manual phase of operation. Line 3470 fetches the Port 0 status, and the next line isolates the MAN bit. If the MAN bit is set, as picked up by the conditional instruction in line 3500, the entire MAN subroutine begins from scratch. But if the conditional is not satisfied, the user has evidently turned OFF the MAN switch. It is then time to return to the program that called MAN in the first place.

THE STALL-CHECKING SUBROUTINE

The stall-checking subroutine, STCHK, simply determines whether or not one or both motors is stalled. It takes no action regarding a *response* to a stall. The STALL subroutine does that.

The flowchart for STCHK appears in Fig. 12-5. Basically, it looks at the one-bit signal from the left-motor speed monitor and determines from that whether or not the left motor is running. Whether or not the left motor is supposed to be running is not relevant at this point in the routine. If it so

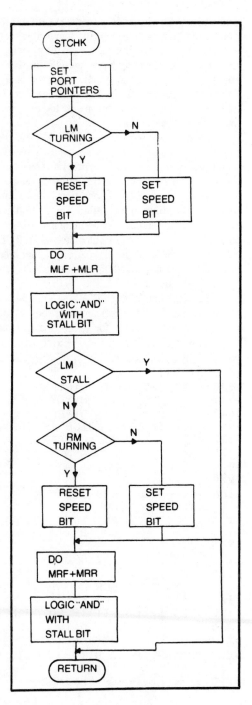

Fig. 12-5. Stall-checking subroutine flowchart.

401

happens that the left motor is running, a speed bit is reset to zero. If that motor is not running, however, the speed bit is set to logic 1.

Then that speed bit is saved and the system goes through a series of operations that ultimately Exclusive-OR the MLF and MLR bits picked up from the Port 0 inputs and outputs. As in the case of the system described in the last part of Chapter 11, the motor is supposed to be running if either MLF or MLR, but not both, are at logic 1. That's the purpose of the Exclusive-OR operation.

The result of that Exclusive-OR is ANDed with the speed bit. So if the motor is supposed to be turning and, indeed, it is, the LM STALL conditional is *not* satisfied. The program consequently turns to the task of checking out the right motor. But if the left motor is supposed to be turning and it is not, the LM STALL conditional is then satisfied. The system then returns to the program that called it in the first place. Having just the left motor stalled is adequate for defining a motor-stall condition; there is no point in checking out the right motor as well. The checks on the right motor follow the same general pattern of operations, ultimately returning to the calling program with information spelling out the fact that the right motor is or is not stalled.

The tricky part of the STCHK subroutine is figuring out whether or not the signals from the optical speed sensors are actually changing at a rate considered adequate for defining a turning situation. Here is the program listing in an Intel 8080A/8085 format:

```
4000 ;BEGINNING OF STALL-CHECK SUBROUTINE
4010 STCHK:                      ;
4020 ;SET THE PORT POINTERS
4030             LXI  H,PORT0     ;HL TO PORT0
4040             LXI  D,PORT1     ;DE TO PORT1
4050 ;SEE IF LEFT MOTOR IS TURNING
4060             MVI  C,0FFH      ;SET INTERVAL
4070             MOV  A,M         ;GET SPEEDS
4080             ANI  10H         ;ISOLATE LM
4090                              ;SPEED BIT
4100             MOV  B,A         ;SAVE IN B4
4110 CYC2:       MOV  A,M         ;GET SPEEDS
4120             ANI  10H         ;ISO. LMSP
4130             CMP  B           ;CMP WITH
4140                              ;BEFORE
4150             JNZ  SPOK1       ;JMP IF
4160                              ;DIFFERENT
4170             DCR  C           ;
4180             JNZ  CYC2        ;CONTINUE
4190 ;SET SPEED BIT IF NO CHANGE BY NOW
4200             MVI  B,1H        ;
4210             JMP  NEXT2       ;
4220 ;RESET SPEED BIT -- LM IS RUNNING
4230 SPOK1:      MVI  B,0H        ;
4240 ;EXCLUSIVE-OR MLF,MLR
```

```
4250 NEXT2:      MOV A,M        ;GET COMM
4260             ANI 1H         ;ISO. MLF
4270             MOV C,A        ;SAVE IN C0
4280             MOV A,M        ;GET COMM
4290             ANI 2H         ;ISO. MLR
4300             RRC            ;MLR TO A0
4310             XRA C          ;EX-OR WITH
4320                            ;MLF
4330 ;AND WITH LM SPEED BIT SAVED IN B0
4340             ANA B          ;
4350 ;RETURN TO CALLING PROGRAM IF LEFT
4360 ;MOTOR IS STALLED
4370             RNZ            ;
4380 ;ELSE CHECK OUT THE RIGHT MOTOR
4390             MVI C,0FFH     ;SET TIME
4400             MOV A,M        ;GET SPEED
4410                            ;BITS
4420             ANI 20H        ;ISO. MRSP
4430             MOV B,A        ;SAVE IN B5
4440 CYC3:       MOV A,M        ;GET SPEED
4450                            ;BITS
4460             ANI 20H        ;ISO. MRSP
4470             CMP B          ;CMP WITH
4480                            ;BEFORE
4490 ;JUMP IF THERE IS A DIFFERENCE
4500             JNZ SPOK2      ;
4510             DCR C          ;
4520             JNZ CYC3       ;JMP IF NO
4530                            ;CHANGE
4540 ;SET SPEED BIT IF STILL NO CHANGE
4550             MVI B,4H       ;
4560             JMP NEXT3      ;
4570 ;RESET SPEED BIT IF RT. MOTOR IS RUNNING
4580 SPOK2:      MVI B,0H       ;
4590 ;EXCLUSIVE-OR MRF,MRR
4600 NEXT3:      MOV A,M        ;GET COMM
4610             ANI 4H         ;ISO. MRF
4620             MOV C,A        ;SAVE IN C2
4630             MOV A,M        ;GET COMM
4640             ANI 8H         ;ISO. MRR
4650             RRC            ;MRR TO A2
4660             XRA C          ;EX-OR WITH
4670                            ;MRF
4680             RLC            ;TO A3
4690             RLC            ;TO A4
4700             RLC            ;TO A5
4710 ;AND WITH RM SPEED BIT SAVED IN B5
4720             ANA B          ;
4730 ;RETURN TO CALLING PROGRAM
4740             RET
```

403

```
4750      ;Z=1 IF RM IS STALLED
4760      ;Z=0 IF NEITHER IS STALLED
4770  ;END OF STALL-CHECK SUBROUTINE
4780  ;
4790  ;USER MUST DEFINE THESE
4800      ;PORT0
4810      ;PORT1
4820      ;STCHK
4830  ;ASSEMBLER WILL DEFINE THESE
4840      ;CYC2
4850      ;SPOK1
4860      ;NEXT2
4870      ;CYC3
4880      ;SPOK2
4890      ;NEXT3
4900  ;
4910  ;
```

Virtually all operations related to STCHK concern I/O Port 0, so the first step in the program is to point to that port with the HL register pair. The process of determining whether or not the left motor is turning is analogous to debouncing the manual control switches described in connection with the MAN subroutine. The purpose is different here, but the process is almost identical.

Line 4060 sets a timing interval in register C. Then the instruction in line 4070 fetches the Port 0 input status. Line 4080 isolates the speed-monitoring signal for the left motor, and the instruction in line 4100 then saves the result in register B.

The program next picks up the same information and isolates the left-motor speed-sensing bit. Line 4130 then compares the result with that already saved in register B. If they are *not* the same, it means the motor must have moved during that interval. The conditional instruction in line 4150 is thus satisfied and operations go down to SPOK1.

If the comparison instruction in line 4130 shows there is no difference between the two successive readings of the MLSP bit, however, the motor has not turned during the short interval between lines 4100 and 4130. That does not mean the motor is not turning at all; it simply means it did not turn very much during the short comparison time. The timing interval is thus decremented, and the operation goes back up to CYC2 to check the motor speed status again.

The speed status is compared with the original one 256 times, as determined by the initial value set into register C in line 4060. If during that time there are no changes in the motor speed status, you can bet the motor isn't turning, and the system responds by putting a logic 1 in the B register. That is the SET SPEED BIT operation specified in the flowchart in Fig. 12-5.

Recall that the system jumped out of the checking loop when it found a difference between two successive readings. It jumped down to SPOK1; at

404

that point, the B register is cleared to zero. This is the RESET SPEED BIT operation appearing on the flowchart.

In either case, program operations resume at line 4250, with the speed bit saved in register B. Line 4250 begins the Exclusive-OR phase of the job. The instruction at that line fetches the data from Port 0, and the next line isolates the MLF bit. That bit value is saved in register C by the instruction in line 4270, and the system proceeds to pick up the Port 0 information again, isolating the MLR bit in line 4290.

Line 4300 shifts that bit one place to the right, lining it up with the MLF bit saved in register C. The instruction in line 4310 then does an Exclusive-OR on the MLF and MLR bits as specified on the flowchart. Line 4340 ANDs the result with the speed bit saved earlier in register B. At this point, the Z flag is set to 1 or 0 (depending on whether the left motor is stalled or not stalled, respectively).

If the left motor is stalled, as picked up by the conditional instruction in line 4370, the program returns with Z=1. Otherwise, operations proceed to line 4390 to check out the status of the right motor.

The operation of the right-motor portion of the subroutine is practically identical to the left-motor operations just described. The right-motor bits are in different bit positions, though, which means it is necessary to shift them around somewhat to get the right sort of results.

The subroutine concludes at line 4740, where it returns to the calling program. It returns with Z=1 if the right motor is stalled or Z=0 if neither motor is stalled. Remember that it returned to the calling program, with Z=1, at an earlier time if the left motor was stalled.

THE STALL RESPONSE SUBROUTINE

The stall-checking subroutine just described always returns to the calling program with the Z flag carrying a bit that indicates whether or not the machine is stalled. And if that subroutine returns with Z=1, it means at least one of the motors is stalled. It is thus appropriate to call a subroutine intended to get the machine out of the stalling situation. That is the task of the STALL routine described here.

The basic idea is to use the current motion code as the lower-order bit for addressing the response memory. You have already set the address of the starting point of that memory when defining label MEM in the main program. This STALL routine simply sums the current motion code with the MEM address to come up with the actual address for the prescribed response.

Figure 12-6 illustrates the process in a flowchart fashion. The first step is to set the MEM pointer to the first address in the response memory and point to Port 0. After that, the program fetches the motion code being executed as the stall condition occurs, adds the motion code to the MEM address, fetches the prescribed response residing in that memory location, and then outputs it to the motors via Port 0.

The system then executes a time delay operation intended to give the machine a reasonable opportunity to get away from the stall condition. It is running the response motion code through that timing interval. The time delay interval should be on the order of 1 or 2 seconds, but you can modify that to suit your own situation.

After running the response motion code through the time delay interval, the system then calls the stall-checking subroutine, STCHK. The STALL subroutine calls another subroutine, thus marking the first step of nested subroutines used in this particular program.

The idea is to see whether or not the response really works. Usually it will, and STCHK will return to STALL with a Z-flag value of 0. But such machines have a nasty habit of getting themselves tangled up in fairly complicated stall situations (into corners, between pieces of furniture, and so on), so there must be some provision for handling sequences of responses necessary for maneuvering out of tricky places.

If, indeed, the system finds it is *not* stalled after executing the prescribed response motion code and trial time delay, the program immediately returns to the one that called it. This happens whenever the STALL conditional in the flowchart in Fig. 12-6 is *not* satisfied. The subroutine is thus finished for a while.

If the machine is still stalled, however, the STALL conditional is satisfied. Rather than going back immediately to the beginning of the subroutine and picking another response motion code, the program first checks to see whether or not the user has switched over to manual control. That is the purpose of the MANUAL conditional in the flowchart.

Here, the subroutine checks the status of the MAN switch on the user's control panel. If it turns out that the user has decided to give the machine some outside help with the current stall situation, the MANUAL conditional is satisfied. The routine in turn calls the MAN subroutine described earlier in this section.

Recall that the system remains in the MAN subroutine as long as the MAN switch is set to its ON position. Thus the user has full control over the machine as long as necessary or desired. Upon concluding the MAN subroutine, by setting the MAN switch to its OFF position, the STALL routine concludes, returning to the calling program immediately. See Fig. 12-6.

But if the user has not turned ON the MAN switch after this subroutine does a stall check, the MANUAL conditional is *not* satisfied. Consequently, the system returns to the beginning of STALL to select another response motion code from the response memory. The new response code is bound to be different from the original one for the simple reason that the current motion code is different from the earlier one.

The machine thus has the potential for eventually executing all possible response motion codes. If one of them does not clear up the stall situation—and the user doesn't intervene from the manual control

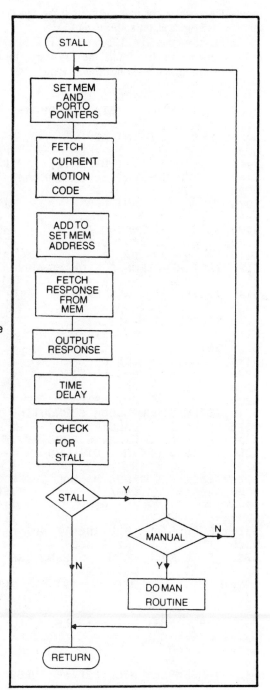

Fig. 12-6. Stall routine flowchart.

407

panel—the machine continues executing the sequence of motion codes until it is free (or, alas, the batteries die).

In practice, the STALL subroutine is a superbly effective stall-response mechanism. It is, in fact, the very heart of the operating system.

Here is the program listing in Intel mnemonics:

```
5000 ;BEGINNING OF STALL RESPONSE SUBR
5010 STALL:                    ;
5020 ;SET THE POINTERS
5030           LXI H,MEM        ;HL TO MEM
5040           LXI D,PORTO      ;DE TO PORTO
5050 ;FETCH CURRENT MOTOR COMMAND
5060           LDAX D           ;
5070           ANI OFH          ;ISO. MOTOR
5080                            ;COMMAND
5090           MOV C,A          ;MOVE TO C
5100           MVI B,OH         ;ZERO B
5110           DAD B            ;ADD TO HL
5120 ;FETCH RESPONSE FROM MEMORY
5130           MOV A,M          ;
5140 ;OUTPUT THE RESPONSE
5150           STAX D           ;
5160 ;DO THE TIME DELAY
5170           MVI B,OFH        ;
5180 SETC:     MVI C,FFH        ;
5190 CNTC1:    DCR C            ;
5200           JNZ CNTC1        ;
5210           DCR B            ;
5220           JNZ SETC         ;
5230 ;CALL STALL-CHECK SUBROUTINE
5235           CALL STCHK       ;
5240 ;RETURN TO CALLING PROGRAM IF THE
5250 ;MOTORS ARE NOT STALLED
5260           RZ               ;
5270 ;ELSE CHECK FOR MANUAL OVERRIDE
5280           LDAX D           ;GET STATUS
5290           ANI 80H          ;ISO. MANUAL
5300                            ;BIT
5310 ;REDO STALL ROUTINE IF NOT MANUAL
5320           JNZ STALL        ;
5330 ;ELSE CALL MANUAL SUBROUTINE
5340           CALL MAN         ;
5350           RET              ;RETURN TO
5360                            ;PROGRAM
5370 ;
5380 ;USER MUST DEFINE THESE
5390      ;STALL
5400      ;MEM
5410      ;PORTO
5420 ;ASSEMBLER WILL DEFINE THESE
```

```
5430        ; SETC
5440        ; CNTC1
5450 ;
5460 ;
```

The pointers relevant to STALL are set in lines 5030 and 5040. The HL pair points to the beginning of the response memory at address MEM, while the DE register pair is set up to point to Port 0. Port 1 is not used through the STALL subroutine. Line 5060 fetches the current motion code from Port 0. Then the next two instructions isolate the four motor command bits and save them in register C.

The B register is cleared to zero in line 5100. Then the content of the BC register pair (a two-byte version of the current motion code) is added to the content of the HL register pair (the starting point of the response memory). As a result of that double-register adding operation in line 5110, the HL pair carries the appropriate address for the prescribed response— the response appropriate for the motion code being executed the moment the stall condition occurs.

The response is moved to the accumulator by the instruction in line 5130. It is then loaded into the Port 0 output by the instruction in line 5150. The machine thus begins executing the prescribed response for the stall situation at hand.

The time delay operation occupies lines 5170 through 5220. Registers B and C are set to reflect the length of the time delay. If it turns out that you need a longer time delay for your particular machine, simply increase the hexadecimal value moved into register B at line 5170. Of course, if you need a shorter time delay, simply reduce the hexadecimal value at that line.

The instruction in line 5235 calls the stall-checking subroutine. After checking for a continuing stall situation, the process picks up with the conditional instruction in line 5260. If the STCHK subroutine returns here with a Z-flag set to 1, a stall condition still exists, and the program defaults to line 5280. Otherwise, the stall condition must have been remedied during the time delay interval. The conditional return instruction in line 5260 concludes the STALL subroutine, returning operations to the program that called it in the first place.

If a stall condition still exists, however, the instruction in line 5280 fetches the byte at Port 0, and the next instruction isolates the MAN bit from the user's control panel. If that bit is not set (if the user has not set the MAN switch to its ON position), the conditional instruction in line 5320 returns operations all the way back to the *beginning* of the STALL subroutine.

But if the machine is still stalled and the user has decided to take over manual control, the program defaults to line 5340, which calls the MAN subroutine for doing manual motion-code operations. Upon returning to line 5350, the program returns to the one that originally called it, which is the purpose of the unconditional return instruction in line 5350.

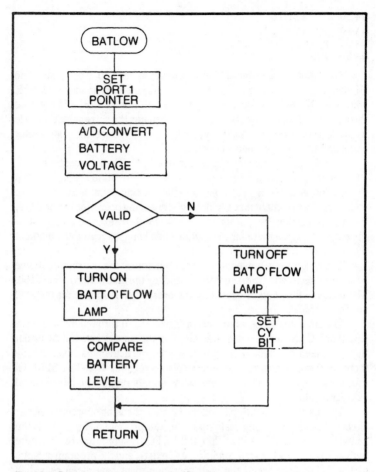

Fig. 12-7. Battery-checking subroutine flowchart.

CHECKING THE BATTERY VOLTAGE LEVEL

BATLOW is a relatively simple subroutine that does an A/D conversion on the main battery voltage level of the machine, returning to the calling program with a CY flag bit that is set to 1 or 0, depending on the results of the A/D conversion and battery-level check. See the flowchart for this subroutine in Fig. 12-7.

The first step in the operation sets a pointer to Port 1. This is the only port involved in the battery voltage-checking scheme. After that, the program does a four-bit A/D conversion on the battery voltage level.

The VALID conditional on the flowchart is used only for detecting an overflow of the A/D conversion operation. It does not check for a low battery level at this point—just the validity of the A/D conversion.

410

If it should happen that the A/D conversion process turns up an invalid check, the VALID conditional on the flowchart is *not* satisfied. The program responds by turning on the INVAL BATT CONVERSION warning lamp and setting the CY flag bit to 1. The reasoning behind this operation will become apparent later in this discussion. In any case, the subroutine does an unconditional return instruction to the program that called it.

Whenever the VALID conditional is satisfied, the A/D conversion evidently was a good one. The program thus turns off the INVAL BATT CONVERSION warning lamp, compares the actual battery voltage level with some prescribed low-level trigger voltage and then returns to the calling program.

Here is the Intel program listing:

```
6000 ;BEGINNING OF LOW BATTERY CHECK
6010 BATLOW:                      ;
6020 ;SET PORT POINTER
6030            LXI H,PORT1       ;
6040            MVI C,0H          ;ZERO A/D
6050                              ;COUNTER
6060 CYC4:      MOV M,C           ;OUTPUT A/D
6070                              ;COUNT
6080            MOV A,M           ;GET PORT 1
6090            ANI 1H            ;ISO. CORR0
6100            JZ ADOK1          ;JMP IF DONE
6110            MVI A,0EH         ;SET MAX A/D
6120            CMP C             ;
6130            JC ADINV1         ;JMP IF BAD
6140            INR CC            ;INCR A/D
6150            JMP CYC4          ;DO AGAIN
6160 ADINV1:    MVI M 10H         ;SET O'FLOW
6170                              ;LAMP BIT
6180            STC               ;SET CY=1
6185            RET               ;RETURN
6190 ;RETURN TO CALLING PROGRAM WITH CY=1
6200            RET               ;
6210 ADOK1:     MVI M,00H         ;RESET O'FLOW
6220                              ;LAMP BIT
6230 ;CHECK BATTERY VOLTAGE LEVEL
6240            MVI A,0CH         ;SET LEVEL
6250            CMP C             ;CMP WITH A/D
6260 ;RETURN TO CALLING PROGRAM
6270            RET               ;
6280     ;CY=0 IF BATT IS LOW
6290     ;CY=1 IF BATT OK OR BAD A/D
6300 ;
6310 ;USER MUST DEFINE THESE
6320     ;BATLOW
6330     ;PORT1
6340 ;ASSEMBLER WILL DEFINE THESE
```

```
6350        ;CYC4
6360        ;ANINV1
6370        ;ADOK1
6380   ;
6390   ;
```

Line 6030 sets the HL pair to the memory-mapped address for Port 1. Line 6040 marks the beginning of the A/D operation, and it zeros the A/D counter, register C. The next line outputs the current A/D count to Port 1. Line 6080 fetches the input status of Port 1, and the line following that one isolates the CORR0 bit—the correction bit for the A/D conversion process on the battery level monitor circuit.

The CORR0 bit is compared with a 1 in line 6090 to see whether or not the A/D conversion is done. If so, the conditional instruction in line 6100 takes operations down to label ADOK1 in line 6160. Otherwise, the system checks for a possible overflow of the A/D counter. If an overflow exists, operations go to ADINV1. If everything is progressing smoothly, though, line 6140 increments the A/D counter and returns operations to CYC4 to check out the process once again.

ADINV1 simply turns on the INVAL BATT CONVERSION lamp and sets the CY bit in the flag register. The reason for the latter operation will be apparent in a moment. The program executes the instructions beginning at ADOK1 if the conversion is a good one, and it responds by resetting the INVAL BATT CONVERSION lamp in line 6210; it turns off that lamp.

By the time operations reach line 6240, a good A/D conversion operation is completed, and the next step is to check the converted value against some minimum tolerable operating level. In this case, the criteria is hexadecimal 0C. The values are compared in line 6250, and the subroutine returns to the calling program, carrying a CY flag set to 0 if the battery level is below par, or a logic-1 level is the battery is OK or a bad A/D conversion occurred.

The status of the CY flag bit thus reflects the results of the BATLOW subroutine. The CY flag is set to logic 1 if the A/D conversion overflows because you really don't want the creature to begin searching for its battery charger on the basis of a bad A/D check. You might not agree with this somewhat arbitrary decision. If you don't, replace the carry-setting instruction in line 6180 with XRA A. That instruction will clear the CY flag bit, ultimately causing the machine to begin searching for the nest whenever a bad A/D conversion on the battery voltage happens to occur.

THE NEST SEARCH AND BATTERY RECHARGE SUBROUTINE

The purpose of this subroutine is to give the machine an opportunity to seek out its own nest (battery charger) and soak up life-giving electrical energy as long as it is necessary to do so. It is a rather extensive subroutine, mainly because so many things can go wrong between the time

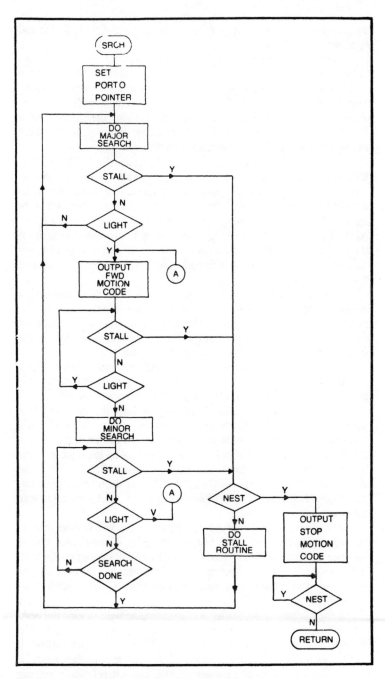

Fig. 12-8. Nest-search and battery recharging flowchart.

413

the machine first notices it is time to search for the nest, find it and recharge its batteries.

The procedure begins with the usual port-pointing operation. It then moves to DO MAJOR SEARCH. What is a major search? The point of the major search is to locate the bright light that is presumably attached to the nest and battery charger. All DO MAJOR SEARCH involves is making the machine do a FORWARD RIGHT TURN motion code, running it around in a circle with the hope that a phototransistor attached to the front mainframe assembly will pick up the light on the nest.

Specifically, then, DO MAJOR SEARCH, means OUTPUT FWD RT. Of course, there is always a chance the machine will blunder into an obstacle while executing this major search pattern, so it is necessary to do a STALL checking routine. If, indeed, a stall condition occurs while the machine is doing a major search, the flowchart shows a jump to a NEST conditional. The NEST conditional tests the nature of the stall, looking for the possibility that the machine is stalled against the nest and battery charger. That phase of the subroutine will be covered later in this discussion.

If a stall does not occur during the time the machine is executing a major search pattern, the next step is to see whether or not the phototransistor is picking up the light from the nest. That is the purpose of the LIGHT conditional on the flowchart. If the machine does not see the light, it loops back to check for a stall condition again.

The machine is thus locked into an operational loop, checking for a stall condition and looking for the light of the nest. In the meantime, it is doing that FORWARD RIGHT, major search motion code.

Suppose the machine finally senses the light on the nest. That satisfies the LIGHT conditional, and operations go to OUTPUT FWD MOTION CODE. Because the phototransistor is mounted on the front of the mainframe assembly, it follows that it can pick up the light from the nest only when the machine is oriented so that it is facing it. Thus the desirability of outputting a FORWARD motion code under that circumstances is obvious. The machine responds by running straight toward the nest.

But, again, there is always the chance that it will blunder into an obstacle along the way. That justifies the STALL conditional that is inserted after the OUTPUT FWD MOTION CODE operation. If a stall occurs while the machine is running toward the nest, one can hope it is the nest, itself, that is causing the stall condition. Operations thus drop down to the NEST conditional.

If the machine is not stalled while running straight toward the nest, the system checks to see whether or not the light is still in view. The machine you see, might not be running exactly straight toward the nest. If that's the case, it will lose sight of the light somewhere along the way. If not, it continues looping between STALL and LIGHT while running toward the nest. If it does lose track of the light along the way, however, it

414

shouldn't go into a major search pattern right away.

The machine probably did a lot of work to find the light in the first place. It would therefore be rather inefficient to risk throwing it way off the track at this time by making it do a wide, circular search pattern. Instead, the machine does a MINOR SEARCH routine. This one makes the machine move a bit to the right and then to the left, oscillating back and forth with the intent of relocating the nest light. This minor search routine is executed a reasonable number of times, as prescribed by the program. If, during that time, the machine becomes stalled, the STALL conditional sends operations over to the NEST conditional. Or if it happens to pick up the light again, the LIGHT conditional sends operations back up to the point where the system outputs the FORWARD motion code. In that case, the machine has evidently managed to relocate the light during the minor search operation.

The minor search should not be carried out indefinitely. Something might have happened, such as a serious stall condition, to throw the machine way off track. The SEARCH DONE conditional determines how many times the machine will go through the left-and-right, minor search operations without encounting an obstacle or finding the light again.

If it turns out that the machine doesn't find the light while doing the minor search, the flowchart shows operations returning all the way back to the point where the machine is instructed to begin a major search. In other words, it gives up on the minor search and begins all over again, doing a wide right turn to pick up the light from the nest.

Every time the machine encounters a stall condition while searching for the nest, there is a chance that it is stalling against the nest, itself. That, in fact, is the point of the whole routine. So the NEST conditional asks the question: *Am I stalled against the nest or am I against some other object?*

If the object causing the stall condition is *not* the nest, the system does the STALL subroutine described earlier in this chapter. The main point is to get the machine away from the obstacle. Recall, however, that running the STALL subroutine always gives the user a chance to take over manual control of the motion codes. That's handy at this point, because the little machine is "hungry" and is going to die if it doesn't find some "food" shortly. As soon as it blunders into an obstacle, it gives you a chance to take charge of things and steer it to the nest via the control panel. If the MAN switch isn't turned on during the STALL subroutine, the poor machine labors its way out of the stall condition and returns to the beginning of the program to DO MAJOR SEARCH. That starts it all over again from scratch.

Now suppose the NEST conditional is satisfied after a stall condition is picked up. The machine, in other words, finds it has blundered right into the nest, which is where it wants to be. The result is a STOP motion code, and the machine rests at the nest, soaking up energy from the battery charger.

The program is written such that the machine distinguishes the nest from other objects because contacts with the nest and the live battery charger cause recharge current to flow into the main battery. That's how the machine knew it ran into the nest in the first place.

So now it rests at the nest as long as recharge current flows. Once the battery is recharged, very little recharge current flows, and the second NEST conditional returns operations to the main calling program. If you refer to the flowchart for the main program (Fig. 12-3), you will see that the system leaves this NEST SEARCH subroutine and returns to a point where a FORWARD motion code is loaded into the Port 0 output.

The machine will push right against the nest in this case. Because the battery charging operation is done, however, the machine will interpret that contact as a foreign object to be avoided. It consequently fetches a motion code aimed at getting away from the object—the nest, itself, in this case. At this point, the machine is going its merry way around the room, blundering away from obstacles, including the nest, until the battery runs critically low again.

Consult the flowchart in Fig. 12-8 while working your way through the program listing for this rather extensive SRCH subroutine:

```
7000  ;BEGINNING OF NEST SEARCH
7010  SRCH:                      ;
7020  ;SET PORT 0 POINTER
7030          LXI H,PORT0        ;
7040  ;START MAJOR SEARCH PATTERN
7050  STA1:   MVI M,4H           ;DO FWD RT
7060          CALL STCHK         ;CHECK FOR
7070                             ;A STALL
7080          JNZ NCHK           ;JMP IF SO
7090  ;CHECK FOR NEST LIGHT
7100          MOV A,M            ;GET STAT
7110          ANI 40H            ;ISO. LIGHT
7120          JZ STA1            ;JMP IF NO
7130                             ;LIGHT
7140  ;ELSE GO FAST FORWARD
7150  FFWD:   MVI M,5H           ;
7160  STA2:   CALL STCHK         ;CHECK FOR
7170                             ;A STALL
7180          JNZ NCHK           ;JMP IF SO
7190  ;ELSE CHECK FOR NEST LIGHT
7200          MOV A,M            ;GET STAT
7210          ANI 40H            ;ISO. LIGHT
7220          JNZ STA2           ;JMP IF
7230                             ;LIGHT
7240  ;ELSE BEGIN MINOR SEARCH
7250          MVI C,8H
7260  MISRCH: MVI M,4H           ;DO FWD RT
7270          MVI B,8H           ;SET TIME
7280  SCYC1:  PUSH B             ;SAVE BC
```

```
7290                  PUSH H              ;SAVE HL
7300  ;CHECK FOR A STALL
7310                  CALL STCHK          ;
7320                  JNZ NCHK            ;JMP IF SO
7330  ;ELSE RESUME MINOR SEARCH
7340                  POP H               ;GET HL
7350                  POP B               ;GET BC
7360  ;CHECK FOR NEST LIGHT
7370                  MOV A,M             ;GET STAT
7380                  ANI 40H             ;ISO. LIGHT
7390                  JZ  FFWD            ;JMP IF
7400                                      ;LIGHT
7410                  DCR B               ;
7420                  JNZ SCYC1           ;DO AGAIN
7430  ;ELSE DO SECOND PHASE OF MINOR SEARCH
7440                  MVI M,1H            ;DO FWD LT
7450                  MVI B,8H            ;
7460  SCYC2:          PUSH B              ;SAVE BC
7470                  PUSH H              ;SAVE HL
7480  ;CHECK FOR A STALL CONDITION
7490                  CALL STCHK          ;
7500                  JNZ NCHK            ;JMP IF SO
7510                  POP H               ;GET HL
7520                  POP B               ;GET BC
7530  ;CHECK FOR NEST LIGHT
7540                  MOV M,A             ;GET STAT
7550                  ANI 40H             ;ISO. LIGHT
7560                  JNZ FFWD            ;JMP IF LIGHT
7570                  DCR B               ;
7580                  JNZ SCYC2           ;
7590                  DCR C               ;
7600                  JNZ MISRCH:         ;
7610                  JMP SRCH            ;GO BACK TO
7620                                      ;BEGINNING
7630  ;CHECK FOR NEST CONTACT
7640  NCHK:           LXI H,PORT1         ;HL TO PORT1
7650                  MVI C,0H            ;ZERO A/D
7660                                      ;COUNTER
7670  CYC5:           MOV M,C             ;OUTPUT A/D
7680                                      ;COUNT
7690                  MOV A,M             ;GET STAT
7700                  ANI 2H              ;ISO. CORR1
7710                  JZ  ADOK2           ;JMP IF DONE
7720                  MVI A,0EH           ;SET MAX
7730                                      ;A/D COUNT
7740                  CMP C               ;COMP WITH
7750                                      ;A/D COUNT
7760                  JC  ADINV2          ;JMP IF OK
7770                  INR C               ;INC A/D
7780                                      ;COUNT
```

417

```
7790              JMP CYC5        ;DO AGAIN
7800 ADINV2:      MVI M,20H       ;SET INVAL
7810                              ;LAMP
7820              JMP NSTOP       ;
7830 ADOK2:       MVI M,0H        ;RESET INVAL
7840                              ;LAMP
7850              MVI A,2H        ;SET MIN
7860                              ;CURRENT
7870              CMP C           ;COMP WITH
7880                              ;CURRENT
7890              JC NSTOP        ;STOP IF
7900                              ;CURR FLOW
7910 ;ELSE DO THE STALL ROUTINE
7920              CALL STALL      ;
7930 ;AND BEGIN MAJOR SEARCH AGAIN
7940              JMP SRCH        ;
7950 ;STOP MOVING AT NEST
7960 NSTOP:       XRA A           ;SET STOP
7970              LDA PORTO       ;OUTPUT STOP
7980 ;MONITOR RECHARGE CURRENT
7990 NOK:         MOV C,0H        ;ZERO A/D
8000                              ;COUNTER
8010 CYC6:        MOV M,C         ;OUTPUT A/D
8020                              ;COUNT
8030              MOV A,M         ;GET STAT
8040              ANI 2H          ;ISO. CORR1
8050              JZ ADOK 3       ;JMP IF OK
8060              MVI A,0EH       ;SET MAX A/D
8070              CMP C           ;COM WITH
8080                              ;A/D COUNT
8090              JC ADINV3       ;JMP IF
8100                              ;0'FLOW
8110              INR C           ;INC A/D COUNT
8120              JMP CYC6        ;DO AGAIN
8130 ADINV3:      MVI M,20H       ;SET INVAL
8140                              ;LAMP
8150              JMP NOK         ;DO A/D AGAIN
8160 ADOK3:       MVI M,0H        ;RESET INVAL
8170                              ;LAMP
8180 ;SET FULL-CHARGE CRITERION
8190              MVI A,2H        ;
8200              CMP C           ;COMP WITH
8210                              ;ACTUAL
8220 ;RETURN IF CHARGE IS DONE
8230              RNC             ;
8240 ;ELSE MONITOR CHARGE CURRENT AGAIN
8250              JMP NOK         ;
8260 ;END OF NEST SEARCH SUBROUTINE
8270 ;USER MUST DEFINE THESE
8280      ;SRCH
```

```
8290      ;PORT0
8300      ;PORT1
8310  ;ASSEMBLER WILL DEFINE THESE
8320      ;STA1
8330      ;FFWD
8340      ;STA2
8350      ;MISRCH
8360      ;SCYC1
8370      ;SCYC2
8380      ;NCHK
8390      ;CYC5
8400      ;ADINV2
8410      ;ADOK2
8420      ;NSTOP
8430      ;NOK
8440      ;CYC6
8450      ;ADINV3
8460      ;ADOK3
8470  ;
8480  ;
```

The nest-searching operations begin by setting the HL register pair to the memory-mapped address for Port 0. This allows the numerous data exchanges to take place with Port 0, using memory-related instructions.

Line 7050, labeled STA1, outputs the MAJOR SEARCH motion code, setting the pattern of motion to one characterized by doing a FORWARD WITH RIGHT TURN. Line 7060 then calls the STCHK subroutine to see whether or not the machine is stalled. If the machine is stalled, that subroutine returns with the Z flag set to logic 1, and the conditional jump instruction in line 7080 sends operations down to NCHK in line 7640.

If STCHK returns with Z=0, however, the program defaults to line 7100 to begin checking for the appearance of the nest light. Line 7100 fetches the data at Port 0, and the nest line isolates the LIGHT bit. If the nest light is not in sight, the instruction in line 7120 returns operations to STA1 in line 7050. That completes the MAJOR SEARCH loop, including the stall and light checks.

When the machine does indeed pick up the nest light, operations default to FFWD in line 7150, immediately outputting a FORWARD motion code. Then it is time to repeat the same sequence of stall and light checks. See the instructions in lines 7160 through 7220. The only way out of that loop is a stall or detection of the light. In the former case, operations go down to NCHK in line 7640. In the latter case, operations default to line 7250, which is the beginning of the MINOR SEARCH routine.

The MINOR SEARCH routine begins by setting the register C counter to hexadecimal 8. This number dictates the number of left-and-right cycles the machine will execute before giving up the minor search and resorting to starting all over again with MAJOR SEARCH. Immediately after setting that counter, the system outputs a motion code that drives the machine forward with a right turn.

In line 7250, another counter register, register B, is initialized. This one determines how long the machine will execute that forward right turn before changing to a forward left turn.

You might want to make line 7250 for future reference. When using a microprocessor system different from mine, you might find the count too large or too small to be of any use. If that number is set just right, the oscillations between left and right motions during this minor search phase should each carry the machine no more than 45 degrees from its original path. You will probably have to adjust that initial value for register B to suit the characteristics of your own system.

After getting the search started, it is time to check for a stall condition. Unfortunately, the STCHK subroutine uses some of the counting registers for its own purposes. Thus, the contents must first be saved in the stack. That is the purpose of the PUSH instructions in lines 7280 and 7290.

Once the machine determines whether or not it is stalled, the conditional jump instruction in line 7320 responds by sending operations down to NCHK (if a stall condition exists) or by letting the machine continue its minor search mode of operation (if no stall condition exists).

If there are no stall condition, lines 7340 and 7350 restore the counts in registers BC and HL and then begin a set of instructions aimed at checking for the nest lamp. The instruction in line 7370 fetches the Port 0 input, and the nest instruction isolates the LIGHT bit.

If the LIGHT bit is set, indicating the machine is picking up the light at the nest, the conditional instruction in line 7390 sends operations back up to FFWD, where the machine is instructed to run straight ahead. Otherwise, the B register is decremented to count off some minor search time. If that particular phase is not done, operations return to SCYC1 in line 7280 to begin another stall check.

Register B decrementing to zero marks the end of the right-motion phase of the minor search. The instruction in line 7440 sets up the motors for doing FORWARD WITH LEFT TURN, and the second phase of the minor search begins.

The sequence of instructions for this left-turn phase is identical to the right-turn phase just described. Unless there is contact with an obstacle or the machine senses the presence of the light from the nest, that phase winds up with register B decrementing to zero in line 7590.

Register C, the minor-search cycle counter, is finally decremented. Unless it is decremented to zero, the program begins another forward-right phase at MISRCH. Register C counts the number of complete, left-right search cycles. Register B determines how long the machine executes the left-turn and right-turn operations within each cycle. So once register C decrements to zero, as detected by line 7600, the program abandons the minor search and jumps back to the beginning of the entire subroutine where the major search begins.

Label NCHK at line 7640 marks the beginning of the nest-checking phase. The program enters this phase only when the system has picked up a stall condition. The first step in line 7640 is to use the HL register pair as a Port 1 pointer.

The basic idea of NCHK is to do an A/D conversion on the battery recharge current level. So the instructions zero the A/D counter (line (7650), fetch the Port 1 data and isolate the CORR1 bit (lines 7690 and 7700), and use a conditional jump instruction to see whether or not the A/D conversion process is done (line 7710). If the conversion is done, operations pick up at ADOK2 in line 7830. Otherwise, the instructions in lines 7720 through 7760 check for an A/D overflow and jump the operations to ADOK2 if an overflow has occurred. Such A/D conversion algorithms have already been described in detail several times, so there is no need to dwell on the point any further here.

The important feature of the conversion is that it jumps NSTOP in line 7960 under two different conditions: If the A/D conversion is an invalid one (as determined by the instruction in line 7800) or if the recharge current level just converted is above a threshold level determined by the hexadecimal byte moved into the accumulator at line 7850. Failing to meet either of the machine-stopping conditions, the system calls the STALL subroutine from the instruction in line 7920 and then returns to the very beginning of the whole SEARCH routine.

The NSTOP portion of the routine, beginning at line 7960, is executed whenever the machine has found the nest or gets an invalid A/D check. The instruction in that line zeros the accumulator. The next instruction outputs those zeros to the motor control circuit. That stops the machine in its tracks.

The instructions between lines 7990 and 8160 do another A/D conversion on the battery recharge current level. Every time the system picks up a good A/D conversion, it compares the actual recharge current value with the full-charge criterion entered at line 8190. If the recharge current value is less than that specified criterion, the subroutine returns to the call program. Otherwise, it jumps back to NOK, beginning the A/D operation all over again.

Examples of Alpha-Class and Beta-Class Robots

13

True robots, as described in Chapter 1, are machine capable of adapting to unforeseen circumstances in their environments. And since the possibility of encountering unforeseen circumstances exists, the machine—and true *robot*—must be able to devise its own methods for dealing with such circumstances. The parabots represented in Chapters 11 and 12 are incapable of making independent decisions, so they do not qualify as true robots.

This chapter offers examples of true robots. The Alpha-Class machine presented first deals with its environment in a purely random fashion. If you are not acquainted with Alpha-Class machine behavior, you might be surprised at how well the machine gets along without your help. Refer to TAB book No. 1241, *How to Build Your Own Self-Programming Robot* and No. 1191, *Robot Intelligence...with experiments.*

The Beta-Class robot uses the Alpha mechanism, but includes some self-programmed memory that enables it to remember and recall responses executed under similar circumstances in the past. Its behavior is purely random at first. As the robot builds up a repertoire of successful responses, its behavior becomes more efficient and rational.

You will see that the hardware is identical for both classes of robots. The program listings are rather different, however.

I/O PORT AND PERIPHERAL CONFIGURATION

Figure 13-1 shows the block diagram for both the Alpha-Class and Beta-Class machines described in this chapter. If you have studied the material in Chapters 11 and 12, you won't find much here that is new. Only the four-bit binary counter, connected to the Port 1 inputs, is new. That is the random motion code generator—something unique to self-programming robots.

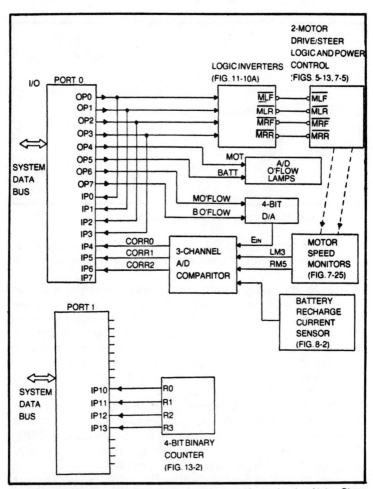

Fig. 13-1. I/O Port configuration and peripheral hardware for the Alpha-Class and Beta-Class machines.

I/O Port 0 is a busy one. The four lower-order input and output bits handle the motion codes for a standard two-motor drive/steer system. This is a simple on/off type of motor control scheme. If you want to add speed control as well, you will have to shift around the port configurations to make room for the speed bits. Then rewrite the software. The random-number scheme at Port 1 will also have to be expanded to a six-bit binary counter. All this adds up to a fine project for any experimenter who has the motivation and know-how to carry it off. Otherwise, you can be content with the simpler scheme as it is presented here.

The system monitors three internal parameters: the speeds for the two motors and the battery recharge current. The motor speed parameters

423

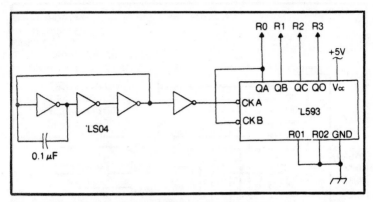

Fig. 13-2. Circuit for the random number generator function.

are used for detecting stall conditions that arise whenever the robot encounters an obstacle in its environment. That is its only contact-sensing mechanism.

The battery recharge current sensor plays two important roles. For one, it senses a full-charge condition while resting at a battery-charger nest assembly. That same parameter, however, is also used for distinguishing the battery-charger nest from any other sort of obstacle it encounters in its world. Run into the nest, and battery recharge current flows. But run into anything else, and recharge current does not flow. The battery recharge parameter is thus a vital part of the ability of a robot to function independently in their environments.

These three sensory parameters are represented as analog voltage levels, so they must be translated into a binary format through an A/D conversion process. The correction bits for this operation (CORR0, CORR1 and CORR2) represent the left motor speed (LMS), right motor speed (RMS) and battery recharge current, respectively.

The two bits from Port 0 going to the D/A converter imply a four-bit A/D conversion. A high-resolution conversion is not needed.

As mentioned earlier, Port 1 is used only for fetching a four-bit number—a number used as a random response motion code. The circuit is detailed for you in Fig. 13-2. It is a simple, free-running, high-speed counter. Because it is not synchronized to any part of the microprocessor system, the numbers fetched from it will appear to be random in nature.

PROGRAM FLOWCHART FOR THE ALPHA-CLASS ROBOT

Figure 13-3 is the overall flowchart for the Alpha-Class robot represented in this chapter. It follows the general format for simple Alpha-Class intelligence.

After setting the Port 0 pointer, the system fetches a four-bit, random motion code from Port 1. It then immediately outputs that motion code to the motor assembly.

424

So when you first fire up this machine, you have no way of knowing in advance what it will do first. It might run in circles or run straight backward or forward. You can only be sure it will start moving, because the software eliminates the possibility of outputting any stop codes at this point.

After fetching and beginning to execute some sort of motion code, the program runs a time delay. The time delay is not relevant while the machine is running freely, but it is executed anyway. The delay will be important when the machine is trying to get away from a stall situation.

In any case, the program does the time delay. Then it checks the motors for a stall condition. If a stall condition does *not* exist, the program merely checks for a stall again and again and again. All the while, it is executing the motion code it picked up earlier in the program.

This goes on until the robot runs into something, satisfying the STALL conditional. At that point, the robot must determine whether that object is its nest or something else.

So it checks for battery recharge current from the nest assembly at CHECK NEST CURRENT. If it turns out that battery recharge current is flowing, the CURRENT conditional is satisfied and the program outputs a stop motion code. The robot, in other words, comes to a stop only when it senses battery current flowing from the battery charger. It just sits there, soaking up energy.

All the time it is resting at the battery charger, it continuously rechecks the nest current level. That is the loop between CHECK NEST CURRENT and OUTPUT STOP MOTION CODE. Once the robot finds that the recharge current has fallen below a prescribed level, the CURRENT conditional is no longer satisfied, and the program loops all the way back up to FETCH RANDOM MOTION CODE.

An encounter with the nest and battery charger thus causes the robot to stop in its tracks and soak up electrical energy until the battery is fully charged. At that moment, it begins treating the nest as any other sort of obstacle, taking action to get away from it.

If the machine encounters an obstacle, as determined by satisfying the STALL conditional, it checks for nest current as described earlier. If no nest current is flowing, the CURRENT conditional sends operations back up to FETCH RANDOM MOTION CODE.

The system picks a motion code, checks to make sure it isn't a stop code, outputs it to the motor assembly and then does the TIME DELAY operation. This is where the time delay is important. The machine has run into something and responded by executing a random motion code. It must have some time to try that motion. If it works, the program goes into the CHECK FOR STALL loop. If it doesn't work, operations eventually get down to the CURRENT conditional and then back up to FETCH RANDOM MOTION CODE again.

This particular robot will not default to a forward-running motion code whenever it is free of a contact situation. Instead, it runs the motion code that successfully freed it from the obstacle, and it continues running that motion code until it encounters another obstacle. The software could be rewritten so that the robot defaults to a forward motion code whenever it is free, but that is left to experimenters who wish to use the problem as an exercise in microprocessor programming.

The machine thus runs according to randomly generated motion codes. It responds to contacts with obstacles (other than the nest) by executing as many other random motion codes as necessary for freeing itself from the obstacle. Upon making contact with the nest, the robot can identify it as the nest only as long as the battery requires recharging. When the battery is fully charged, the robot treats the nest as any other kind of obstacle in the room.

Incidentally, the robot finds the nest by chance. There is no built-in mechanism for locating and tracking down the nest and battery charger when they are needed. The Alpha-Class robot is such an active creature, however, that it rarely exhausts its main battery before making contact with the nest—assuming, of course, that the system has a battery with adequate ampere-hour specifications.

PROGRAM LISTINGS FOR THE ALPHA-CLASS ROBOT

This section includes two different program listings for executing the flowchart just described. One is in an Intel 8080A/8085 format and the other uses Z80 mnemonics. The following discussion for these listings applies to both of them, and you can consult the listing that suits your own taste and experience.

The operational part of the listing begins at line 1020, where the HL register pair is set to the Port 0 address. Immediately after that, the FETCH label marks the beginning of normal operations. If your microprocessor system requires a specific instruction for setting the location of the stack, you should enter that statement between lines 1020 and 1030.

The steps in the FETCH routine are to get a random number from Port 1 (line 1050), isolate the random bits, save them in register B, fetch the data at Port 0 (line 1070), clear the motion code bits, replace them with the random motion code from register B, and output the new motion code (line 1120). Lines 1040 through 1120 thus take care of the operations involved in FETCH RANDOM MOTION CODE and OUTPUT MOTION CODE designated on the flowchart in Fig. 13-3.

Lines 1140 through 1200 take care of the TIME DELAY operation. If you find that the amount of time delay is too short for your own system, you can lengthen it by increasing the value of the hexadecimal number loaded into register B by the instruction in line 1140. Likewise, you can shorten the TIME DELAY by reducing the value of that number.

In any event, TIME DELAY is finished by the time the program reaches the instruction in line 1260. That line, carrying label STA1, sets up the system for doing an A/D conversion on CORR0, the left motor

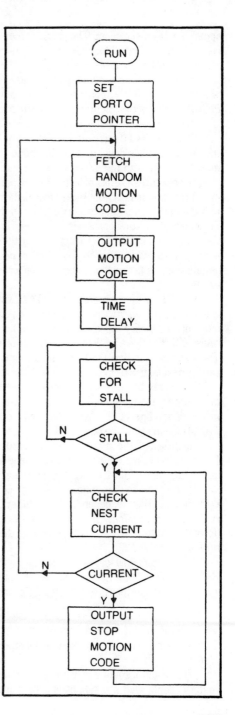

Fig. 13-3. Flowchart for the Alpha-Class machine.

427

speed sensing mechanism. The program then calls ADCON, a subroutine responsible for carrying out the A/D conversion. Details of that subroutine are discussed later in this section.

As noted on the program listing, the system returns from ADCON with the carry flag bit equal to 0 if the conversion is a good one. The digitized value of the left motor speed is then carried back in register C. If the A/D conversion overflows, however, ADCON returns with CY=1.

So if there is a 0 carry bit, as detected by the instruction in line 1300, the program jumps down to ADOK1 in line 1380. Otherwise, something must be wrong with the A/D conversion of the left motor speed, and the program executes a series of instructions from line 1310 through 1350.

That set of instructions picks up the data from Port 0 and modifies it so that the data then output to Port 0 turns on the MOTOR O'FLOW lamp. Since the information concerning the speed of the left motor is defective, the instruction in line 1350 aborts any further motor-related operations, sending the program down to a point where nest battery checks begin. The assumption, as far as the system is concerned, is that a stall condition exists.

But if the program picks up a good A/D value for the left motor speed, it executes the instructions beginning at ADOK1 in line 1380. At that moment, the binary representation of the left motor speed is residing in register C (left there by the recent execution of the ADCON subroutine). The instruction in line 1380 defines a minimum acceptable motor speed, and the next instruction compares it with the actual motor speed.

If it turns out that the actual motor speed is less than the designated minimum acceptable speed, the CY bit is reset to zero, and the conditional instruction in line 1410 causes operations to skip down to NEST1 in line 1460. But if the actual motor speed is equal to or greater than the minimum acceptable speed, the instruction in line 1430 sets a turning flag in register D to a logic-1 level. NEXT1 is called whenever the motor is not running, and all that instruction does is set that turning flag register to zero.

By the time the program reaches line 1480, therefore, register D is carrying a 1 or 0 value, depending on whether the system finds the motor is running or not running. The instructions beginning at NEST2 and running down through line 1510 simply make sure the MOTOR O'FLOW lamp is turned off.

The system now knows whether or not the left motor is turning. The next major step is to find out whether or not that motor is *supposed* to be turning. The basic idea is to solve the following Boolean expression:

$$MLF \oplus MLR \oplus LMS$$

where MLF and MLR are the actual command bits for the left motor, and LMS is the bit in register D that tells whether or not the motor is actually turning.

The result of solving that Boolean expression is a logic-0 level if the motor is doing what it is supposed to be doing; *i.e.*, if it is supposed to be

428

stopped and it is, indeed, stopped, or if it is supposed to be running and it is actually running. A logic-1 level comes out of the expression if something is wrong; i.e., if the motor is supposed to be running and it is not, or if the motor is supposed to be stopped and the wheel is turning for some reason.

The conditional instruction in line 1670 detects the result of calculating the Boolean expression. In essence, that instruction aborts any further stall checks if the left motor is stalled, jumping down to BTCHK to begin nest battery checks. If the Boolean expression determines that the left motor is actually doing what it is supposed to do, however, operations pick up with the beginning of the right motor checks at line 1710.

The sequence of instructions for checking the right motor is practically identical to the left motor operations just described. The only difference is that it is necessary to do some bit shifting here in order to align them for proper execution of the Boolean expression:

$$MRF \oplus MRR \oplus RMS.$$

For instance, if the A/D conversion is a bad one, the instruction in line 1800 aborts any further checks on the right motor system and jumps the program to the point where it begins checking the nest current. By the time the right motor operations reach line 2120, the Boolean expression has been solved.

If the right motor is stalled, the conditional instruction in line 2120 carries operations to the battery checking routine. Otherwise, the unconditional jump in line 2140 returns operations to STA1, thus beginning the entire stall checking routine from the start. This particular operation is represented on the flowchart (Fig. 13-3) as the return line from STALL to CHECK FOR STALL.

The process of checking for nest current begins at line 2190. The program first sets up the system for doing an A/D conversion on CORR2, the battery recharge current parameter. Then it calls the ADCON, A/D conversion subroutine.

ADCON returns to line 2230, carrying a CY=0 if the conversion was a good one or a CY=1 if the A/D counter overflowed. If the carry flag bit is 0, the instruction in line 2230 jumps operations down to ADOK3 in line 2310. Otherwise, the program executes a series of instructions (lines 2240 through 2350) that turns on the BATTERY O'FLOW lamp and bypasses any further nest checks.

If the A/D conversion is good, though, the instruction in line 3210 sets the criterion for a minimum recharge current level, and the next instruction compares that criterion with the actual recharge current now carried in the C register (the ADCON subroutine put it there earlier in the program). If the actual current level is then less than or equal to the criterion for minimum recharge current, the carry flag bit is reset to 0 and operations return to FETCH. The actual battery recharge current, in other words, is either very low or nonexistent. The flowchart shows this return to FETCH as a line looping up from CURRENT to FETCH RANDOM MOTION CODE.

If the battery check shows that there is, indeed, some reasonable amount of current flowing into the battery, the program picks up at line 2370. The instruction in that line fetches the contents of Port 0, the next instruction zeros the motor motion code bits, and the instruction in line 2400 outputs this stop code to the motor assembly. With the robot thus stopped at the nest and soaking up energy from the battery charger, the instruction in line 2440 forces the system to continuously cycle through the battery current-checking routines until something happens to lower the current level.

That marks the end of the main program. The A/D conversion subroutine occupies the last operational segment of the program listing. It begins at line 2480 by clearing the A/D counter, register C, to zero. Operations after that follow the basic A/D software routine already described a number of times in Chapters 11 and 12.

8080A/8085 Program Listing

```
ALPHA-CLASS ROBOT PROGRAM LISTING
(8080A/8085 VERSION)
1000 START:                     ;
1010 ;SET PORT 0 POINTER
1020              LXI H,PORT0    ;
1030 ;FETCH RANDOM MOTION CODE
1040 FETCH:       LDA PORT1      ;GET RANDOM
1050              ANI 0FH        ;ISO. BITS
1060              MOV B,A        ;SAVE IN B
1070              MOV A,M        ;GET PORT 0
1080              ANI 0F0H       ;CLEAR MOT BITS
1090              ORA B          ;PUT IN NEW
1100                             ;MOTION CODE
1110 ;OUTPUT MOTION CODE
1120              MOV M,A        ;
1130 ;BEGIN TIME DELAY
1140              MOV B,0FH      ;SET B COUNT
1150 SETC:        MOV C,0FFH     ;SET C COUNT
1160 CNTC:        DCR C          ;COUNT DOWN
1170              JNZ CNTC       ;JMP IF NOT
1180                             ;DONE
1190              DCR B          ;COUNT DOWN
1200              JNZ SETC       ;JMP IF NOT
1210                             ;DONE
1220 ;END OF TIME DELAY
1230 ;BEGIN CHECKING FOR POSSIBLE STALL
1240 ;GET SPEED STATUS OF THE LEFT MOTOR
1250 ;SET FOR A/D CONVERSION ON CORR0
1260 STA1:        MVI B,10H      ;
1270              CALL ADCON     ;
1280 ;RETURNS WITH CY=0 IF CONVERSION
1290 ;IS GOOD, ELSE CY=1
1300              JNC ADOK1      ;JMP IF OK
```

```
1310              MOV A,M          ;GET PORT 0
1320              ORI 10H          ;SET O'FLOW
1330              MOV M,A          ;OUTPUT TO
1340                               ;PORT 0
1350              JMP BTCHK        ;JMP TO BATT.
1360                               ;CHECK
1370 ;CHECK LEFT MOTOR SPEED
1380 ADOK1:       MVI A,1H         ;SETMIN.SPEED
1390              CMP C            ;CMP WITH A/D
1400                               ;SPEED
1410              JNC NEXT1        ;JMP IF LM IS
1420                               ;NOT TURNING
1430              MVI D,1H         ;SET TURNING
1440                               ;FLAG
1450              JMP NEXT2        ;
1460 NEXT1:       MVI D,0H         ;SET NOT TURN-
1470                               ;ING FLAG
1480 NEXT2:       MOV A,M          ;GET PORT 0
1490              ANI 0EFH         ;RESET O'FLOW
1500                               ;BIT
1510              MOV M,A          ;OUTPUT TO
1520                               ;PORT 0
1530 ;GET MLF COMMAND BIT
1540              MOV A,M          ;GET PORT 0
1550              ANI 1H           ;ISO. MLF
1560              MOV B,A          ;SAVE MLF IN
1570                               ;B0
1580 ;GET MLR COMMAND BIT
1590              MOV A,M          ;GET PORT 0
1600              ANI 2H           ;ISO. MLR
1610              RRC              ;MLR TO A0
1620 ;EXCLUSIVE-OR MLF,MLR,LMS
1630              XRA B            ;
1640              XRA D            ;
1650                               ;
1660 ;IF LEFT MOTOR IS STALLED, Z=0
1670              JNZ  BTCHK       ;JMP IF SO
1680 ;ELSE DO RIGHT MOTOR SPEED CHECKS
1690 ;GET SPEED STATUS OF RIGHT MOTOR
1700 ;SET FOR A/D CONVERSION ON CORR1
1710              MVI B,20H        ;
1720              CALL ADCON       ;
1730 ;RETURNS WITH CY=0 IF CONVERSION
1740 ;IS GOOD, ELSE CY=1
1750              JNC ADOK2        ;JMP IF OK
1760              MOV A,M          ;GET PORT 0
1770              ORI 10H          ;SET O'FLOW
1780              MOV M,A          ;OUTPUT TO
1790                               ;PORT 0
1800              JMP BTCHK        ;JMP TO BATT.
1810                               ;CHECK
```

431

```
1820 ADOK2:      MVI A,1H       ;SET MIN. SPEED
1830            CMP C          ;CMP WITH A/D
1840            JNC NEXT3      ;JMP IF RM IS
1850                           ;NOT TURNING
1860            MVI D,1H       ;SET TURNING
1870                           ;FLAG
1880            JMP NEXT4
1890 NEXT3:     MVI D,0H       ;SET NOT TURN-
1900                           ;ING FLAG
1910 NEXT4:     MOV A,M        ;GET PORT 0
1920            ANI 0EFH       ;RESET O'FLOW
1930                           ;BIT
1940            MOV M,A        ;OUTPUT TO
1950                           ;PORT 0
1960 ;GET MRF COMMAND BIT
1970            MOV A,M        ;GET PORT 0
1980            ANI 4H         ;ISO. MRF
1990            MOV B,A        ;SAVE MRF
2000                           ;IN B2
2010 ;GET MRR COMMAND BIT
2020            MOV A,M        ;GET PORT 0
2030            ANI 8H         ;ISO. MRR
2040            RRC            ;MRR TO A2
2050 ;EXCLUSIVE-OR MRF,MRR,RMS
2060            XRA B          ;
2070            RRC            ;TO A1
2080            RRC            ;TO A0
2090            XRA D          ;
2100                           ;
2110 ;IF RIGHT MOTOR IS STALLED, Z=0
2120            JNZ BTCHK      ;JMP IF SO
2130 ;ELSE GO BACK TO DO STALL CHECKS AGAIN
2140            JMP STA1       ;
2150 ;END OF STALL CHECKING ROUTINE
2160 ;
2170 ;BEGIN CHECKING NEST RECHARGE CURRENT
2180 ;SET FOR A/D CONVERSION ON CORR2
2190 BTCHK:     MVI B,40H      ;
2200            CALL ADCON     ;
2210 ;RETURNS WITH CY=0 IF CONVERSION
2220 ;IS GOOD, ELSE CY=1
2230            JNC ADOK3      ;JMP IF OK
2240            MOV A,M        ;GET PORT 0
2250            ORI 20H        ;SET O'FLOW
2260                           ;BIT
2270            MOV M,A        ;OUTPUT TO
2280                           ;PORT 0
2290 ;ABORT NEST CURRENT TEST IF BAD A/D
2300            JMP FETCH      ;
2310 ADOK3:     MVI A,1H       ;SET MIN. I
2320            CMP C          ;CMP WITH A/D
```

```
2330  ;IF NEST CURRENT FLOWING, CY=1
2340  ;IF NOT AT NEST OR NOT HUNGRY, CY=0
2350            JNC FETCH        ;JMP IF CY=0
2360  ;SET AND OUTPUT STOP MOTION CODE
2370            MOV A,M          ;GET PORT 0
2380            ANI 0F0H         ;ZERO MOTOR
2390                             ;BITS
2400            MOV M,A          ;OUTPUT TO
2410                             ;PORT 0
2420                             ;
2430  ;CHECK NEST CURRENT AGAIN
2440            JMP BTCHK        ;
2450  ;END OF NEST CURRENT CHECKING ROUTINE
2460  ;
2470  ;A/D CONVERSION SUBROUTINE
2480  ADCON:   MVI C,0H         ;ZERO A/D
2490                             ;COUNTER
2500  ADCNT:   MOV A,M          ;GET PORT 0
2510            ANI 3FH          ;ISO. A/D BITS
2520            ORA C            ;FIT IN A/D
2530                             ;COUNT
2540            MOV M,A          ;OUTPUT TO
2550                             ;PORT 0
2560            MOV A,M          ;GET PORT 0
2570            ANA B            ;ISO. CORR BIT
2580  ;RETURN IF DONE -- CY=0
2590            RNZ              ;
2600            MVI A,3H         ;SET MAX. A/D
2610            CMP C            ;CMP WITH A/D
2620  ;RETURN IF INVALID -- CY=1
2630            RC               ;
2640            INR C            ;
2650            JMP ADCNT        ;DO AGAIN
2660  ;END OF A/D CONVERSION SUBROUTINE
2670  ;
2680  ;
2690  ;USER MUST DEFINE THESE ADDRESSES
2700       ;START
2710       ;PORT0
2720       ;PORT1
2730  ;ASSEMBLER WILL DEFINE THESE
2740       ;FETCH
2750       ;SETC
2760       ;CNTC
2770       ;STA1
2780       ;ADOK1
2790       ;NEXT1
2800       ;NEXT2
2810       ;ADOK2
2820       ;NEXT3
2830       ;NEXT4
```

```
2840        ;BTCHK
2850        ;ADCON
2860        ;ADCNT
2870 ;
2880 ;
```

Z80 Program Listing

```
ALPHA-CLASS ROBOT PROGRAM LISTING
(Z80 VERSION)
1000 START:                      ;
1010 ;SET PORT 0 POINTER
1020         LD HL,PORT0          ;
1030 ;FETCH RANDOM MOTION CODE
1040 FETCH:  LD A,(PORT1)         ;GET RANDOM
1050         AND 0FH              ;ISO. BITS
1060         LD B,A               ;SAVE IN B
1070         LD A,(HL)            ;GET PORT 0
1080         AND 0F0H             ;CLEAR MOT BITS
1090         OR B                 ;PUT IN NEW
1100                              ;MOTION CODE
1110 ;OUTPUT MOTION CODE
1120         LD (HL),A            ;
1130 ;BEGIN TIME DELAY
1140         LD B,0FH             ;SET B COUNT
1150 SETC:   LD C,0FFH            ;SET C COUNT
1160 CNTC:   DEC C                ;COUNT DOWN
1170         JP NZ,CNTC           ;JMP IF NOT
1180                              ;DONE
1190         DEC B                ;COUNT DOWN
1200         JP NZ,SETC           ;JMP IF NOT
1210                              ;DONE
1220 ;END OF TIME DELAY
1230 ;BEGIN CHECKING FOR POSSIBLE STALL
1240 ;GET SPEED STATUS OF THE LEFT MOTOR
1250 ;SET FOR A/D CONVERSION ON CORR0
1260 STA1:   LD B,10H             ;
1270         CALL ADCON           ;
1280 ;RETURNS WITH CY=0 IF CONVERSION
1290 ;IS GOOD, ELSE CY=1
1300         JP NC,ADOK1          ;JMP IF OK
1310         LD A,(HL)            ;GET PORT 0
1320         OR 10H               ;SET O'FLOW
1330         LD (HL),A            ;OUTPUT TO
1340                              ;PORT 0
1350         JP BTCHK             ;JMP TO BATT.
1360                              ;CHECK
1370 ;CHECK LEFT MOTOR SPEED
1380 ADOK1:  LD A,1H              ;SET MIN. SPEED
1390         CP C                 ;CMP WITH A/D
1400                              ;SPEED
```

434

```
1410              JP NC,NEXT1     ;JMP IF LM IS
1420                              ;NOT TURNING
1430              LD D,1H         ;SET TURNING
1440                              ;FLAG
1450              JP NEXT2        ;
1460 NEXT1:       LD D,0H         ;SET NOT TURN-
1470                              ;ING FLAG
1480 NEXT2:       LD A,(HL)       ;GET PORT 0
1490              AND 0EFH        ;RESET O'FLOW
1500                              ;BIT
1510              LD (HL),A       ;OUTPUT TO
1520                              ;PORT 0
1530 ;GET MLF COMMAND BIT
1540              LD A,(HL)       ;GET PORT 0
1550              AND 1H          ;ISO. MLF
1560              LD B,A          ;SAVE MLF IN
1570                              ;B0
1580 ;GET MLR COMMAND BIT
1590              LD A,(HL)       ;GET PORT 0
1600              AND 2H          ;ISO. MLR
1610              RRCA            ;MLR TO A0
1620 ;EXCLUSIVE-OR MLF,MLR,LMS
1630              XOR B           ;
1640              XOR D           ;
1650                              ;
1660 ;IF LEFT MOTOR IS STALLED, Z=0
1670              JP NZ,BTCHK     ;JMP IF SO
1680 ;ELSE DO RIGHT MOTOR SPEED CHECKS
1690 ;GET SPEED STATUS OF RIGHT MOTOR
1700 ;SET FOR A/D CONVERSION ON CORR1
1710              LD B,20H        ;
1720              CALL ADCON      ;
1730 ;RETURNS WITH CY=0 IF CONVERSION
1740 ;IS GOOD, ELSE CY=1
1750              JP NC,ADOK2     ;JMP IF OK
1760              LD A,(HL)       ;GET PORT 0
1770              OR 10H          ;SET O'FLOW
1780              LD (HL),A       ;OUTPUT TO
1790                              ;PORT 0
1800              JP BTCHK        ;JMP TO BATT.
1810                              ;CHECK
1820 ADOK2:       LD A,1H         ;SET MIN. SPEED
1830              CP C            ;CMP WITH A/D
1840              JP NC,NEXT3     ;JMP IF RM IS
1850                              ;NOT TURNING
1860              LD D,1H         ;SET TURNING
1870                              ;FLAG
1880              JP NEXT4        ;
1890 NEXT3:       LD D,0H         ;SET NOT TURN-
1900                              ;ING FLAG
```

```
1910 NEXT4:     LD A,(HL)        ;GET PORT 0
1920            AND 0EFH         ;RESET O'FLOW
1930                             ;BIT
1940            LD (HL),A        ;OUTPUT TO
1950                             ;PORT 0
1960 ;GET MRF COMMAND BIT
1970            LD A,(HL)        ;GET PORT 0
1980            AND 4H           ;ISO. MRF
1990            LD B,A           ;SAVE MRF
2000                             ;IN B2
2010 ;GET MRR COMMAND BIT
2020            LD A,(HL)        ;GET PORT 0
2030            AND 8H           ;ISO. MRR
2040            RRCA             ;MRR TO A2
2050 ;EXCLUSIVE-OR MRF,MRR,RMS
2060            XOR B            ;
2070            RRCA             ;TO A1
2080            RRCA             ;TO A0
2090            XOR D            ;
2100                             ;
2110 ;IF RIGHT MOTOR IS STALLED, Z=0
2120            JP NZ,BTCHK      ;JMP IF SO
2130 ;ELSE GO BACK TO DO STALL CHECKS AGAIN
2140            JP STA1          ;
2150 ;END OF STALL CHECKING ROUTINE
2160 ;
2170 ;BEGIN CHECKING NEST RECHARGE CURRENT
2180 ;SET FOR A/D CONVERSION ON CORR2
2190 BTCHK:     LD B,40H         ;
2200            CALL ADCON       ;
2210 ;RETURNS WITH CY=0 IF CONVERSION
2220 ;IS GOOD, ELSE CY=1
2230            JP NC,ADOK3      ;JMP IF OK
2240            LD A,(HL)        ;GET PORT 0
2250            OR 20H           ;SET O'FLOW
2260                             ;BIT
2270            LD (HL),A        ;OUTPUT TO
2280                             ;PORT 0
2290 ;ABORT NEST CURRENT TEST IF BAD A/D
2300            JP FETCH         ;
2310 ADOK3:     LD A,1H          ;SET MIN. I
2320            CP C             ;CMP WITH A/D
2330 ;IF NEST CURRENT FLOWING, CY=1
2340 ;IF NOT AT NEST OR NOT HUNGRY, CY=0
2350            JP NC,FETCH      ;JMP IF CY=0
2360 ;SET AND OUTPUT STOP MOTION CODE
2370            LD A,(HL)        ;GET PORT 0
2380            AND 0F0H         ;ZERO MOTOR
2390                             ;BITS
2400            LD (HL),A        ;OUTPUT TO
2410                             ;PORT 0
```

```
2420                                    ;
2430 ;CHECK NEST CURRENT AGAIN
2440              JP BTCHK             ;
2450 ;END OF NEST CURRENT CHECKING ROUTINE
2460 ;
2470 A/D CONVERSION SUBROUTINE
2480 ADCON:       LD C,0H              ;ZERO A/D
2490                                    ;COUNTER
2500 ADCNT:       LD A,(HL)            ;GET PORT 0
2510              AND 3FH              ;ISO. A/D BITS
2520              OR C                 ;FIT IN A/D
2530                                    ;COUNT
2540              LD (HL),A            ;OUTPUT TO
2550                                    ;PORT 0
2560              LD A,(HL)            ;GET PORT 0
2570              AND B                ;ISO. CORR BIT
2580 ;RETURN IF DONE -- CY=0
2590              RET Z                ;
2600              LD A,3H              ;SET MAX. A/D
2610              CP C                 ;CMP WITH A/D
2620 ;RETURN IF INVALID -- CY=1
2630              RET C                ;
2640              INC C                ;
2650              JP ADCNT             ;DO AGAIN
2660 ;END OF A/D CONVERSION SUBROUTINE
2670 ;
2680 ;
2690 ;USER MUST DEFINE THESE ADDRESSES
2700     ;START
2710     ;PORT0
2720     ;PORT1
2730 ;ASSEMBLER WILL DEFINE THESE
2740     ;FETCH
2750     ;SETC
2760     ;CNTC
2770     ;STA1
2780     ;ADOK1
2790     ;NEXT1
2800     ;NEXT2
2810     ;ADOK2
2820     ;NEXT3
2830     ;NEXT4
2840     ;BTCHK
2850     ;ADCON
2860     ;ADCNT
2870 ;
2880 ;
```

PROGRAM FLOWCHART FOR A BETA-CLASS ROBOT

One of the most attractive features of using microprocessor systems is that it is often possible to make radical changes in the performance of a

437

machine without having to change a single portion of the hardware. It is certainly easier—usually less costly as well—to change the content of some program memory than to buy some new parts and assemble them into a working system.

A case in point is making a change from an Alpha-Class to a Beta-Class machine. As presented here, there are absolutely no hardware changes. The port configuration and peripheral circuits shown in Figs. 13-1 and 13-2 work equally well for the Alpha-Class and Beta-Class mechanisms.

All that changes here is the software, or the programming. The revised flowchart is shown in Fig. 13-4.

The main difference between an Alpha mechanism and a Beta mechanism is that the latter has provisions for remembering responses to certain circumstances previously encountered in the environment. That means the Beta flowchart should reflect some operations aimed at self-programming a specified section of memory.

To make sure the machine begins with a completely blank response memory, the first operation on the flowchart calls for clearing that response memory. This operation is carried out just one time, and, of course, at the very beginning of the life of the machine. In a sense, it is "born" knowing nothing at all about how to respond to its environment.

After clearing out the response memory, the flowchart specifies some operations for setting the program stack and Port 0 pointer. Then it is time to fetch a random motion code and output it to the motor assembly. The robot thus makes its first spastic reaction to being born.

So the robot does something initially. There's no telling what it will do, but it will do something, because the random-number routine is written to exclude STOP motion codes.

With the first motion code being executed, the next step is to check for a stall condition. This stall-checking loop is identical to the one described for the Alpha-Class program presented earlier in this chapter.

Once a stall condition is detected, the STALL conditional is no longer satisfied, and the system begins checking for nest conditions. Has the machine stalled against the nest? If so, it is a good time to stop and soak up some electrical energy. If the STALL conditional is satisfied, the system is detecting battery recharge current from the nest. The program responds by outputting the STOP code. This is not a response from the self-programmed response memory. Rather, it is a "natural, built-in reflex" that stops the machine at the nest. It is the only situation that allows the program to output the STOP motion code.

After executing the OUTPUT STOP MOTION CODE operation on the flowchart, the system loops back up to CHECK NEST CURRENT. As long as battery recharge current continues to flow, the program remains in that loop, checking for recharge current and outputting the STOP motion code.

The recharge current level will eventually fall below some point prescribed in the program, however. When that happens, the CURRENT

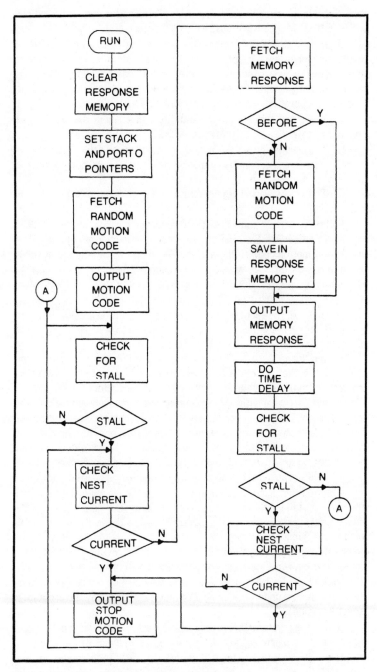

Fig. 13-4. Flowchart for the Beta-Class machine.

conditional is no longer satisfied, and its N alternative carries operations to FETCH MEMORY RESPONSE.

That same CURRENT conditional is not satisfied under a different set of circumstances, too. Suppose the machine blunders into some obstacle other than the nest assembly. Certainly that obstacle will not supply battery recharging current, so the CURRENT conditional is not satisfied. Therefore, operations go to FETCH MEMORY RESPONSE.

So there are three kinds of situations that cause the CURRENT conditional to lead to FETCH MEMORY RESPONSE:

☐ The machine blunders into an obstacle that is not the nest.

☐ The machine blunders into the nest but is not "hungry" at the moment.

☐ The machine has been feeding at the nest, but the recharge cycle is done.

Failing to satisfy that CURRENT conditional brings up the FETCH MEMORY RESPONSE operation. The response memory is addressed by the current motion code of the machine. Whatever it is doing the moment the stall occurs dictates where in the response memory it looks for an appropriate response. The details of this particular matter will be clearer when you have a chance to study the actual program listing.

At any rate, the robot fetches a memory response and then checks to see whether or not it has every encountered and solved that particular problem before. If not, the data in the response memory is still zero: — the "responses" loaded into the response memory at the very beginning of the program. It is the task of the BEFORE conditional to determine whether or not the situation has been encountered and solved before.

If that particular situation has never been encountered before, the BEFORE conditional is *not* satisfied. The machine doesn't have the foggiest idea how to deal with it, so it fetches a random motion code and saves it in the response memory. Even if the randomly generated response is destined to fail, it is still inserted into the response memory at this point. You'll see how it is replaced with a workable response later on.

On the other hand, it might turn out that the machine has dealt with the current stall condition in the past. That being the case, the BEFORE conditional *is* satisfied and the program bypasses the fetching and saving operations, jumping down to a point where it outputs the earlier response to the motor control system. The OUTPUT MEMORY RESPONSE is thus executed, whether the response saved in memory is a new or old one.

DO TIME DELAY gives the machine a chance to try the memory response motion code. The amount of time delay is set by the program to several seconds.

Then the system effectively checks for a stall or nest-contact situation. This particular set of operations is identical to the one used earlier in the program. The exit points go to different places on the flowchart, though.

Remember that the system is now trying to find out whether or not the response code saved in memory is a good one. First, suppose the remembered response is a good one. In that case, the STALL conditional is not satisfied, the operations jump all the way back to the first part of the program to the original CHECK FOR STALL operation. The machine thus continues executing the remembered response until another stall condition arises. This machine, unlike the one described in Chapter 12, does not automatically default to a FORWARD motion code after getting away from a stall situation.

Now return to that second STALL conditional and assume it *is* satisfied. That means the motion code from the response memory does not work or the machine has blundered into its nest assembly while trying the response.

To distinguish these two possibilities, the program does another CHECK NEST CURRENT and CURRENT conditional. If the machine is picking up battery recharge current, the second CURRENT conditional is satisfied. The robot aborts the entire response-checking routine, outputs a STOP motion code and rests at the nest until the battery is recharged.

But here is the interesting situation. Suppose that second CURRENT conditional is *not* satisfied, which means the response called from the response memory is not working. It makes no difference whether it is a new response or an old one that worked before, the program still returns to the point where it picks a new random response code, saves it in the response memory, and goes through the entire response-checking routine again.

The system will loop through that code-selecting and testing routine until it comes up with a workable response to the stall condition at hand, or until the machine blunders into its nest. The response memory is continuously modified until the machine gets away from the stall condition. At best, the machine will leave the situation being better prepared to deal with it in the future. The SAVE IN RESPONSE MEMORY operation is wholly responsible for retaining the sequence of motion codes necessary for solving a given stall condition.

It is important to point out that this system, unlike some simpler types of artificial intelligence, is capable of revising its memory as it experiences more complicated elements in its environment. A remembered response that worked at one time in the past might not work very well under slightly different conditions in the future; it is therefore revised as the occasion arises. The system, in a manner of speaking, is self-perfecting as a function of first-hand experience with the real environment.

A complete discussion of Beta behavior—the sort of behavior exhibited by this program—is beyond the scope of this book. You can learn more about the matter from my earlier books cited in the opening paragraphs of this chapter.

PROGRAM LISTINGS FOR THE BETA-CLASS ROBOT

This chapter concludes with program listings for the Beta-Class robot. The first is written in Intel 8080A/8085 mnemonics, and the second uses Z-80 mnemonics.

The listings are rather extensive but employ a number of techniques described for the Alpha-Class program. Rather than describing the entire listing from beginning to end, the following discussion first specifies sections of the program that can be directly related to the flowchart in Fig. 13-4. Then it describes some of the details peculiar to the Beta-Class mechanism. Programming details that are identical to those used in the Alpha-Class program can be gleaned from the discussion of that particular system presented earlier in this chapter.

The first step on the flowchart is CLEAR RESPONSE MEMORY, which occupies lines 1020 through 1080. The user must specify a 16-byte section of RAM for this response memory, beginning at address MEM. The routine simply loads zeros from the accumulator into 16 successive memory locations, beginning, of course, from address MEM.

The second step, SET STACK AND PORT 0 POINTERS, is done by instructions in lines 1110 and 1120. Lines 1150 through 1210 do the FETCH RANDOM MOTION CODE operation, looping around as many times as necessary to get a motion code that does not result in stopping both motors simultaneously.

The single instruction in line 1240 does the OUTPUT MOTION CODE operation. The flowchart specifies two operations labeled CHECK FOR STALL. The oprations are identical and called a subroutine, labeled STCHK. In the first instance, STCHK is called by the instruction in line 1270. In the second case (when the system is trying out a response from the response memory), the routine is called in line 1820. The STCHK subroutine, itself, is specified in program lines 2040 through 3050.

As described for the Alpha-Class program, this STCHK subroutine returns with a 1 in the accumulator if something is wrong: at least one of the motors is stalled or an A/D overflow occurs while doing a conversion on the actual motor speed. But if everything is in order, STCHK returns with a 0 value in the accumulator.

The jump instruction in line 1870 is executed if the STALL conditional on the flowchart is satisfied. That means an overflow or stall condition exists. Otherwise, the unconditional jump instruction in line 1890 carries operations back up to the point where the system does another CHECK FOR STALL

The flowchart calls for doing a CHECK NEST CURRENT at two different places: first, during the initial stall/nest checking routine; then again, while trying out a response called up from memory. On the program listings, the CHECK NEST CURRENT operations are called as the BTCHK subroutine in lines 1350 and 1910. The BTCHK subroutine, itself, occupies lines 3100 through 3340.

BTCHK returns to the calling program with the CY bit in the flag register set to 1 if the machine is not at the nest or recharge is done. Otherwise, it returns with CY=0 if the machine is at the nest and consuming battery recharge current.

The conditional jump instruction in line 1390 responds to the nest-checking operation, doing the job of the CURRENT conditional in the flowchart. If that instruction in line 1390 is satisfied, it carries the program to FETCH MEMORY RESPONSE. If not satisfied, operations go to OUTPUT STOP MOTION CODE. This operation is carried out by the instruction in line 1410. The unconditional jump instruction in line 1420 executes the loop back up to CHECK NEST CURRENT.

The flowchart operation, FETCH MEMORY RESPONSE, is carried out by lines 1440 through 1530. The basic idea is to add the current motion code from Port 0 to the starting address of the response memory, MEM. Line 1530 is the one that actually gets the data from the response memory into the accumulator.

The next operation on the flowchart is the BEFORE conditional, which is handled by the instructions in lines 1550 and 1560. The response pulled from memory, you see, if compared with zero in line 1550. If the machine has not experienced the situation before, the comparison will turn up a zero result. If the robot has already put a response there, the comparison operation will turn up something other than a zero response.

The conditional jump instruction in line 1560 therefore sends operations down to the OUTPUT MEMORY RESPONSE operation on the flowchart—if some response is indeed found in the memory. Otherwise, the system does FETCH RANDOM MOTION CODE again (see program lines 1580 through 1640). Then it does SAVE IN RESPONSE MEMORY via the instruction in line 1660.

OUTPUT MEMORY RESPONSE is done by line 1700, after resetting the HL register pair to Port 0 again by line 1680. The DO TIME DELAY operation on the flowchart is executed by lines 1730 through 1790.

The remaining operations involved in the code-checking routine involve the stall-checking and nest-checking subroutines already described. The main program ends at line 1970.

8080/8085 Program Listing

```
BETA-CLASS ROBOT LISTING

( 8080A/8085 VERSION )


1000 STRT:                          ;START PROG.
1010 ;CLEAR RESPONSE MEMORY
1020            LXI  H,MEM          ;POINT TO
1030            XRA  A              ;ZERO ACCUM.
1040            MVI  B,10H          ;ZERO COUNT
```

443

```
1050 CLR1:      MOV M,A          ;CLEAR DATA
1060           INX H            ;NEW ADDR
1070           DCR B            ;COUNT
1080           JNZ CLR1         ;JMP IF NOT
1090                            ;DONE
1100 ;ELSE SET STACK AND PORT 0 POINTERS
1110           LXI SP,TOPS      ;SET STACK
1120           LXI H,PORTO      ;POINT TO
1130                            ;PORT 0
1140 ;FETCH RANDOM MOTION CODE
1150 FETCH1:    LDA PORT1        ;GET PORT 1
1160           ANI 0FH          ;ISO. RANDOM
1170                            ;BITS
1180           CPI 0H           ;CHECK FOR 0
1190           JZ FETCH1        ;JMP IF SO
1192           CPI 3H           ;
1194           JZ FETCH1        ;
1196           CPI 0CH          ;
1198           JZ FETCH1        ;
1200           CPI 0FH          ;CHECK FOR F
1210           JZ FETCH1        ;JMP IF SO
1220 ;VALID MOTION CODE NOW IN ACCUMULATOR
1230 ;OUTPUT MOTION CODE
1240           MOV M,A          ;OUTPUT TO
1250                            ;PORT 0
1260 ;CHECK FOR STALL
1270 STA1:      CALL STCHK       ;
1280 ;RETURNS WITH 1 IN ACCUMULATOR IF
1290 ;A/D O'FLOW OR STALL CONDITION
1300 ;RETURNS WITH 0 IN ACCUMULATOR IF
1310 ;BOTH MOTORS ARE RUNNING PROPERLY
1320           JNZ NCHK         ;JMP IF BAD
1330           JMP STA1         ;REPEAT IF OK
1340 ;CHECK NEST CURRENT
1350 NCHK:      CALL BTCHK       ;
1360 ;RETURNS WITH CY=1 IF AT NEST,
1370 ;RETURNS WITH CY=0 IF NOT AT NEST,
1380 ;A/D O'FLOW OR RECHARGE DONE
1390           JNC FETCH2       ;JMP IF NO
1400 ;OUTPUT STOP MOTION CODE
1410 OSTP:      MVI M,0H         ;
1420           JMP NCHK         ;REPEAT
1430 ;FETCH MEMORY RESPONSE
1440 FETCH2:    MVI A,M          ;GET PORT 0
1450           ANI 0FH          ;ISO. MOTOR
1460                            ;COMMAND
1470           MVI B,0          ;ZERO B REG.
1480           MOV C,A          ;CURRENT
1490                            ;CODE TO C
1500           LXI H,MEM        ;POINT TO
```

444

```
1510                                   ;START OF MEM
1520              DAD B                 ;SET ADDR
1530              MOV A,M               ;GET RESPONSE
1540 ;SEE IF BEFORE
1550              CPI OH                ;
1560              JNZ MEMOUT            ;JMP IF SO
1570 ;ELSE FETCH NEW RANDOM MOTION CODE
1580 FETCH3:      LDA PORT1             ;GET PORT 1
1590              ANI OFH               ;ISO. RANDOM
1600                                    ;BITS
1610              CPI OH                ;CHECK FOR O
1620              JZ FETCH3             ;JMP IF SO
1622              CPI 3H                ;
1624              JZ FETCH3             ;
1626              CPI OCH               ;
1628              JZ FETCH3
1630              CPI OFH               ;CHECK FOR F
1640              JZ FETCH3             ;JMP IF F
1650 ;SAVE RESPONSE IN MEMORY
1660              MOV M,A               ;
1670 ;OUTPUT MEMORY RESPONSE
1680 MEMOUT:      LXI H,PORTO           ;SET PORT 0
1690                                    ;POINTER
1700              MOV M,A               ;OUTPUT TO
1710                                    ;PORT 0
1720 ;DO TIME DELAY
1730              MVI B,OFH             ;SET B COUNT
1740 SETC:        MVI C,OFFH            ;SET C COUNT
1750 CNTC:        DCR C                 ;COUNT C
1760              JNZ CNTC              ;AGAIN IF NOT
1770                                    ;DONE
1780              DCR B                 ;COUNT B
1790              JNZ SETC              ;JMP IF NOT
1800                                    ;DONE
1810 ;CHECK FOR STALL AGAIN
1820              CALL STCHK            ;
1830 ;RETURNS WITH 1 IN ACCUMULATOR IF
1840 ;A/D O'FLOW OR STALL CONDITION
1850 ;RETURNS WITH 0 IN ACCUMULATOR IF
1860 ;BOTH MOTORS ARE RUNNING PROPERLY
1870              JNZ NCHK2             ;JMP IF BAD
1880 ;ELSE RETURN TO FIRST STALL CHECK
1890              JMP STA1              ;
1900 ;CHECK NEST CURRENT AGAIN
1910              CALL BTCHK            ;
1920 ;RETURNS WITH CY=1 IF AT NEST
1930 ;RETURNS WITH CY=0 IF NOT AT NEST,
1940 ;OR A/D O'FLOW
1950              JNC FETCH2            ;JMP TO GET
1960                                    ;ANOTHER CODE
```

```
1970              JMP OSTP          ;ELSE JMP TO
1980                                ;OUTPUT STOP
1990                                ;CODE
2000 ;END OF MAIN PROGRAM
2010 ;
2020 ;
2030 ;BEGINNING OF STALL CHECKING SUBR.
2040 STCHK:      MVI B,10H         ;SET FOR LM
2050                                ;A/D CONVERT
2060 ;DO A/D CONVERSION ON CORRO
2070              CALL ADCON        ;
2080 ;RETURNS WITH CY=0 IF CONVERSION IS GOOD,
2090 ;ELSE CY=1
2100              JNC ADOK1         ;JMP IF OK
2110 ;ELSE TURN ON THE MOT O'FLOW LAMP AND
2120 ;ABORT THE MOTOR CHECKS
2130              MOV A,M           ;GET PORT 0
2140              ORI 10H           ;SET O'FLOW
2150                                ;LAMP BIT
2160              MOV M,A           ;OUTPUT TO
2170                                ;PORT 0
2180              STC               ;SET CY=1
2190              RET               ;RETURN TO
2200                                ;MAIN PROG.
2210 ;CHECK LEFT MOTOR SPEED
2220 ADOK1:      MVI A,1H          ;SET MIN. SPEED
2230              CMP C             ;CMP WITH A/D
2240                                ;SPEED
2250              JNC NEXT1         ;JMP IF LM IS
2260                                ;NOT TURNING
2270              MVI D,10H         ;SET TURNING
2280                                ;FLAG
2290              JMP NEXT2         ;
2300 NEXT1:      MVI D,0H          ;SET NOT-TURN-
2310                                ;ING FLAG
2320 ;TURN OFF THE MOTOR O'FLOW LAMP
2330 NEXT2:      MOV A,M           ;GET PORT 0
2340              ANI 0EFH          ;CLEAR O'FLOW
2350                                ;LAMP BIT
2360              MOV M,A           ;OUTPUT TO
2370                                ;PORT 0
2380 ;GET MLF COMMAND BIT
2390              MOV A,M           ;GET PORT 0
2400              ANI 1H            ;ISO. MLF
2410              MOV B,A           ;SAVE MLF IN
2420                                ;B0
2430 ;GET MLR COMMAND BIT
2440              MOV A,M           ;GET PORT 0
2450              ANI 2H            ;ISO. MLR
2460              RRC               ;MLR TO A0
```

446

```
2470 ;EXCLUSIVE-OR MLF,MLR,LMS
2480              XRA B            ;
2490              XRA D            ;
2500 ;1 IS IN ACCUMULATOR IF MOTOR IS STALLED
2510 ;ELSE 0 IS IN ACCUMULATOR
2520 ;RETURN TO MAIN PROGRAM IF STALLED
2530              RNZ              ;
2540 ;ELSE DO RIGHT MOTOR CHECKS
2550              MVI B,20H        ;SET FOR RM
2560                               ;A/D CONVERT
2570 ;DO A/D CONVERSION ON CORR1
2580              CALL ADCON       ;
2590 ;RETURNS WITH CY=0 IF CONVERSION IS GOOD.
2600 ;ELSE CY=1
2610              JNC ADOK2        ;JMP IF OK
2620 ;ELSE TURN ON THE MOT O'FLOW LAMP AND
2630 ;ABORT THE MOTOR CHECKS
2640              MOV A,M          ;GET PORT 0
2650              ORI 10H          ;SET O'FLOW
2660                               ;LAMP BIT
2670              MOV M,A          ;OUTPUT TO
2680                               ;PORT 0
2690              STC              ;SET CY=1
2700              RET              ;RETURN TO
2710                               ;MAIN PROG.
2720 ;CHECK RIGHT MOTOR SPEED
2730 ADOK2:       MVI A,1H         ;SET MIN. SPEED
2740              CMP C            ;CMP WITH A/D
2750              JNC NEXT3        ;JMP IF RM IS
2760                               ;NOT TURNING
2770              MVI D,1H         ;SET TURNING
2780                               ;FLAG
2790              JMP NEXT4        ;
2800 NEXT3:       MIV D,0H         ;SET NOT-TURN-
2810                               ;ING FLAG
2820 ;TURN OFF THE MOTOR O'FLOW LAMP
2830 NEXT4:       MOV A,M          ;GET PORT 0
2840              ANI 0EFH         ;CLEAR O'FLOW
2850                               ;LAMP BIT
2860              MOV M,A          ;OUTPUT TO
2870                               ;PORT 0
2880 ;GET MRF COMMAND BIT
2890              MOV A,M          ;GET PORT 0
2900              ANI 4H           ;ISO. MRF
2910              MOV B,A          ;SAVE MRF IN
2920                               ;B2
2930 ;GET MRR COMMAND BIT
2940              MOV A,M          ;GET PORT 0
2950              ANI 8H           ;ISO. MRR
2960              RRC              ;MRR TO A2
```

```
2970  ;EXCLUSIVE-OR MRF,MRR,RMS
2980            XRA B           ;
2990            RRC             ;TO A1
3000            RRC             ;TO A0
3010            XRA D           ;
3020  ;RETURN TO MAIN PROGRAM --
3030  ;1 IS IN ACCUMULATOR IF MOTOR IS STALLED
3040  ;ELSE 0 IS IN ACCUMULATOR
3050            RET             ;
3060  ;END OF STALL CHECKING SUBROUTINE
3070  ;
3080  ;
3090  ;BEGINNING OF NEST CHECKING SUBR.
3100  BTCHK:    MVI B,40H       ;SET FOR BATT
3110                            ;A/D CONVERT
3120  ;DO A/D CONVERSION ON CORR2
3130            CALL ADCON      ;
3140  ;RETURNS WITH CY=0 IF CONVERSION IS GOOD
3150  ;ELSE CY=1
3160            JNC ADOK3       ;JMP IF OK
3170  ;ELSE TURN ON THE BATT O'FLOW LAMP
3180  ;AND ABORT FURTHER NEST CHECKS
3190            MOV A,M         ;GET PORT 0
3200            ORI 20H         ;SET O'FLOW
3210                            ;LAMP BIT
3220            MOV M,A         ;OUTPUT TO
3230                            ;PORT 0
3240            RET             ;RETURN TO
3250                            ;MAIN PROG.
3260  ;CHECK NEST CURRENT LEVEL
3270  ADOK3:    MVI A,1H        ;SET MIN. I
3280                            ;LEVEL
3290            CMP C           ;CMP WITH A/D
3300                            ;LEVEL
3310  ;RETURN TO MAIN PROGRAM --
3320  ;CY=1 IF NOT AT NEST OR RECHARGE IS DONE
3330  ;CY=0 IF AT NEST AND TAKING CURRENT
3340            RET             ;
3350  ;END OF NEST CHECKING SUBROUTINE
3360  ;
3370  ;
3380  ;BEGINNING OF A/D CONVERSION SUBROUTINE
3390  ADCON:    MVI C,0H        ;ZERO A/D
3400                            ;COUNTER
3410  ADCNT:    MOV A,M         ;GET PORT 0
3420            ANI 3FH         ;ISO. A/D BITS
3430            ORA C           ;FIT IN A/D
3440                            ;COUNT
3450            MOV M,A         ;OUTPUT TO
3460                            ;PORT 0
```

```
3470                 MOV A,M          ;GET PORT 0
3480                 ANA B            ;ISO. CORR BIT
3490 ;RETURN IF DONE -- CY=0
3500                 RNZ              ;
3510                 MVI A,3H         ;SET MAX. A/D
3520                 CMP C            ;CMP WITH A/D
3530 ;RETURN IF O'FLOW -- CY=1
3540                 RC               ;
3550                 INR C            ;
3560                 JMP ADCNT        ;DO MORE
3570 ;END OF A/D CONVERSION SUBROUTINE
3580 ;
3590 ;
3600 ;USER MUST DEFINE THESE ADDRESSES
3610      ;STRT
3620      ;MEM
3630      ;PORT0
3640      ;PORT1
3650 ;ASSEMBLER WILL DEFINE THESE
3660      ;CLR1
3670      ;FETCH1
3680      ;STA1
3690      ;NCHK
3700      ;OSTP
3710      ;FETCH2
3720      ;FETCH3
3730      ;MEMOUT
3740      ;SETC
3750      ;CNTC
3760      ;STCHK
3770      ;ADOK1
3780      ;NEXT1
3790      ;NEXT2
3800      ;ADOK2
3810      ;NEXT3
3820      ;NEXT4
3830      ;BTCHK
3840      ;ADOK3
3850      ;ADCON
3860      ;ADCNT
```

Z80 Program Listing

BETA-CLASS ROBOT LISTING

(Z80 VERSION)

```
1000 STRT:                           ;START PROG.
1010 ;CLEAR RESPONSE MEMORY
1020                 LD HL,MEM        ;POINT TO
```

```
1030              XOR A           ;ZERO ACCUM.
1040              LD B,10H        ;ZERO COUNT
1050 CLR1:        LD (HL),A       ;CLEAR DATA
1060              INC HL          ;NEW ADDR
1070              DEC B           ;COUNT
1080              JP NZ,CLR1      ;JMP IF NOT
1090                              ;DONE
1100 ;ELSE SET STACK AND PORT 0 POINTERS
1110              LD SP,TOPS      ;SET STACK
1120              LD HL,PORT0     ;POINT TO
1130                              ;PORT 0
1140 ;FETCH RANDOM MOTION CODE
1150 FETCH1:      LD A,(PORT1)    ;GET PORT 1
1160              AND 0FH         ;ISO. RANDOM
1170                              ;BITS
1180              CP 0H           ;CHECK FOR 0
1190              JP Z,FETCH1     ;JMP IF SO
1192              CP 3H           ;
1194              JP Z,FETCH1     ;
1196              CP 0CH          ;
1198              JP Z,FETCH1     ;
1200              CP 0FH          ;CHECK FOR F
1210              JZ FETCH1       ;JMP IF SO
1220 ;VALID MOTION CODE NOW IN ACCUMULATOR
1230 ;OUTPUT MOTION CODE
1240              LD (HL),A       ;OUTPUT TO
1250                              ;PORT 0
1260 ;CHECK FOR STALL
1270 STA1:        CALL STCHK      ;
1280 ;RETURNS WITH 1 IN ACCUMULATOR IF
1290 ;A/D 0'FLOW OR STALL CONDITION
1300 ;RETURNS WITH 0 IN ACCUMULATOR IF
1310 ;BOTH MOTORS ARE RUNNING PROPERLY
1320              JP NZ,NCHK      ;JMP IF BAD
1330              JP STA1         ;REPEAT IF OK
1340 ;CHECK NEST CURRENT
1350 NCHK:        CALL BTCHK      ;
1360 ;RETURNS WITH CY=1 IF AT NEST,
1370 ;RETURNS WITH CY=0 IF NOT AT NEST,
1380 ;A/D 0'FLOW OR RECHARGE DONE
1390              JP NC,FETCH2    ;JMP IF NO
1400 ;OUTPUT STOP MOTION CODE
1410 OSTP:        LD (HL),0H      ;
1420              JP NCHK         ;REPEAT
1430 ;FETCH MEMORY RESPONSE
1440 FETCH2:      LD A,(HL)       ;GET PORT 0
1450              AND 0FH         ;ISO. MOTOR
1460                              ;COMMAND
1470              LD B,0          ;ZERO B REG.
1480              LD C,A          ;CURRENT
1490                              ;CODE TO C
```

```
1500            LD HL,MEM       ;POINT TO
1510                            ;START OF MEM
1520            ADD HL,BC       ;SET ADDR
1530            LD A,(HL)       ;GET RESPONSE
1540  ;SEE IF BEFORE
1550            CP 0H           ;
1560            JP NZ,MEMOUT    ;JMP IF SO
1570  ;ELSE FETCH NEW RANDOM MOTION CODE
1580  FETCH3:   LD A,(PORT1)    ;GET PORT 1
1590            AND 0FH         ;ISO. RANDOM
1600                            ;BITS
1610            CP 0H           ;CHECK FOR 0
1620            JP Z,FETCH3     ;JMP IF SO
1622            CP 3H           ;
1624            JP Z,FETCH3     ;
1626            CP 0CH          ;
1628            JP Z,FETCH3     ;
1630            CP 0FH          ;CHECK FOR F
1640            JP Z,FETCH3     ;JMP IF F
1650  ;SAVE RESPONSE IN MEMORY
1660            LD (HL),A       ;
1670  ;OUTPUT MEMORY RESPONSE
1680  MEMOUT:   LD HL,PORT0     ;SET PORT 0
1690                            ;POINTER
1700            LD (HL),A       ;OUTPUT TO
1710                            ;PORT 0
1720  ;DO TIME DELAY
1730            LD B,0FH        ;SET B COUNT
1740  SETC:     LD C,0FFH       ;SET C COUNT
1750  CNTC:     DEC C           ;COUNT C
1760            JP NZ,CNTC      ;AGAIN IF NOT
1770                            ;DONE
1780            DEC B           ;COUNT B
1790            JP NZ,SETC      ;JMP IF NOT
1800                            ;DONE
1810  ;CHECK FOR STALL AGAIN
1820            CALL STCHK      ;
1830  ;RETURNS WITH 1 IN ACCUMULATOR IF
1840  ;A/D O'FLOW OR STALL CONDITION
1850  ;RETURNS WITH 0 IN ACCUMULATOR IF
1860  ;BOTH MOTORS ARE RUNNING PROPERLY
1870            JP NZ,NCHK2     ;JMP IF BAD
1880  ;ELSE RETURN TO FIRST STALL CHECK
1890            JP STA1         ;
1900  ;CHECK NEST CURRENT AGAIN
1910            CALL BTCHK      ;
1920  ;RETURNS WITH CY=1 IF AT NEST
1930  ;RETURNS WITH CY=0 IF NOT AT NEST,
1940  ;OR A/D O'FLOW
1950            JP NC,FETCH2    ;JMP TO GET
```

```
1960                                    ;ANOTHER CODE
1970              JP OSTP               ;ELSE JMP TO
1980                                    ;OUTPUT STOP
1990                                    ;CODE
2000 ;END OF MAIN PROGRAM
2010 ;
2020 ;
2030 ;BEGINNING OF STALL CHECKING SUBR.
2040 STCHK:       LD B,10H              ;SET FOR LM
2050                                    ;A/D CONVERT
2060 ;DO A/D CONVERSION ON CORRO
2070              CALL ADCON            ;
2080 ;RETURNS WITH CY=0 IF CONVERSION IS GOOD,
2090 ;ELSE CY=1
2100              JP NC,ADOK1           ;JMP IF OK
2110 ;ELSE TURN ON THE MOT O'FLOW LAMP AND
2120 ;ABORT THE MOTOR CHECKS
2130              LD A,(HL)             ;GET PORT 0
2140              OR 10H                ;SET O'FLOW
2150                                    ;LAMP BIT
2160              LD (HL),A             ;OUTPUT TO
2170                                    ;PORT 0
2180              SCF                   ;SET CY=1
2190              RET                   ;RETURN TO
2200                                    ;MAIN PROG.
2210 ;CHECK LEFT MOTOR SPEED
2220 ADOK1:       LD A,1H               ;SET MIN. SPEED
2230              CP C                  ;CMP WITH A/D
2240                                    ;SPEED
2250              JP NC,NEXT1           ;JMP IF LM IS
2260                                    ;NOT TURNING
2270              LD D,10H              ;SET TURNING
2280                                    ;FLAG
2290              JP NEXT2              ;
2300 NEXT1:       LD D,0H               ;SET NOT-TURN-
2310                                    ;ING FLAG
2320 ;TURN OFF THE MOTOR O'FLOW LAMP
2330 NEXT2:       LD A,(HL)             ;GET PORT 0
2340              AND 0EFH              ;CLEAR O'FLOW
2350                                    ;LAMP BIT
2360              LD (HL),A             ;OUTPUT TO
2370                                    ;PORT 0
2380 ;GET MLF COMMAND BIT
2390              LD A,(HL)             ;GET PORT 0
2400              AND 1H                ;ISO. MLF
2410              LD B,A                ;SAVE MLF IN
2420                                    ;B0
2430 ;GET MLR COMMAND BIT
2440              LD A,(HL)             ;GET PORT 0
2450              AND 2H                ;ISO. MLR
```

```
2460                RRC A              ;MLR TO A0
2470 ;EXCLUSIVE-OR MLF,MLR,LMS
2480                XOR B              ;
2490                XOR D              ;
2500 ;1 IS IN ACCUMULATOR IF MOTOR IS STALLED
2510 ;ELSE 0 IS IN ACCUMULATOR
2520 ;RETURN TO MAIN PROGRAM IF STALLED
2530                RET NZ             ;
2540 ;ELSE DO RIGHT MOTOR CHECKS
2550                LD B,20H           ;SET FOR RM
2560                                   ;A/D CONVERT
2570 ;DO A/D CONVERSION ON CORR1
2580                CALL ADCON         ;
2590 ;RETURNS WITH CY=0 IF CONVERSION IS GOOD.
2600 ;ELSE CY=1
2610                JP NC,ADOK2        ;JMP IF OK
2620 ;ELSE TURN ON THE MOT O'FLOW LAMP AND
2630 ;ABORT THE MOTOR CHECKS
2640                LD A,(HL)          ;GET PORT 0
2650                OR 10H             ;SET O'FLOW
2660                                   ;LAMP BIT
2670                LD (HL),A          ;OUTPUT TO
2680                                   ;PORT 0
2690                SCF                ;SET CY=1
2700                RET                ;RETURN TO
2710                                   ;MAIN PROG.
2720 ;CHECK RIGHT MOTOR SPEED
2730 ADOK2:         LD A,1H            ;SET MIN. SPEED
2740                CP C               ;CMP WITH A/D
2750                JP NC,NEXT3        ;JMP IF RM IS
2760                                   ;NOT TURNING
2770                LD D,1H            ;SET TURNING
2780                                   ;FLAG
2790                JP NEXT4           ;
2800 NEXT3:         LD D,0H            ;SET NOT-TURN-
2810                                   ;ING FLAG
2820 ;TURN OFF THE MOTOR O'FLOW LAMP
2830 NEXT4:         LD A,(HL)          ;GET PORT 0
2840                AND 0EFH           ;CLEAR O'FLOW
2850                                   ;LAMP BIT
2860                LD (HL),A          ;OUTPUT TO
2870                                   ;PORT 0
2880 ;GET MRF COMMAND BIT
2890                LD A,(HL)          ;GET PORT 0
2900                AND 4H             ;ISO. MRF
2910                LD B,A             ;SAVE MRF IN
2920                                   ;B2
2930 ;GET MRR COMMAND BIT
2940                LD A,(HL)          ;GET PORT 0
2950                AND 8H             ;ISO. MRR
```

```
2960                RRC A              ;MRR TO A2
2970  ;EXCLUSIVE-OR MRF,MRR,RMS
2980                XRA B              ;
2990                RRC A              ;TO A1
3000                RRC A              ;TO A0
3010                XOR D              ;
3020  ;RETURN TO MAIN PROGRAM --
3030  ;1 IS IN ACCUMULATOR IF MOTOR IS STALLED
3040  ;ELSE 0 IS IN ACCUMULATOR
3050                RET                ;
3060  ;END OF STALL CHECKING SUBROUTINE
3070  ;
3080  ;
3090  ;BEGINNING OF NEST CHECKING SUBR.
3100  BTCHK:        LD B,40H           ;SET FOR BATT
3110                                   ;A/D CONVERT
3120  ;DO A/D CONVERSION ON CORR2
3130                CALL ADCON         ;
3140  ;RETURNS WITH CY=0 IF CONVERSION IS GOOD
3150  ;ELSE CY=1
3160                JP NC,ADOK3        ;JMP IF OK
3170  ;ELSE TURN ON THE BATT O'FLOW LAMP
3180  ;AND ABORT FURTHER NEST CHECKS
3190                LD A,(HL)          ;GET PORT 0
3200                OR 20H             ;SET O'FLOW
3210                                   ;LAMP BIT
3220                LD (HL),A          ;OUTPUT TO
3230                                   ;PORT 0
3240                RET                ;RETURN TO
3250                                   ;MAIN PROG.
3260  ;CHECK NEST CURRENT LEVEL
3270  ADOK3:        LD A,1H            ;SET MIN. I
3280                                   ;LEVEL
3290                CP C               ;CMP WITH A/D
3300                                   ;LEVEL
3310  ;RETURN TO MAIN PROGRAM --
3320  ;CY=1 IF NOT AT NEST OR RECHARGE IS DONE
3330  ;CY=0 IF AT NEST AND TAKING CURRENT
3340                RET                ;
3350  ;END OF NEST CHECKING SUBROUTINE
3360  ;
3370  ;
3380  ;BEGINNING OF A/D CONVERSION SUBROUTINE
3390  ADCON:        LD C,0H            ;ZERO A/D
3400                                   ;COUNTER
3410  ADCNT:        LD A,(HL)          ;GET PORT 0
3420                AND 3FH            ;ISO. A/D BITS
3430                OR C               ;FIT IN A/D
3440                                   ;COUNT
3450                LD (HL),A          ;OUTPUT TO
```

HEX	D7	D6	D5	D4	D3	D2	D1	D0	LEFT MOTOR	RIGHT MOTOR
					BINARY					
C0	1	1	0	0	0	0	0	0	FWD SSC	STOP
C1	1	1	0	0	0	0	0	1	FWD SSC	REV SS9
C2	1	1	0	0	0	0	1	0	FWD SSC	REV SSA
C3	1	1	0	0	0	0	1	1	FWD SSC	REV SSB
C4	1	1	0	0	0	1	0	0	FWD SSC	REV SSC
C5	1	1	0	0	0	1	0	1	FWD SSC	REV SSD
C6	1	1	0	0	0	1	1	0	FWD SSC	REV SSE
C7	1	1	0	0	0	1	1	1	FWD SSC	REV SSF
C8	1	1	0	0	1	0	0	0	FWD SSC	STOP
C9	1	1	0	0	1	0	0	1	FWD SSC	FWD SS9
CA	1	1	0	0	1	0	1	0	FWD SSC	FWD SSA
CB	1	1	0	0	1	0	1	1	FWD SSC	FWD SSB
CC	1	1	0	0	1	1	0	0	FWD SSC	FWD SSC
CD	1	1	0	0	1	1	0	1	FWD SSC	FWD SSD
CE	1	1	0	0	1	1	1	0	FWD SSC	FWD SSE
CF	1	1	0	0	1	1	1	1	FWD SSC	FWD SSF
D0	1	1	0	1	0	0	0	0	FWD SSD	STOP
D1	1	1	0	1	0	0	0	1	FWD SSD	REV SS9
D2	1	1	0	1	0	0	1	0	FWD SSD	REV SSA
D3	1	1	0	1	0	0	1	1	FWD SSD	REV SSB
D4	1	1	0	1	0	1	0	0	FWD SSD	REV SSC
D5	1	1	0	1	0	1	0	1	FWD SSD	REV SSD
D6	1	1	0	1	0	1	1	0	FWD SSD	REV SSE
D7	1	1	0	1	0	1	1	1	FWD SSD	REV SSF
D8	1	1	0	1	1	0	0	0	FWD SSD	STOP
D9	1	1	0	1	1	0	0	1	FWD SSD	FWD SS9
DA	1	1	0	1	1	0	1	0	FWD SSD	FWD SSA
DB	1	1	0	1	1	0	1	1	FWD SSD	FWD SSB
DC	1	1	0	1	1	1	0	0	FWD SSD	FWD SSC
DD	1	1	0	1	1	1	0	1	FWD SSD	FWD SSD
DE	1	1	0	1	1	1	1	0	FWD SSD	FWD SSE
DF	1	1	0	1	1	1	1	1	FWD SSD	FWD SSF

INPUT CODES   MOTOR RESPONSES

HEX	D7	D6	D5	D4	D3	D2	D1	D0	LEFT MOTOR	RIGHT MOTOR
					BINARY					
E0	1	1	1	0	0	0	0	0	FWD SSE	STOP
E1	1	1	1	0	0	0	0	1	FWD SSE	REV SS9
E2	1	1	1	0	0	0	1	0	FWD SSE	REV SSA
E3	1	1	1	0	0	0	1	1	FWD SSE	REV SSB
E4	1	1	1	0	0	1	0	0	FWD SSE	REV SSC
E5	1	1	1	0	0	1	0	1	FWD SSE	REV SSD
E6	1	1	1	0	0	1	1	0	FWD SSE	REV SSE
E7	1	1	1	0	0	1	1	1	FWD SSE	REV SSF
E8	1	1	1	0	1	0	0	0	FWD SSE	STOP
E9	1	1	1	0	1	0	0	1	FWD SSE	FWD SS9
EA	1	1	1	0	1	0	1	0	FWD SSE	FWD SSA
EB	1	1	1	0	1	0	1	1	FWD SSE	FWD SSB
EC	1	1	1	0	1	1	0	0	FWD SSE	FWD SSC
ED	1	1	1	0	1	1	0	1	FWD SSE	FWD SSD
EE	1	1	1	0	1	1	1	0	FWD SSE	FWD SSE
EF	1	1	1	0	1	1	1	1	FWD SSE	FWD SSF
F0	1	1	1	1	0	0	0	0	FWD SSF	STOP
F1	1	1	1	1	0	0	0	1	FWD SSF	REV SS9
F2	1	1	1	1	0	0	1	0	FWD SSF	REV SSA
F3	1	1	1	1	0	0	1	1	FWD SSF	REV SSB
F4	1	1	1	1	0	1	0	0	FWD SSF	REV SSC
F5	1	1	1	1	0	1	0	1	FWD SSF	REV SSD
F6	1	1	1	1	0	1	1	0	FWD SSF	REV SSE
F7	1	1	1	1	0	1	1	1	FWD SSF	REV SSF
F8	1	1	1	1	1	0	0	0	FWD SSF	STOP
F9	1	1	1	1	1	0	0	1	FWD SSF	FWD SS9
FA	1	1	1	1	1	0	1	0	FWD SSF	FWD SSA
FB	1	1	1	1	1	0	1	1	FWD SSF	FWD SSB
FC	1	1	1	1	1	1	0	0	FWD SSF	FWD SSC
FD	1	1	1	1	1	1	0	1	FWD SSF	FWD SSD
FE	1	1	1	1	1	1	1	0	FWD SSF	FWD SSE
FF	1	1	1	1	1	1	1	1	FWD SSF	FWD SSF

INPUT CODES   MOTOR RESPONSES

Index

Edited by Raymond A. Collins